Southeast Asian Water Environment 1

Southeast Asian Water Environment 1

Selected papers from The First International Symposium on Southeast Asian Water Environment (Biodiversity and Water Environment), Bangkok, Thailand, October 2003

Edited by Shinichiro Ohgaki
with Kensuke Fukushi, Hiroyuki Katayama,
Satoshi Takizawa and Chongrak Polprasert

IWA Publishing

LONDON • SEATTLE

Published by IWA Publishing, Alliance House, 12 Caxton Street, London SW1H 0QS, UK

Telephone: +44 (0) 20 7654 5500; Fax: +44 (0) 20 7654 5555; Email: publications@iwap.co.uk
Web: www.iwapublishing.com

First published 2006

Printed by The Alden Group, Oxford, UK

British Library Cataloguing in Publication Data
A CIP catalogue record for this book is available from the British Library

Library of Congress Cataloging- in-Publication Data
A catalog record for this book is available from the Library of Congress

ISBN: 1843390981

Contents

Preface

Southeast Asia, a region of the developing world, has been undergoing rapid commercial and industrial development over the past half century, which continues to bring economic stability and prosperity to the inhabitants. The combined factors of population growth, urbanization, and industrialization continue to wield pressures on the natural resources and the environment. At the same time, globalization is another momentous challenge for the region.

Southeast Asian megacities are some of the most dynamic and diverse regions, which are playing a key role in achieving global sustainability. Twelve out of nineteen megacities of the world are located in the region and, of these, six are ranked as the top 10 largest cities. Most of them are the fastest growing cities and some of them are the most highly polluted cities in the world. In these cities, highly developed areas with high skylines coexist with poorly developing areas where large economic growth is expected. The poor quality of the living standards of the latter creates many environmental problems that are major threats to the inhabitants of the region.

The Southeast Asian environment has been degraded by the release of industrial and domestic wastes, agricultural and aquacultural chemicals, and pollutants from automobiles. Although there is apparent public awareness that damage to the environment may have serious consequences for future generations, the economic well-being of the region is proving to mitigate against protection of natural resources. Southeast Asia, the monsoon Asia, has quasi-common natural and cultural conditions in most of the countries. It suffers from water-related disasters, Tsunami, floods, typhoons, etc. In order to deal with these issues an integrated approach from the inhibitants, Government and researchers is essential.

The environmental threats arising from the increasing population, over use of natural resources, industrialization, urbanization, and natural disasters are intimidating in pursuing sustainable development of the region. Japan has experienced severe environmental pollution during the early stages of its commercial and industrial development, some of which still remains unresolved. These experiences should be communicated to the global community with the goal of preventing such pollution in other developing regions, like Southeast Asia. However, literal transfer of this knowledge is not pragmatic, since the patchwork-like development makes the situation even more complicated than that experienced. Networking of the knowledge would be meaningful in the related countries; in particular, it would work most effectively when experts on these issues are integrated. We have been trying hard to build an academic network among the researchers of related academic fields of different regions including Southeast Asian region, starting from a preparative workshop held in 2002 in Tokyo, Japan, and an international symposium held in 2003 in Bangkok, Thailand. These were initiated to provide an opportunity for Southeast Asian researchers to exchange information and to get benefits from the accumulated wisdom of researchers from Japan and other countries. I would like to note the contribution of the Japanese Government through the Special Coordination Fund for Promoting Science and Technology, the Ministry of Education, Culture, Sports, Science and Technology for this initiative.

Since the symposium held in Thailand was very informative, we have taken an initiative to record the exchanged knowledge in the symposium and to deliver it worldwide. This book is the fruit of our effort. Among the discussed topics in the symposium the following were emphasized: (1.) Environmental and ecological management, (2.) Dialogue between environmental science and policy making, (3.) Emerging issues and environmental technologies. Some important articles from these topics presented in the symposium were selected and have been included in this book. Our efforts would be completely successful if this book proves to be useful in assisting researchers, policy makers, NGOs and NPOs of the different parts of the world to deal with their environmental issues. It would give me great pleasure if this book contributes to establishing global sustainability in any respect.

Shinichiro Ohgaki
Professor
Department of Urban Engineering
The University of Tokyo

Southeast Asian Water Environment Scenario at Present and the Future Perspective

C. Polprasert and K. Sharma

School of Environment, Resources and Development, Asian Institute of Technology, P.O. Box 4 Khlong Luang, Pathumthani, Thailand 12120.
(E-mail: *chongrak@ait.ac.th; kusums@ait.ac.th*)

Abstract

The Southeast Asian (SEA) region is endowed with vast water resources but challenges are ahead in harnessing the resource. There are problems in water management in urban and rural parts of SEA countries, especially providing safe and sustainable water supply and sanitation facilities. These issues of SEA water environment need to be properly addressed with new strategies. In harnessing water resources the concept of integrated water resources management and interstate river basin management is vital. To meet the Millennium Development Goal (MDG) by all SEA countries, water supply and sanitation sector development is a top priority. Since progress made so far is not sufficient, new strategies might include a concept of urban water district for holistic urban water management. Development of waste minimization technologies and application of appropriate technologies that underline principles of sustainable development might be able to break the chain of water pollution and reduce the stress on water resources. In the new strategies, involvement of all sectors including the community, the government and private sector is vital for the success of the development.

1. Introduction

Water, essential for human beings and ecosystems, is used for domestic, industrial, and agricultural purposes. The importance of water is illustrated by the staggering proportions of the problems faced by those who do not have access to safe, secure supplies of freshwater. It is an unfortunate situation that sector demand for water in the Southeast Asian (SEA) countries is being accompanied by water scarcity on the one hand and deteriorating quality on the other.

Although the SEA region is endowed with massive water resources, the demand for water is increasing in all sectors. The irrigation sector is the highest consumer of water, about three-quarters of the total available water resource. The region is facing problems in supplying quality water to the demanded quantity. The serious problem of maintaining the water environment is in the urban setting, where supply of water in quantity and quality is scarce and the city product, i.e. municipal wastewater and industrial wastewater, is exacerbating the water pollution problem. Agricultural runoff is another non-point source contributing pollution to surface water systems.

To meet the Millennium Development Goal (MDG) of halving by 2015 the proportion of people living in extreme poverty or unable to have access to safe drinking water and adequate sanitation, the SEA countries should prioritize the water supply and sanitation sector development. The challenging issues of water supply and sanitation are not only related to scarcity of financial resources, but also to development of appropriate technologies that underline the principle of sustainable development and hygiene education/awareness.

This paper gives an insight into the present water environment situation in SEA countries. A strategic approach to address the challenges in water management and improving the water supply and sanitation sector in the region is presented.

2. Water environment scenario in Southeast Asian countries

2.1 Water availability and scarcity in the Southeast Asian region

The traditional water demand for domestic consumption, industrial processes, and agricultural use in the region is increasing. The global campaign of 'safe drinking water for all' has certainly increased the water demand. While there is increasing pressure on the water resources by industrialization and urbanization to maintain the economic growth, the more expansion of irrigable land for food production to feed the ever increasing population is increasing water demand.

Fortunately, the SEA Region is endowed with large quantities of water resources except Singapore, which has to depend on neighboring countries. The per capita annual internal renewable water resources in 2000 is shown in Table 1. The Lao PDR, with the smallest population among other SEA countries, has the largest water resources capacity; however, harnessing this resource is a challenge to the country. It can be inferred that, if these rates of fresh water withdrawal continue, Thailand, Philippines and Vietnam are going to face a water stressed situation in the near future.

Table 1 Internal renewable water resources and annual fresh water withdrawal in SEA countries (Source: UNDP, 2000)

Country	Internal Renewable Water Resources (m^3/capita/year)*	Annual Fresh Water Withdrawal (m^3/capita/year)**
Camobdia	10,795	66
Indonesia	13380	407
Laos	35,049	259
Malaysia	26,074	633
Myanmar	19,306	101
Philippines	6,305	811
Thailand	3,420	596
Vietnam	4,591	416

* For year 2000
** For any year 1987-1997

In the Philippines, the country's available water supply and demand shows that from a total national perspective, the country's water resources are adequate to meet the demand up to 2005; the dependable water supply: water demand ratio in the year 2000 is 1:1.6, but if the groundwater safe yield is included, this ratio becomes 1:1.9 (Arellano, 2002).

2.2 Water quality situation in Southeast Asian countries

The economic growth of the region has shadowed the environmental impact, particularly of the receiving water bodies. The water quality situation in SEA countries is highly variable, reflecting social, economic, and physical factors as well as the state of development. And while not all countries are facing a crisis of water shortage, all have, to some extent, serious problems associated with degraded water quality. Most of the small and medium industries located near water bodies, often discharge untreated wastewaters to nearby rivers, thus contaminating the water sources for public use. Landfill leachates can normally find their way through the soils to contaminate ground water in poorly designed landfill sites. Water pollution has affected human health. Inadequate water supply and poor sanitation has affected human health. Use of polluted water for drinking and cooking has led to health problems such as diarrhea, hepatitis and occasional outbreaks of typhoid and cholera.

The surface water quality of the lower part of Chao Phraya River flowing across the Bangkok city area is classified in Class IV, which needs special treatment process for consumption and is not suitable for direct agricultural use (Bangkok SoE, 2001; UNEP, 2001). Indonesia is experiencing similar problems of surface water pollution. Textile industries located along the banks of Citarum River and Ciliwung River are the major pollution contributors of BOD and color to degrade river water quality. While, the major polluters of Ciujung River are pulp and paper industries, the textile and pulp & paper industries are the major polluters of Cisadane and Celeunngsi rivers (Wisjnoprato, 2001).

Most of the monitored rivers in Vietnam are found to be polluted with substances like N and P, from 2 to nearly 200 times compared with water resource of category A and from 2 to 20 times in comparison with water resource category B. Organic pollution in Sai Gon River, Vam Co Dong River and canals is very serious while it is rather severe in Dong Nai River. Quality of water in rivers in the central region is relatively good (SoE Vietnam, 2001; UNEP, 2001)

Table 2 Water supply and sanitation situation in SEA Countries (Source: UNDP, 2004)

Country	Population (in million)	Access to Facilities (%)	
		Sustainable Water Supply	Sustainable Sanitation
Cambodia	13.1	30	17
Indonesia	212.1	78	55
Lao PDR	5.3	37	30
Malaysia	22.2	92	95
Myanmar	47.7	72	64
Philippines	75.7	87	83
Singapore	4.0	100	100
Thailand	62.8	80	96
Vietnam	78.1	77	47

Efforts have been made by governments to minimize water pollution in recent years. Thailand has established the National Environmental Act in 1992, and the number of municipal wastewater treatment plants has substantially increased. After completion of construction/upgrading work, the total treatment capacity throughout the country is expected to be 2.8 million m^3/day. However, based on the current population, the existing treatment capacity can handle only about 20% of the total municipal wastewater (Polprasert *et at.*, 2001).

2.3 Water supply and sanitation coverage in Asian countries

The present situation of sustainable water supply and sanitation coverage in the SEA region in 2000 is presented in Table 2. The accessibility of the water supply and sanitation services in Lao PDR and Cambodia is remote. Other SEA countries still need much effort to increase coverage of sanitation services. The table also shows that about two-thirds of Cambodians and Laotians do not have access to safe and sustainable drinking water and about three-quarters of the same people do not have access to sustainable sanitation facilities. Singapore is the only country which has achieved the target of sustainable water supply and safe sanitation for all. The Lao PDR, with the highest quantity of water resource available among SEA countries, has not been able to exploit the resource to provide water supply to its people. Therefore, it is imperative that from the experience of Lao PDR, water resources management needs to give priority to the sustainable water supply.

3. Southeast Asian water environment issues

1. Water management
Appropriate exploitation of water resources in the SEA countries is a big challenge, as problems are seen in harnessing available resources. National water resources management policies should take into account integrated water resources management giving top priority to water supply and sanitation.

2. Accessibility to the water supply and sanitation services
SEA Region is heading towards industrialization and modernization at the cost of the environment. There is more demand for water supply and waste management in new urban areas. The naturally available water resource has not been successfully tapped to provide water supply facilities to the people. In some countries the safe and sustainable water supply coverage is one in three and sustainable sanitation coverage is one in four. Thus efforts are needed to upgrade these facilities and meet the MDGs (Millennium Development Goals).

3. Application of appropriate technologies
The technology to be applied for the treatment of water supply and sanitation systems depends on the level of affluence of the people demanding the services. The form of appropriate technology may differ depending on specific situations. For example, an improved pit latrine might be a hygienically appropriate technology in rural areas of poor incomes, but certainly middle-class families in the same area will prefer a better facility e.g. pour-flush toilets. It is thus desirable to try to establish technology that enhances the sustainability of water supply and sanitation. The water supply and sanitation sector have to be developed to achieve functional, financial and environmental sustainability.

4. Financial and regulatory framework
Many of the least economically developed SEA countries are facing problems in maintaining the water supply and sanitation facilities, as the people consider water as nature's gift. Improved water supply should be considered as an economic commodity. For the effective development of the water supply and sanitation sectors, services must be charged with appropriate criteria. The people must be made aware of the fact that sector services need investment. In rural areas, the water supply and sanitation sector may be owned by the community to properly operate and maintain the facilities. The habitual tendency shows that

as long as the water supply and sanitation services are free, the consumption pattern is high. The charging system would generate revenue for the sustainability of the project and the service providers are bound to perform effectively.

4. Future Strategies

To properly address the problems of water management, maintaining the water quality and the water supply and sanitation sector development, SEA countries need to develop national strategies. In view of the water environment issues a set of strategies are given here:

1. Integrated water resources management

As the fresh water withdrawal is increasing to expand irrigation coverage for increasing agricultural products, the water resources become more stressed. As discussed earlier, some countries in SEA could face water scarcity in such a situation. Thus new technologies in irrigation systems, such as drip irrigation, could be an important strategy in water management. Consequently, interstate regional river basin management and integrated water resources management will be necessary in the future.

It is essential that the formulation and implementation of time-bound targets for the provision of water supply and sanitation for the forthcoming years be carried out within the framework of an integrated approach to water resources management. Priority attention should be given to the provision of water supply and sanitation within the context of on going efforts made by national governments and the international community towards poverty alleviation.

2. Developing urban water district

Urban Water District implies an integrated approach in planning and implementing all water-related urban activities and components, including water and wastewater management, solid and hazardous waste, economic and social factors, as well as community and governance (Ujang and Buckley, 2001).

3. Innovation on appropriate technologies and waste minimization

The technology to be applied for the treatment of water supply and sanitation systems depends on the level of affluence of the people demanding the services. The form of appropriate technology may differ depending on specific situations. For example, an improved pit latrine might be a hygienically appropriate technology in rural areas of poor incomes, but certainly middle-class families in the same area will prefer better facilities e.g. pour-flush toilets. It is thus desirable to try to establish technology that enhances the sustainability of water supply and sanitation. The water supply and sanitation sector have to be developed to achieve functional, financial and environmental sustainability.

To improve the sanitary conditions in urban areas, some innovative concepts must be adopted. As is well known, the more the water usage, the more the chance of polluting another water source. Water usage must be done in such a way that the chain of transmission of water-borne diseases is broken down. The options that can be applied for sustainable water supply and sanitation facilities are: waste minimization, reuse and recycling. Waste minimization is the process that reduces the degree of potential pollution in the receiving environment. Source reduction, in plant recycling etc. are the methods of waste minimization. In urban sanitation, proper treatment of wastewater before discharging it to the receiving water bodies reduces the pollution level, thus providing the opportunity for downstream use. Reuse of wastewater and septage (septic tank sludge) could break the chain of water pollution. Septage and wastewater can be reused as nutrients for agricultural practice and land reclamation. This also reduces the stress on water resources for irrigation purpose. Recycling is now a promising option in the urban water supply to reduce fresh water withdrawal. Grey waters are now being recycled in Japan after simple but appropriate treatment in buildings for secondary purposes such as flushing toilets, gardening, cleaning. Public water supplies are used only for drinking, bathing and culinary purposes.

4. Partnership and empowerment

The involvement of all sectors is the key to success for the development of water supply and sanitation sectors. Past experiences in the 1990s indicate the importance of the involvement of the community in all phases of water supply and sanitation project development, as well as the involvement of the private sector in financing the project. In several cases, the role of the government has now been keenly addressed as the facilitator not the implementer.

5. References

Arellano, F. A. (2002) The Philippines: Assessing Quality and Quantity, In: *IWA Year Book 2002,* International Water Association, UK, pp:15-16.

Polprasert, C., Koottatep, T. and Visvanathan, C. (2001) Water pollution in Thailand: Appropriate treatment technologies for urban and rural areas, *Proc. IWA Conf. on Water and Wastewater for Developing Countries*. Kuala Lumpur, Malaysia, Vol. 1, pp:64-71.

Bangkok SoE (2001) *Bangkok State of the Environment*, United Nations Environment Program, Nairobi, 2001.

SoE Vietnam (2001) State of the Environment Vietnam 2001, United Nations Environment Program, Nairobi.

UNDP (2000) *Human Development Report*, United Nations Development Program, New Yourk, 2000.

UNDP (2004) *Human Development Report*, United Nations Development Program, New Yourk, 2004.

Wisjnoprato (2001) Wastewater Management in Indonesia: Problems on river pollution, *Proc. IWA Conf. on Water and Wastewater for Developing Countries.* Kuala Lumpur, Malaysia, Vol. 1, pp:56-63.

Ujang, Z. and Buckley, C. (2001) Water and Wastewater in Developing Countries: Present Reality and Strategy for the Future. *Proc. IWA Conf. on Water and Wastewater for Developing Countries.* Kuala Lumpur, Malaysia, Vol. 1, pp:1-9.

Section 1

Environmental Ecological Management

Biodiversity and Fisheries in the Mekong River Basin

D. Coates[1], C. Pongsri[1], O. Poeu[1], U. Suntornratana[1],
N. T. Tung[1] and S. Viravong[1]

1) Assessment of Mekong Fisheries Component
The MRC Programme for Fisheries Management and Cooperation
Mekong River Commission Secretariat
e-mail: chumnarnp@mrcmekong.org

Abstract

The fishery of the Lower Mekong Basin is one of the most abundant river fisheries in the world. It is of vital importance to the 55 million people who live in the basin. Not only do millions of people earn much needed income from catching, preserving and marketing fish and other aquatic products, they also depend on the fishery as their main source of animal protein. Although the Mekong River Commission (MRC)'s research has demonstrated that the fishery is still in good condition, this could change quickly if it is not well managed and developments in other sectors have adverse impacts on aquatic life.

In the report that follows, the authors propose that in the Mekong the importance of the fisheries is the major argument for protecting the biodiversity. They also discuss threats to biodiversity from within and outside the fisheries sector and measures that should be undertaken to sustain biodiversity.

Keywords: biodiversity, inland fishery, Mekong River Basin

1. Introduction

Within the Mekong River system there are flourishing fisheries which exploit a large number of species. Estimates indicate that approximately 120 fish species are commercially traded, although the bulk of the fishery is based on 10-20 species.

There is much current interest in biodiversity and the need to sustain it. Many people, however, often regard fisheries as a threat to biodiversity because of widespread over-exploitation of stocks, the use of destructive fishing gears, large by-catches (killing of unused species) and general mismanagement of resources. But is this true? This paper argues that fisheries are, in fact, not the villain but perhaps the most important ally in the quest for sustaining aquatic biodiversity in the Mekong.

Large tropical river ecosystems have immense value both in terms of high biodiversity and the numbers of people that depend upon that biodiversity for their livelihoods. Unfortunately, all too often, large rivers lose their biodiversity as the environment suffers due to the many demands for water and other resources.

Popular belief is that amongst the world's aquatic environments, it is the sea, and in particular coral reefs, where biodiversity is threatened most. The facts, however, disprove this. The loss of species from freshwater environments is far greater. The IUCN Red List for bony fishes, a database on extinct, threatened and vulnerable species, lists no marine species as extinct, in comparison with 96 freshwater species. Losses in other categories of fish confirm that the threat to biodiversity is much greater in freshwater environments. This is caused by environmental degradation (habitat loss and pollution), not over-exploitation and is generally far worse in freshwaters.

There are some grounds for optimism with respect to freshwater environments, as understanding is growing that freshwater biological resources can be sustained where they are still significant. This is not technically difficult, but it does require awareness and commitment. Serious efforts are also beginning to be taken to rehabilitate degraded rivers, and this is being met with considerable success. Central to this, in developing countries, is the growing appreciation of the importance of freshwater fisheries to the livelihoods of people in rural areas.

2. Materials and methods

A combination of methods were used to understand the biodiversity and fisheries in the Mekong River Basin. These included conventional methods, such as biological studies and marketing surveys, and non-conventional methods, such as desk studies and local ecological knowledge surveys.

3. Results and discussion

3.1 Aquatic biodiversity in the Mekong

There are at least 1200 species of fish, and possibly as many as 1700, living in the Mekong Basin. High diversity is also exhibited by other aquatic animal and plant groups. It is important to note that "fisheries" in the Mekong include other animals and plants as well as fish. Possibly as much as 30 percent of production from the fishery comes from non-fish sources. Although poorly studied, this group of miscellaneous animals and plants is very diverse and important.

In rivers such as the Mekong, which have monsoon climates, tremendous seasonal changes also drastically increase the range and nature of habitats available. These are most marked on the river's floodplain. Every year, the spread of the Mekong's floodwaters influence the basin's aquatic life. This flooding also produces different habitats, in different places, at different times within the year. Changes occur between years due to differences in the timing, extent and duration of flooding. Natural variations in river hydrology, both within and between years, are very important in sustaining ecosystem diversity.

Understanding the role of ecosystem variability (including hydrology) in sustaining the Mekong's rich biodiversity is crucial. Development activities in a river system almost always result in the simplification, or even obliteration, of ecosystem diversity. These disturbances appear to be by far the greatest threat to sustaining aquatic biological resources in the Mekong Basin.

3.2 The fishery resource and its exploitation

Official government statistics still significantly undervalue the Mekong's fishery production in the four Lower Mekong Basin countries. It is between 2.6 and 21 times higher than official statistics suggest. The Lower Mekong Basin produces an estimated two million tonnes of fishery products each year. Of this, less than 10 percent arises from aquaculture.

While finfish dominate the overall catch in the Mekong, other aquatic species are also important, especially with the small-scale fisheries. In Lao PDR, for example, approximately 30 percent of the catch is composed of molluscs, crustaceans, insects, amphibians and reptiles. The region also has a cottage industry producing dried "seaweed" (algae gleaned from rocks in clearer mountain streams).

Throughout the basin, rice fields are another important source of fish and aquatic species, unless these are impacted by excessive pesticide use.

Fisheries vary greatly from region to region, depending on the availability and access to markets. In areas with more abundant resources and greater human population (*e.g.*, the Thai Mekong and central and southern Cambodia), larger commercial fisheries are dominant. In the delta area of Viet Nam, which has the highest population density, the fishery is likely to be the most heavily exploited. In some areas, the fishery takes on industrial proportions, especially in the Tonle Sap in Cambodia.

At peak times, single gears in the bagnet or *dai* fishery in Cambodia and Viet Nam can land up to 0.5 tonnes of fish during a 15 minute period. Scenes here are reminiscent of the days of plenty in ocean fisheries, although catches are highly seasonal and influenced greatly by the lunar cycle. There are up to 68 units of *dai*s located along the Tonle Sap River alone, with large numbers operating also in Viet Nam. Over 40 species can be found in a single catch.

Other spectacular fisheries include the large barrage systems which use fences to channel fish towards various traps or holding pens. This seasonal abundance means that large numbers of people are employed in fisheries and associated activities such as gear making, transportation, fish processing and marketing.

In Cambodia, the best fishing grounds are allocated through a bi-annual auction of fishing lots. Lot "owners" not only have fishing rights, but they protect the supporting environment with security staff. The revenue earned from fishing lots makes the Fisheries Department in Cambodia one of the most prominent agencies in the country. There is much interest in these resource management systems as mechanisms for conserving the environment and biodiversity. Unfortunately, the current fishing lot system is criticised for being socially unjust. The wealthier individuals can afford to purchase lots, whilst the poor are often excluded. Conflicts are exacerbated by an auction system which is less than transparent. The Government of Cambodia has recently attempted to improve this system. This is being achieved, in part, by abolishing some of the lots, and again allowing public access in those areas.

An interesting question then emerges – will the resulting access by poor people result in resource destruction? And, if so, what is the balance between the rights of poor communities and losses to biodiversity? It is argued here that the two considerations (poverty and biodiversity) are in fact closely related. Poor people are stakeholders in sustaining biodiversity too - in fact, they are the most dependent upon it. The solution to these problems rests not so much in access to resources but in empowerment to manage them. Fishing "lots" offer a management system with considerable proven benefits for biodiversity. Many countries are interested in adopting similar approaches to resource management. The requirement, therefore, is to find a way of making the system more just and socially acceptable.

Much of the industrial/commercial fishery in the Mekong is based upon exploitation of fishes that are migrating from flood season feeding grounds to their dry season habitats as flood waters recede. The resulting seasonal peak in catches means that far more fish are caught than local communities can consume. In response, a low-input technology industry has developed which turns surplus fish into fish paste, fish sauce and other products. Production of these products spreads the benefits of seasonal production over the full year, and caters for the demand in seasons when fish are less plentiful. Remarkably, catches along the Tonle Sap River can be so high in the peak season, that when prices drop due to oversupply, fishing actually ceases. This is quite possibly the world's last remaining example of oversupply in a fishery.

Impressive as industrial fisheries are, the fishery is still dominated by individual small-scale operations. Survey results have shown that from 64 to 93 percent of rural households in the Lower Mekong Basin are involved in fisheries, both for their own consumption and for sale.

3.3 The role of aquaculture

Aquaculture is currently dwarfed in comparison to capture fisheries, although it is more prominent in some areas than others. Commercial aquaculture (including industrial production for the luxury and export markets) is increasing. In the delta of Viet Nam, cage culture of pangasiid catfish is a growing industry, as is the culture of snakeheads (*Channa micropeltes*).

Much of the production of higher-value fish relies on the provision of feed made from lower grade fish which are caught from the river and other wetlands. Thus, some aquaculture activities are net consumers of fish (they use up more fish than they produce). Culture and capture are further linked because much of the stock (seed) for commercial aquaculture is taken from the wild. The motivation for aquaculture tends to be profit rather than food.

Smaller-scale aquaculture mixed with farming is particularly prominent in the Thai part of the Mekong Basin, and in the delta in Viet Nam. Not surprisingly, small-scale aquaculture tends to develop best in areas which are away from the main fishing grounds and from the floods.

Fisheries and farming are combined when fish are raised in the waters covering rice fields. Traditionally this practice relies on recruiting wild stocks of fish and other animals. This is a very significant activity in the Mekong. Only recently has production been enhanced through stocking of fish into fields. The intensification of rice farming, particularly the excessive use of pesticides is, however, undermining this additional source of income and food for rural farmers. Modern agricultural practices significantly reduce the biodiversity of products obtained from rice-growing areas.

While aquaculture will clearly play an increasingly important role in the Mekong, there are serious misconceptions about its ability to increase total fish production. The ability of aquaculture to do this will depend upon primarily whether the capture fisheries themselves are sustained. Without appropriate management, capture fisheries are likely to decline much faster than aquaculture can expand, obliterating gains made from fish farming. Another misconception about aquaculture is that it is somehow more environmentally appropriate than catching fish in rivers and other wetlands. Unfortunately, this is not the case.

The aquaculture sector itself is very diverse in the Mekong, but it is exhibiting the usual characteristics that have occurred in agricultural development - that is, simplification of the production system through increasing control over and modification of the environment. Aquaculture can also cause major environmental disturbances. In the Mekong, major impacts of aquaculture include habitat degradation (especially so for coastal shrimp farming), over-exploitation of wild stock for seed and the poor, or sometimes non-existent management of the use of exotic species and biodiversity of native species.

3.4 Threats to aquatic biodiversity in the Mekong

Threats arise from two sources: the impact of fishery activities themselves and the impact arising from outside the fishery sectors. The direct threats to biodiversity posed by the fishery sectors include: the use of destructive fishing methods (explosives, poisons and electrocution); exploitation of fish at vulnerable stages, such as at spawning times; and fishing in sensitive areas such as spawning grounds.

The highly migratory fish stocks are more vulnerable to over harvest. Large numbers of these fish migrate at the same time, making them vulnerable to intensive fishing, and especially to large-scale commercial operations. Examples include the *dais* on the Tonle Sap River in Cambodia. Licenses have a short duration and this encourages many fishermen to over-exploit the fishery. Fish migrations are often associated with spawning, and concentrations of fish at spawning sites also make stocks vulnerable to over-fishing at the critical spawning period.

Unless properly managed, the release of live animals into the wild can have serious impacts upon biodiversity. This happens primarily through aquaculture when animals are either deliberately released into the wild through fish stocking activities, or when they "escape" from fish farms (a common problem when ponds are flooded seasonally). Biodiversity can be adversely affected when exotic (non-native) species compete with native species in the wild.

Caution must also be exercised when using native species. The natural range of native fish can be quite limited within the basin - they do not all occur everywhere. Some species of Mekong native fish are in fact "exotic" (*i.e.*, naturally foreign) to many parts of the basin. In addition, fish stocks can be composed of a number of distinct populations living in different areas, each with significantly different genetic characteristics. This phenomenon is an example of genetic diversity and it is important. It is probable, for example, that several species have different stocks above and below the Khone Falls.

Impacts upon aquatic biodiversity arising from outside the fishery sectors include changes in the environment and habitats in rivers, brought about by the impact of development. These can have significant impact upon biodiversity and include 1) destruction of local spawning grounds or dry season refuges as a result of stream bed dredging, removal or alteration of vegetation, and bank modification; 2) local changes in the quality (*e.g.* pollution) and quantity of water available as a result of storage in dams and abstraction for irrigation; and 3) the construction of barriers (dams, weirs, diversions) which, apart from the local environmental disturbances they might cause, act as physical barriers to fish migrations.

Essentially, these influences undermine ecosystem integrity. They invariably lead to a simplification of the river ecosystem and a significant loss of ecosystem area. The loss of river floodplains is a significant cause for concern. These changes result in significant reductions in both aquatic productivity and biodiversity.

Since industrialisation and urbanisation are still quite limited, by comparison to some areas, these have had little impact on the Mekong so far, except locally. However, as populations grow and development pressures increase, it is likely that aquatic life will be affected. The negative consequences of deforestation, inappropriate agriculture, road construction, hydropower and other forms of development are already evident. In the Mekong, the most pressing problem is one of water quantity, not water quality. A second major concern is the loss of riparian vegetation cover, and, in particular, the rapid loss of flooded forests, which provide a crucial aquatic habitat for fishes.

In 1997, the MRCS listed the impact of fishing on biodiversity as "slight", but the activities of most other sectors as "high" (through their impact on the environment). Most commentators on the state of river fisheries agree that the major threats are environment based.

4. Conclusions

Capture fisheries are not the major threat to the biodiversity in rivers - environmental degradation is.

The important role that river fisheries play in maintaining aquatic biodiversity must be promoted more widely. Emphasising the importance of fisheries to livelihoods and food security provides the strongest and most relevant argument for improved management of the aquatic environment.

There are considerable differences between river fisheries and marine fisheries in their methods of exploitation, participation, spread of economic benefits and impacts on biodiversity. Conclusions drawn from marine fisheries are therefore often not applicable to inland fisheries.

Fishery activities can have negative impacts upon biodiversity. Improved management of exploitation, by moderating the use of unsustainable fishing practices, should centre on the promotion of co-management approaches. Management initiatives need to be prioritised, focusing on those species and habitats under greatest threat.

The fishing "*lot*" system should be properly and scientifically evaluated to determine if it is an effective tool for sustaining biodiversity. Studies should include consideration of whether social equity and sustaining biodiversity are necessarily mutually exclusive. There is considerable interest internationally in this system as a way of improving environmental management.

The use of local ecological knowledge as both a research tool and a mechanism for improving participation in management should be promoted more widely. There should be increased recognition of the importance of this knowledge base for biodiversity-related subjects.

Management measures for the introduction or transfer of exotic species or strains should include consideration of genetic diversity and, in particular, the biodiversity of wild resources. This can be achieved through the development and, more importantly, implementation of workable codes of practice using pre-introduction assessments of the movement of exotic species .

Aquaculture should be managed to avoid significant negative impacts upon biodiversity. Negative impacts can result from the escape of exotic species, and from practices that destroy productive habitats (such as converting mangrove forests to shrimp ponds).

There should be more widespread recognition and use of existing instruments for promoting sustainable management of fisheries, including attention to biodiversity needs. In addition to the 1995 Agreement on Sustainable Development of the Mekong River Basin, these instruments include the Code of Conduct for Responsible Fisheries which FAO prepared in 1995, the Convention on Biological Diversity (in particular, promotion of its implications for fisheries management), and the Convention on Migratory Species.

Reviews, guidelines and promotional materials should, where necessary and appropriate, be written and disseminated in a language and medium that can be easily understood by target audiences (including resource users). This includes references to what biodiversity is, why it is important and how it can be affected, negatively or positively, by management actions.

5. Acknowledgement

The work reported here has come out of the Assessment of Mekong Fisheries (AMF) component of the Mekong River Commission's Fisheries Programme. AMF is based in the host research centres, *i.e.* Department of Fisheries, Cambodia, Living Aquatic Resources Research Centre, Lao PDR, Udon Thani Inland Fisheries Research and Development, Thailand and Research Institute for Aquaculture No. 2, Viet Nam. The on-going support of the managerial, administrative, technical and support staff at these centres for the work of the MRC Fisheries Programme is greatly appreciated.

6. References

Baird I.G., and Mounsouphom B. (1997). Distribution, mortality, diet and conservation of Irrawaddy dolphins (*Orcaella brevirostris* Gray) in Lao PDR, *Asian Marine Biology* 14: pp. 41-48.

Bao, T.Q., Bouakhamvongsa K., Chan S., Chhuon K.C., Phommavong T., Poulsen A.F., Rukawoma P., Suntornratana U., Tien D.V., Tuan T.T., Tung N.T., Valbo-Jorgensen J., Viravong S., and Yoorong N. (2001). Local knowledge in the study of river fish biology:

Experiences from the Mekong, Mekong Development Series No. 1, Mekong River Commission. Phnom Penh, pp. 22.

Balzer, T., Balzer, P. and Pon, S. (2002). Traditional use and availability of aquatic biodiversity in rice-based ecosystems, CD-ROM. FAO, Rome.

Coates, D. (1995a). Inland capture fisheries and enhancement: status, constraints and prospects for food security, Report of the International Conference on the Sustainable Contribution of Fisheries to Food Security, 4-9 December, 1995, Kyoto, Japan, Government of Japan, Document No. KC/ FI/95/TECH/3, pp. 85.

Coates, D. (1995b). Implementation of the EIFAC/ICES code of practice: experiences with the evaluation of international fish transfers into the Sepik River Basin, Papua New Guinea.pp. 160-174, In Philipp D.P., Epifanio J.M., Marsden J.E. and Claussen J.E. (Editors). *Protection of aquatic biodiversity*, Proceedings of the World Fisheries Congress, Theme 3. Oxford and IBH Publishing Co. Pvt. Ltd., New Delhi.

Coates, D. (2001). Biodiversity and fisheries management opportunities in the Mekong River Basin. "Blue millenium - managing global fisheries for biodiversity", GEF-IDRC, 3-7 July 2001, World Fisheries Trust, Victoria, Canada. CD Rom.

Coates, D. (2002). Inland capture fishery statistics of Southeast Asia: current status and information needs, Asia-Pacific Fishery Commission, Bangkok, Thailand, RAP Publication No. 2002/1, FAO, Bangkok. pp. 114.

Dudgeon, D. (1992). Endangered ecosystems: a review of the conservation of tropical Asian Rivers, *Hydrobiologia* 248: pp.167-190.

FAO. (2000). State of world fisheries and aquaculture, www.FAO.org.

MRCS. (1997). Mekong River Basin diagnostic study, Final report, MKG/R.97010, Mekong River Commission, Bangkok.

Rainboth, W. J. (1996). Fishes of the Cambodian Mekong, *FAO species identification field guide for fishery purposes*, FAO, Rome. 27 colour plates, pp. 265.

Sverdrup-Jensen, S. (2002). Fisheries in the lower Mekong Basin: status and perspectives, MRC Technical Paper No. 6, Mekong River Commission, Phnom Penh. pp. 103.

Sjorslev, J. G. (Editor). (2001). A fisheries survey of Luang Prabang province, *Lao PDR*. LARRC Research Report No. 0001, Vientiane, pp. 54.

van Zalinge, N. (2003). Data requirements for fisheries management in the Tonle Sap. pp. 67-74, In: *New approaches for the improvement of inland capture fishery statistics in the Mekong Basin*, RAP publication 2003/01, FAO, Bangkok, pp. 145.

World Commission on Dams. (2000). Dams and development: a new framework for decision-making, Earthscan Publications, London. pp. 404.

Interactions between Water and Economic Development: the Phong Watershed Case Study, Thailand

S. Sthiannopkao and S. Takizawa

The University of Tokyo
Department of Urban Engineering
Graduate School of Engineering
7-3-1, Hongo, Bunkyo-ku,
Tokyo 113-8656 JAPAN
e-mail: suthi@env.t.u-tokyo.ac.jp and suthisuthi@hotmail.com

Abstract

The objective of this paper is to investigate the interactions between water resource development and the national economic development policy in the Phong watershed, located in the northeast of Thailand. The results reveal that the national economic development policy in the northeast of Thailand is to develop water resources as the main strategy aimed at increasing water availability, improving standards of living and river ecosystems. As a result, agricultural sectors, agro-industries and power generation have been aggressively developed. However, economic development has also led to a host of interrelated problems such as water shortages, water pollution and social conflicts concerning water. These kinds of problems have been evident in the Phong watershed since the Ubolratana dam was constructed in 1966 to serve the regional economic development plan.

Keywords: water resource development, national economic policy, Phong watershed

1. Introduction

The Phong watershed is an important water source for the upper northeastern provinces of Thailand. It is a tributary of the Chi River system, which is one of the Mekong's many sub-watersheds. Its watershed covers about 1.5 million hectares, extending into five main provinces of Chaiyaphum, Loei, Khon Kaen, Petchaboon and Nong Bualumphu. According to Khon Kaen University, KKU, (2003), the Phong watershed is divided into three major sub-watersheds: upstream, middle and downstream. The upstream watershed area is again divided according to the four main rivers flowing into the Ubolratana Dam. There are the Choen River, the Phrom River, the Pa Niang River, and the upstream Phong River. The middle watershed area consists entirely of the Ubolratana Dam. The downstream watershed area comprises the Phong River and its tributaries, downstream from the Ubolratana Dam to the Mahasarakham weir on the Chi River and the short river section from Ban Tha Pra by the Chi River to its confluence with the Phong River.

According to the Bank of Thailand (2004), use of natural resources during the economic development of the northeast of Thailand was extravagant. Soil, water, and forest resources have been depleted at staggering rates. More than 1.6 million ha of forests were destroyed during the past 20 years, from approximately 4 million ha in 1975 to only 2.24 million ha in 1995. 0.32 million ha were encroached and turned to paddy fields and 0.64 million ha were turned into crop fields. The KKU (2003) reports that the transformation of land use from forest area into agricultural area for the past 30 years in the Phong watershed has left only 10-15% of the total watershed as natural forest and caused many problems such as flooding, soil erosion and sedimentation, soil salinity, and water quality. Such a situation causes alterations to the watershed in terms of both water quantity and quality leading to an imbalance of water demand and supply within the watershed. An example of this phenomenon took place in 1993 when a water shortage occurred in the downstream area as the Ubolratana Dam received a low amount of water together with inappropriate water management. Such water shortage problems lead to a significant impact on water use for agriculture and industry. Moreover, discharge of effluents from pollution sources within the Phong watershed caused water quality deterioration. The lower Phong River has frequently faced water pollution problems. Water pollution caused adverse effects on aquatic organisms as indicated by the phenomena of fish and shellfish death recorded during 1993-1995.

The statement outlined above indicates that there is a connection between the state's policy of serving the country's economic development by exploitation of its own natural resources and the problems of environmental pollution. This is in accordance with the study of Julal and Rogers (2002) which reported that the cost of remediation of the environmental damage caused by economic development in Thailand in 1990 stood at US$ 820.68 million. This accounted for 0.83% of GDP. In Thailand, water quality has been deteriorating due to human activities, often activities promoted by the state's economic planning. In addition, the impact of water pollution is vast and severe, especially upon the people of the northeast who depend heavily on rivers as a supplier of food, for their daily lives. This paper aims to investigate the interactions of water and human activities under the driving force of national economic development in the Phong watershed.

2. General background

2.1 The northeast economy (www.bot.go.th)

The Northeast economic development: Regional development in a recent two-decade period has raised this area up a level. The average annual economic growth was 6.7%, whereas the average inflation rate was 5.7% and the average annual per capita income increased seven fold from 2,972 baht in 1975 to 24,331 baht in 1995. The Northeast economic expansion was nine fold during these twenty years, from roughly 49,902 million baht in 1975 to 498,601.6 million baht in 1995. It should be noted that the national per capita income increased 14 times and the national economy expanded 7 times. At present, the northeastern economy is greater than in other regions except Bangkok and its vicinity, which accounts for roughly half of the country.

Agriculture: Relatively much development has been achieved in agriculture. There are 182 large, medium, and small water resources in operation in the northeast, totaling a capacity of 3,670 million cubic meters. These water resources are distributing water to approximately 0.48 million ha of agricultural area. Additionally, there is another 0.42 million ha of farmland receiving water from small-scale irrigation projects and power pumping projects. A project to develop the 0.32 million ha of Thung Kula Rong Hai is also underway, along with the Khong-Chi-Mun project, which is a long-term development project. This project, when completed, will be irrigating about 0.80 million ha of farmland, which should result in more developed cropping and husbandry. Farm produce per ha should, in turn, be increased. Crops nowadays are more diversified than what used to be grown which included only rice, cassava, kenaf, and corn. More varieties of vegetables, fruits, perennials, rubber, and grapes for wineries are increasingly being grown on suitable land in the region. Development has also been extended to production and marketing systems, relying more on contract farming. Nevertheless, productivity per ha of major crops is still low when compared to the country's and other countries' productivity.

Industry: Industry is increasingly becoming important. In the past, the northeast industries used to include only rice mills, tapioca mills, and cassava baling factories. At present, labor-intensive factories are becoming more and more common. There are continuing industries and sub-contract production from factories in Bangkok and the eastern seaboard, such as electronics, garments, basic metallurgy, manufacture of farm products, ornaments, transportation equipment, paper, and plastics. Industrial factories increased about 22 times, from 1,908 factories in 1975 to 44,033 factories in 1995, 34,312 being rice mills.

2.2 Land use in the Phong watershed

(1) Upstream sub-watershed

Land use consists of agriculture, forestry, community areas and industry. It encompasses an area of 1,208,900 ha, which is 79.59% of the total area of the Phong watershed. Land use in terms of area is as shown in Table 1.

Table 1. Land use in the upstream sub-watershed

Land use	Area (ha)	Watershed area accounted (%)	Remarks
1.Reservoir	30,830	2.55	-
2.Agriculture	774,295	64.05	Rice, cassava, sugar acne, corn
3.Forest	330,392	27.31	Evergreen and non-evergreen forests
4.Community and household industry	40,020	3.3	Rice noodle production and brick manufacture
5.Industry	-	-	Sugar mills
6.Others	33,366	2.76	Grasslands and abandoned areas

(2) Downstream sub-watershed: Phong river area downstream from the Ubolratana Dam to its confluence with the Chi River

♦ Ubolratana Dam to Nong Wai Irrigation Weir

Land use in this sub-watershed consists of industry, community, agriculture and forestry. These are summarized as follows:

- Industrial area covers 236 ha. There are two large industries within the area, which are the Phoenix Pulp and Paper Mill and the Combined Cycle Thermal Power Plant.
- Significant communities are Ubolratana Sanitary District, Nam Phong Sanitary District and Ubolratana District.
- The agricultural area directly affecting the Phong River is 20,000 ha. Agricultural land consists of rain fed paddy fields, upland crops and other vegetable areas. Typical crops cultivated include rice, cassava, sugarcane, jute, eucalyptus, watermelon and vegetables.
- Much of the forested areas have been transformed into agricultural and industrial areas.

- ♦ Area from Nong Wai Irrigation Weir to Ban Na Piang
  - There are a number of large and medium scale industries as follows: sugar mill, paperboard mill, distillery mill and tapioca mill.
  - Significant communities are Wang Chai Sanitary District, and Amphor Nam Phong
  - Agriculture; the area under irrigation is 31,880 ha. Rice cultivation is carried out in the rainy season. Some rice and vegetables are cultivated in the dry season. Other areas are used for seed production and cantaloop cultivation.
- ♦ Area from Ban Na Piang to Mahasarakham Weir
  - Communities and household industries, such as Khon Kaen Municipality, Thai rice noodle production, meat processing and brick production.
  - Agriculture; the large irrigation area is provided by both the Nong Wai Irrigation Project and also by the Department of Energy Development and Promotion.
  - There is the Mahasarakham Weir Project of the Department of Energy Development and Promotion.
- ♦ Ban Tha Pra to the Chi River Confluence with the Phong River

Water quality in this section generally falls in class 3. The main problem for this water is a high natural chloride content present due to geological characteristics of the area. There is little industrial effluent to adversely affect water quality.

3. Problems identification and discussion

Based on the study of KKU reports, the linkages between water resources and human activities resulting from the national economic development policy are drawn (Figure 1). The concept of Figure 1 can be explained as the followings:

3.1 Water resources development as a strategy for a regional economic development

Human activities, which are mainly based on agriculture, agro-industries, communities and power plants, encourage water use in the Phong watershed. In the case of agriculture, water is needed for the purpose of irrigation in order to increase agricultural production. For industries, communities and power plants, water is required for use as a raw material to produce industrial products, water supply and to generate electricity. It can be seen clearly from Figure 1 that the main target of all human activities using water as a raw material is to serve the demands of end products and services both domestically and internationally. All of these human activities are driven by the national government policies for regional economic development. In order to achieve this target, the Ubolratana Dam was built in 1996. Since then, the Phong watershed has been the focus of intensive development and management initiatives run by different governmental offices in the hope of stimulating regional economic development. Sneddon (2002) reports that some of the planned effects of the dam included an irrigation system covering 50,000 ha and a power station generating 65 GWh of electricity per year. Additional effects, unforeseen by the state authorities, included a sudden influx of families taking advantage of a highly productive, but unmanaged, reservoir fishery.

Moreover, the dam transformed a river characterized by marked seasonal changes in flow rates and flooding patterns into an impounded river characterized by a large reservoir environment, reducing flooding in formerly flood-prone areas, and producing much more constant flow rates in the river throughout the year which created an enabling environment for industrial development in the watershed. According to KKU (2003), there are 6 main industries (pulp and paper, agro-industries and a power plant) using the Phong river water at the average total amount 1,689,600 m^3/month. This is in accordance with Sneddon (2002) who reported that the Thai state had long targeted the northeast as a region amenable to agro-industrial development. Government initiated special incentives for situating a plant in the Phong watershed included land purchasing rights, introduction of skilled alien technicians, and exemptions from import duties for equipment and income taxes.

Furthermore, human activities driven by the economic development plan have also caused changes of land use in the Phong watershed. The agricultural lands increased between 1994-2002 in the provinces of Chaiyaphum, Nong Bua Lumphu, Khon Kaen and Petchaboon at average annual rates of 0.77, 3.28, 4.71, and 10.66%, respectively. During 1988-1999, all of the forest areas in all the provinces of Chaiyaphum, Nong Bua Lumphu, Khon Kaen, Petchaboon and Loei had been decreased at the average annual rate of 0.84, 1.00, 2.34, 3.21, and 2.07%, respectively (Office for Agricultural Economics, 2004) (Figures 2 and 3). Agriculture has long been a dominant economic sector in the northeast region of Thailand. The GPP data from 1993 to 2000 of the main provinces situated within the Phong watershed proves this fact. The agriculture sector has contributed to more than 55% of the provincial economies of Loei, Nong Bua Lumphu and Petchaboon, which is significantly higher than the other two sectors of manufacturing and wholesale (Table 2).

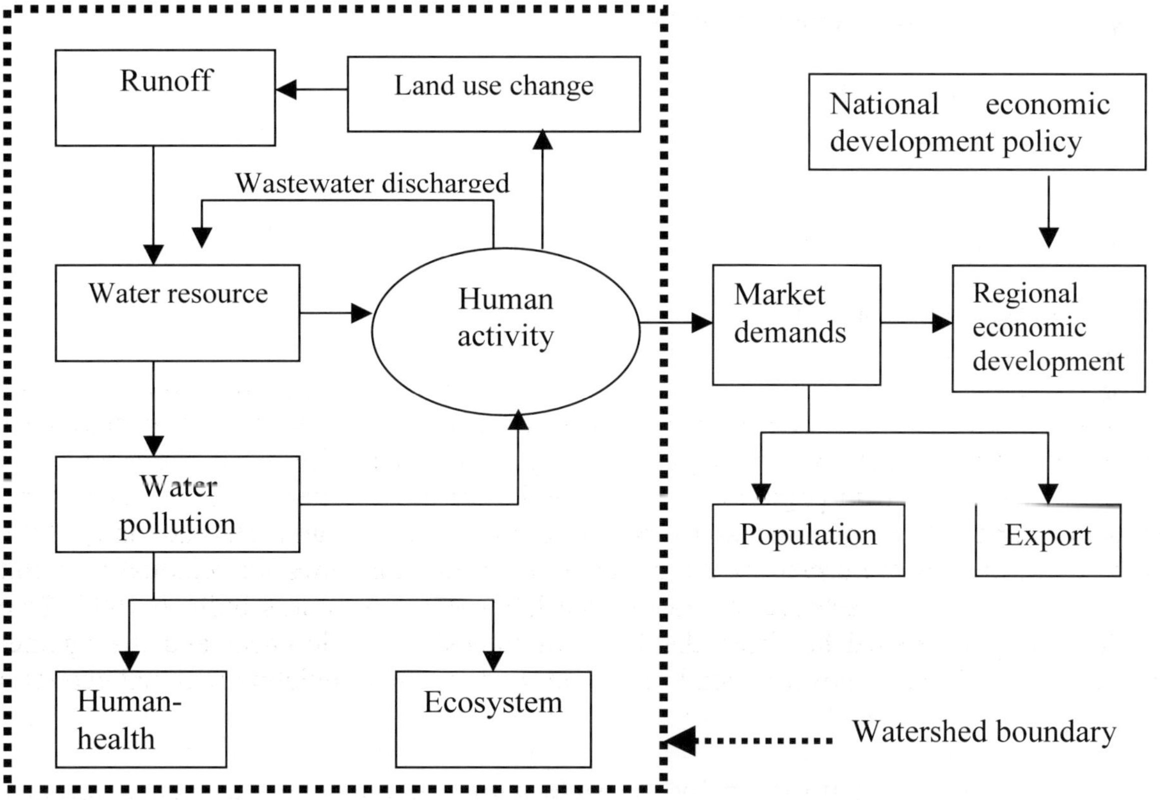

Figure 1. Interactions between water resource and economic development

Table 2. Mean Gross Provincial Products (GPP) during 1993-2000 (million Baht).

Sectors	Chaiyaphum		Khon Kaen		Loei		Nong Bua		Petchaboon	
	(GPP)	(%)	(GPP)	(%)	(GPP)	(%)	(GPP)	(%)	(GPP)	%
Agriculture	6626	41.18	6654	21.12	4041	55.06	2211	57.49	7046	57.20
Manufacture	4214	26.19	15377	48.81	588	8.01	796	20.70	612	4.97
Wholesale and retail	5250	32.63	9475	30.07	2710	36.93	839	21.81	4660	37.83
Total	16090		31506		7339		3846		12318	

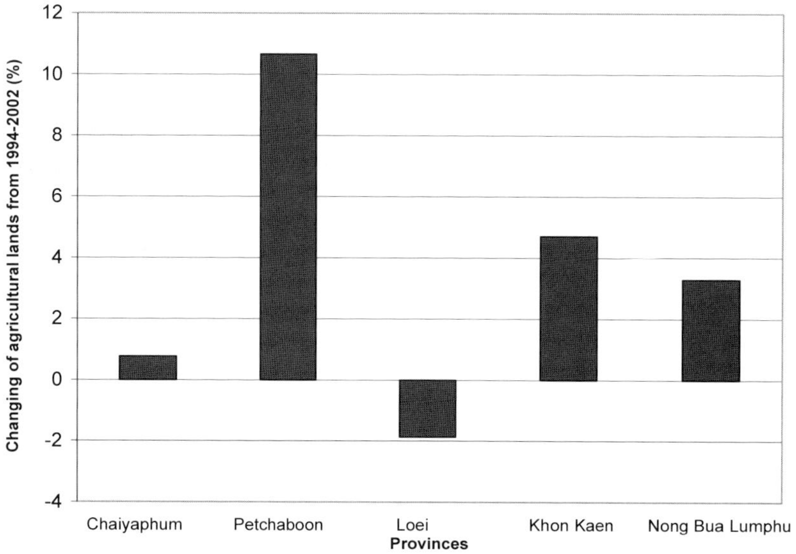

Figure 2. The mean annual rates of increase of agricultural land between 1994 and 2002 (% per year).

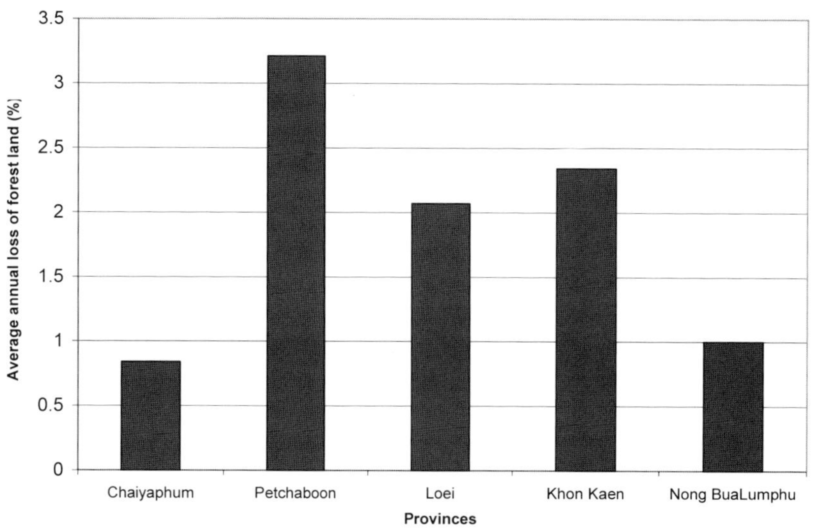

Figure 3. The mean annual rates of decrease of forestlands between 1988 and 1999 (% per year).

3.2 Consequences of economic development

Figure 1 also shows water pollution and land use change as the consequence of human activities and the use of water driven by national economic policy. KKU (2003) states that the upper and lower Phong sub-watersheds have encountered different environmental problems. Deterioration of forest areas is the major problem in the upper Phong sub-watershed while water pollution is the major concern in the lower Phong sub-watershed. Water pollution caused adverse effects on aquatic life as indicated by the phenomena of fish and shellfish death recorded during 1993-1995. Consequently, water pollution will affect both industry and community, as both depend on river water as a raw material and for consumption, respectively. Khon Kaen University (2003) monitored suspended solids in the main rivers of upper Phong watershed and estimated amounts of transported suspended solids in the rivers of Phrom, Choen, upstream Phong, and Pa Niang as 0.06, 0.32, 1.04, 0.19 tons/ha/year, respectively. The Electricity Generation Authority of Thailand (EGAT) reported that the mean annual amount of siltation in the Ubolratana dam was 1.50 million tons/year between 1965-1990. It has decreased the water storage capacity by 32.90 million m^3, which is about 1.4% of its maximum storage capacity of 2,263 million m^3. Mungkarndee and Wirojanagud (2003) report that water pollution in the Phong River has occurred frequently. Pollution has been observed since 1986 and became very serious in 1992 when molasses from a storage tank of a sugar factory leaked into the river and contaminated the water downstream starting from the factory to the Chi and Moon Rivers. Fish is the main source of protein for the northeastern people, with 30.5% of the population consuming fish daily. Catching fish directly from the river is easily done and practiced almost daily. When the amount of fish caught sharply decreases as a result of a molasses leakage, it is necessary to replace this protein from other sources, such as meat or soybeans. The estimated loss of fish in the diet during the 1992 pollution event was at least 271.4-361.8 grams/household/day (calculated for daily fish consumption per household) (KKU, 1995).

4. Conclusions

The economic well being of the northeast region of Thailand has so far exerted demands on regional water resource development. The major economic role of water depends on its relationship with human activities driven by economic development policy such as agriculture, industry, water supply and power generation in order to produce products and services to serve demands both domestically and internationally. The situation of the Phong watershed is well matched to the above statement. The Ubolratana dam was constructed in 1966 together with other reservoirs for stimulating regional economic development. As a result, there is intensive expansion of agricultural areas and of agro-industries along the Phong River. However, water is also an essential part of any ecosystem, in terms of both its quantity and quality. Reducing the availability of water for the natural environment will have devastating effects, as will the pollution from domestic, industrial and agricultural wastewaters. Just as the environment is integrally tied up with the social, health and economic impacts of water use, ensuring environmental sustainability and generation will also have positive effects on these areas.

5. References

Bank of Thailand (2004). The northeast economy: Isaan region and social, economic, and environmental development, www.bot.or.th.

Kazi F. Julal and Peter P.Rogers (2002). Measuring environmental performance in Asia, *Ecological Indicators*, *Volume 2, Issues 1-2, November 2002,* pp. 39-59.

Khon Kaen University (2003). Research on environmental management of the Phong watershed. Khon Kaen University, Thailand.

Khon Kaen University (1995). Study project for preparation of the action plan for the Pong river water quality rehabilitation, Khon Kaen University, Thailand.

Office of Agricultural Economics (2004). Agricultural Statistics of Thailand, www.oae.go.th.

Pinthita Mungkrandee and Wanpen Wirojanagud (2003). Impact of pollution sources on the Phong River water quality, the first international symposium on Southeast Asian Environment, October 23-25, 2003, Bangkok, Thailand.

Sneddon, C. (2002). Water conflicts and river basins: the contradictions of co-management and scale in Northeast Thailand. Society & Natural Resources, 15(8), pp. 725-741.

Development on Expert-Database for Upper Part of the Citarum River Pollution Control

P. Sudjono[1)], A. Yuasa[2)], D. Indira[3)] and E. W. Irianto[4)]

1) Dept. of Environmental Engg., Bandung Institute of Technology
Jalan Ganesha, Bandung, 40132, Indonesia
e-mail: memteq@bdg.centrin.net.id
2) River Basin Research Center, Gifu University
1-1, Yanagido, Gifu, 501-1193, Japan
e-mail: yuasa@green.gifu-u.ac.jp
3) Post Grad. Student in Dept. of Environmental Engg., Bandung Institute of Technology
Jalan Ganesha, Bandung, 40132, Indonesia
e-mail: memteq@bdg.centrin.net.id
4) Research Institute for Water Resources, Indonesia
Jalan Juanda, Bandung, 40135, Indonesia
e-mail: ekowinar@yahoo.com

Abstract

Pollution problems on the upper part of Citarum are so complex that implementation of a pollution control strategy requires structured water data and comprehensive interpretation from experts. In fact, the size of the data on water quantity and water quality is large but is sparse in several institutions. In such a case, development of a system that consists of the structured data and knowledge on river pollution control is initiated from identification on the domain of water related elements. Then a concept of database entity relationships is created, query is set, knowledge is represented in a computer, and finally the system is verified and analyzed. Due to specific conditions of the catchment area, knowledge is acquired from water-management-experts and stakeholders. The knowledge consists of their expectations upon such a database on the point of management and control on river pollution. Finally, the developed database can simply be used by specifying required information from the point of view of industry, river, and administrative boundary. Additionally, the system is also provided with some principle points on river pollution control that have been adopted by local or regional authorities in running daily and long term river pollution control.

Keywords: Citarum river, database, pollution control, water quality

1. Introduction

An expert-database combines database management systems and expert-system environments to solve problems. Data on water quality and sources of pollutants as well as data on physical conditions of a catchment area can be left in a pile of reports in a warehouse. In order to make the data become a useful source of information for decision-makers, the data are structured and linked together in an expert-database.

Experts pay a lot of attention to the Citarum River, the largest river in west Java island covering 6.540 km^2. The upper part of the Citarum River flows from the Wayang mountain to the mouth of the Saguling reservoir, along the middle part there are three reservoirs (Saguling,

Cirata, and Jatiluhur) while the rest is the lower part. The length of each part is about 100 km. In the catchment areas of the upper part, large cities such as Bandung are located. As the activities in the area increase, environment related data are very large and they are worthless without storing them in an appropriate system.

A system to assist decision-makers in managing a catchment area has been discussed (Amien *et al.*, 1996). They stated that the assistance could be in the form of recommendation on fertilizer application, conservation technique, land use management, and crop selection. In another side, a computer-based system for water quality data has been developed and applied in Indonesia since 1994 (Sutriati and Hatmoko, 2001). The program is created by applying MS-Access and it is capable of retrieving and updating water quality data of rivers and reservoirs. Moreover, Gani *et al.* (2001) expect that the formulation of a decision support system for water resources planning in Indonesia should be based on comprehensive inventories of water resources database and available water quality prediction models at the current time. As an alternative to solving the problem and fulfilling the expectations, an expert-database for the upper Citarum river is developed and detailed descriptions on the structure and scheme of the system is presented in the paper.

2. Database structure

The primary problem inherent in modelling any subset of a real river system in a computational medium such as a database is the need to organize real-world knowledge of river systems into structures for efficient storage and retrieval. The data are structured in such a way as to make it possible for users to search the desired data. The type of data to be searched is on water quality monitored at several points along the river, industry as source of pollutants, and physical conditions within the administrative areas. Searching strategy could be in two directions in which the reasoning proceeds, *i.e.* forward and backward reasoning. Forward reasoning process starts from data as initial conditions and works forward toward the effect of a given condition. Inversely, backward reasoning process starts from a given effects and proceeds backward toward the initial state causing the effect. Application of the reasoning processes in reaching the desired goal in the developed expert-database needs to be modeled using real-world information that is defined in a domain element and are schematized in a semantic network.

2.1 Domain element

The most important criteria to consider a domain of a prototype expert-database are small, well bounded, involved in daily tasks of a river water manager, and avoiding interdependency with other systems. A small system, in a practical way, is decided by the size of coverage area, *i.e.* upper part of Citarum River, which flows from Wayang Mountain to the mouth of Saguling reservoir. Well bounded problems in river pollution control are addressed only on retrieving the data through several points of entry that are usually desired by a water manager. River pollution control, at the current stage, consists of several principles that have been adopted by local government. Daily tasks of a water manager concerning the required data could be an important source of information on elements related to river water that should be included in the system. Lastly, the system is built by assuming that data produced by other systems such as water quality prediction models are not required. It means that the developed database is stand alone without requiring extensive integration with other systems. The domain element consists of river, industry, physical conditions of the area, and principles of river pollution management and control. Sampling points are attached in rivers and the data consists of water quantity and water quality as shown in Figure 1. Industry, shown in Figure 2, owns several properties such as locations that cause the condition of the area in Figure 3 and point of discharge that is related to the conditions of the river. Moreover, water pollution should be controlled by institutions as the taxonomy is presented in Figure 4.

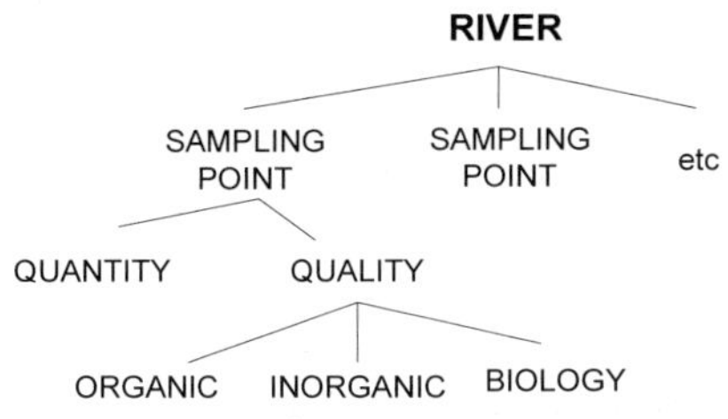

Figure 1. The upper part of Citarum River related to sampling point.

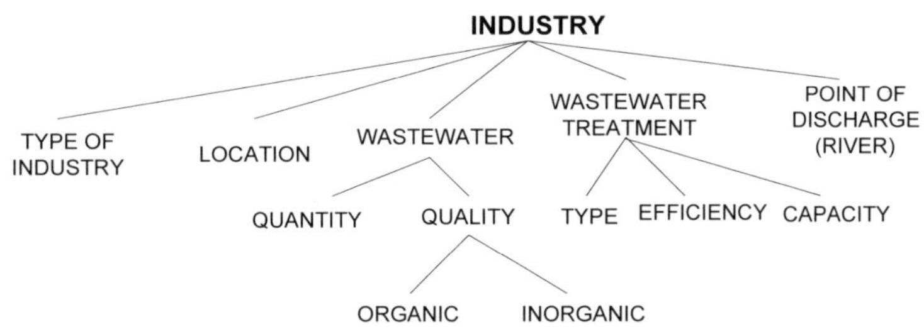

Figure 2. Elements related to industry.

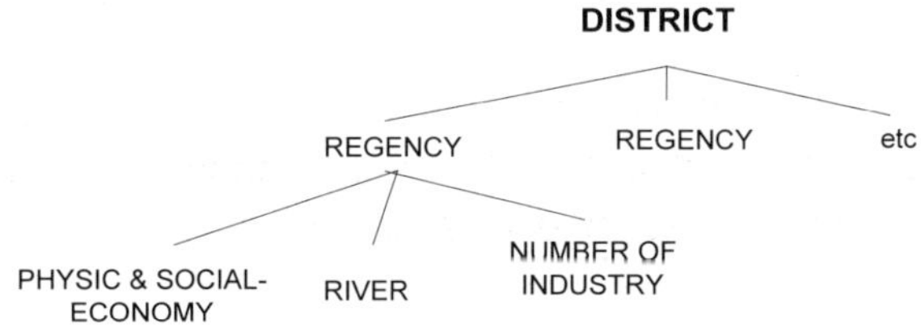

Figure 3. The physical conditions of the catchment.

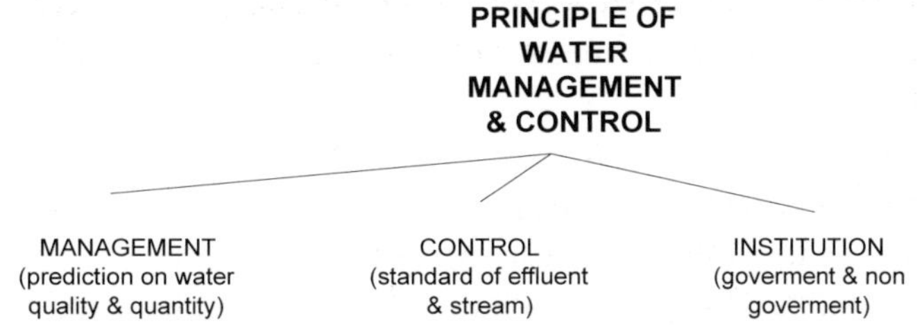

Figure 4. Principle of river pollution control and management.

2.2 Semantic network

The focal point of a semantic network is the representation of relations among elements in a domain to form a network. The main elements in the domain mainly consist of river, industry, physical condition of the area, and principles of river pollution control. The network is presented in Figure 5 showing relationships among elements that represent real-world applications.

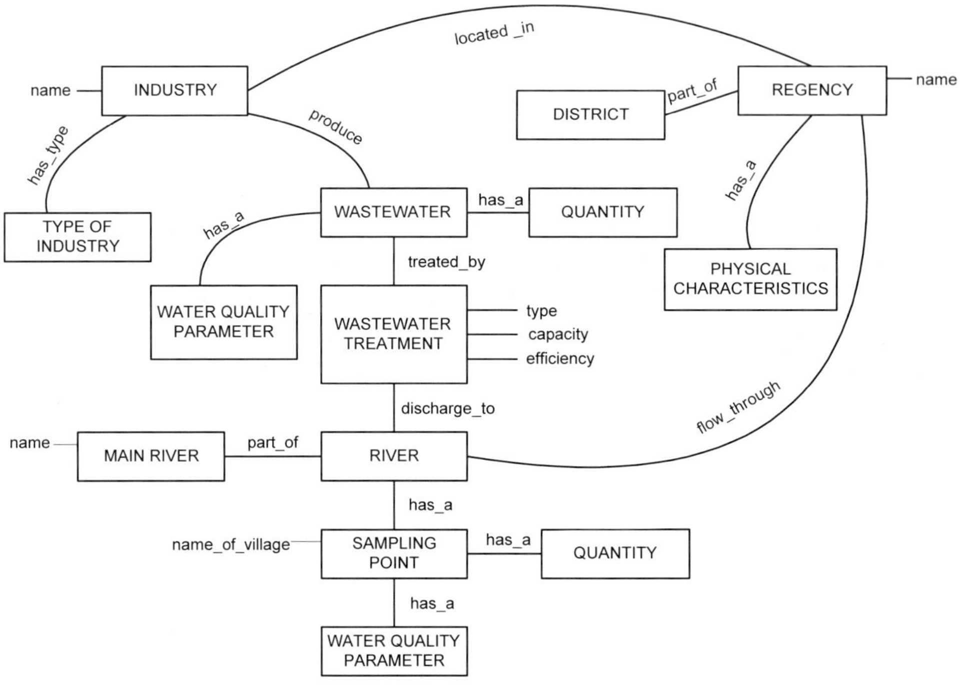

Figure 5. Semantic network for upper part of Citarum river pollution control.

The basic elements of a semantic network are nodes to represent domain elements and links to represent relations between elements. Figure 5 comprehensively shows relationships between elements of the domain and their properties. Binary relations used in the database could be generalization, aggregation, classification, and association (Beynon-Davies, 1991).

Generalization points out similarities between a number of lower-level elements to form high-order elements. Generalization is basically the representation of *is_a* relationships even though inherited properties of an element will be included in the relationships. This kind of abstraction is often used in relation between WATER QUANTITY and its parameters.

Aggregation is a process grouping a number of lower-level elements to form a higher-level element (Smith and Smith, 1977). Aggregation is basically a representation of *is_part_of* relationship. Aggregation may use in relationship between RIVER and MAIN RIVER, also relationship between REGENCY and DISTRICT. Implementation of this kind of abstraction makes tracing of 'river' or 'areas' possible.

Classification is an abstraction of element that is considered as instances of a higher-level element. Usually, classification of elements can use predicate *is_instance_of* relationship. An example of this kind of abstraction is relationship between INDUSTRY and TYPE OF

INDUSTRY. Several types of industry are adopted in the database, these are, textile, pharmacy, paint, steel-works, food, plastic, and electronics. In fact, real application of TYPE OF INDUSTRY is to classify the industries based on products. Unfortunately in the prototype model, the TYPE OF INDUSTRY has no further lower-level elements.

Another important abstraction is association that relationship between member elements is considered a high-level set element (Beynon-Davies, 1991). Further he stated that aggregation may specify attributes of a new element whose value will be of a given type. The *is_member_of* relationship is usually used to express the association concept. Application of this type of abstraction is relationship between INDUSTRY and WASTEWATER, RIVER and REGENCY. Moreover, WASTEWATER TREATMENT PLANT can have association with RIVER where the treated wastewater is discharged.

Advantages of semantic data models can be assessed from economic expression, integrity maintenance, modeling flexibility, and modeling efficiency. Representations of information can be exactly the same as those in the network. The system also maintains database integrity as mentioned by Peckham and Maryanski (1988) that disappearing existence of an entity will scrape the related entity from the system. Furthermore, semantic data model allows users to view data on many different levels of generalization, and network can efficiently be implemented into the system.

3. Query setting for river pollution control

Data are required by users in accordance with problems to be solved at a particular time and location. Even though users are limited to only those who are responsible for river pollution management and control, they consists of several types of expertise. In such case, it is necessary to consult with all of the experts. The expert-database that is provided with expert knowledge should be capable in providing (a). information based on industry and river water body, (b). information based on administrative boundary, and (c). expectation on the river pollution control. The developed database containing the information is schematized in Figure 6.

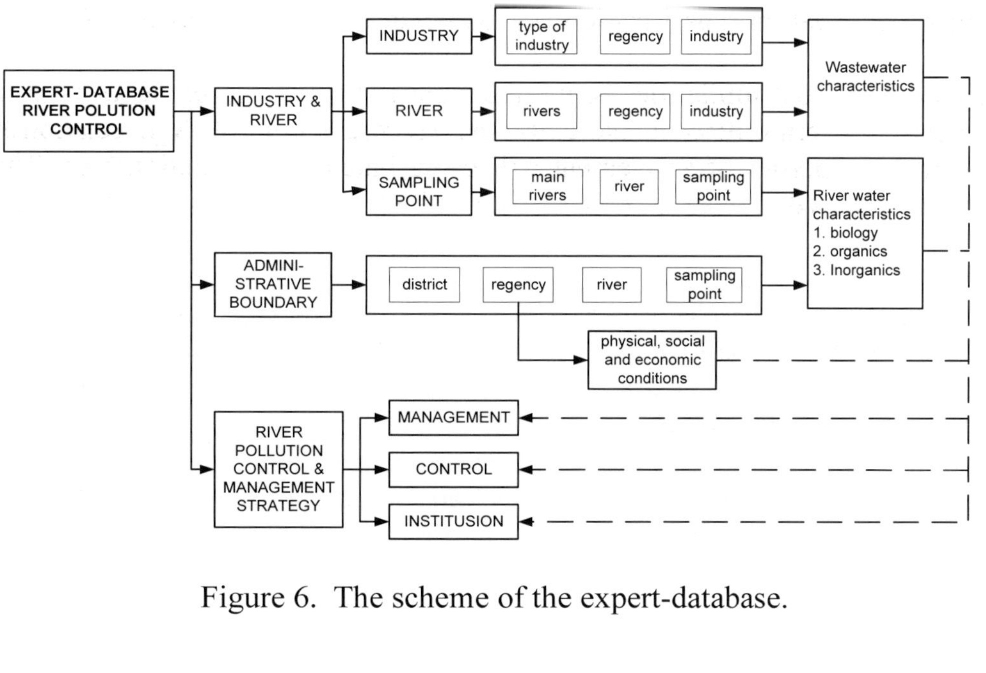

Figure 6. The scheme of the expert-database.

3.1 Information based on industry and river water body
Information can be retrieved through three topics of entry; these are industry, river water body, and sampling point. Based on industrial data, the system will show all types of industry in the upper part of Citarum catchment area. As users choose a type of industry then the system will echo regency where the type of industry is located. Users may choose a regency, then names of the selected type of industry in the regency will be shown. Finally, wastewater characteristics of a chosen industry are shown on the screen.

In some situations, users want to explore the database through a river path. Initially, the system shows all names of rivers in the upper part of Citarum. After a river is chosen, the system shows all regencies where the river flows through. Then users may choose a regency, then the system shows all industries in the regency. When an industry is chosen then wastewater characteristics of the industry and efficiency of the wastewater treatment plant will be shown.

Another desired choice is to find river water quality at a certain sampling point. In this case the system shows all names of rivers in the area. After users chose a river, the system will show sampling points along the river. Finally the system will show characteristics of the river water, *i.e.* physic, biology, organic, and inorganic at the selected sampling point.

3.2 Information based on administrative boundary
Information based on administrative boundary may be useful for users who want to get general figures on physical conditions of a concerned regency. Firstly, the system will show all districts in the upper part of Citarum. After a district is chosen, the system shows all regencies in the district and users may choose a concerned regency to get information about physical, social, and economic conditions of the regency. At the same time all rivers that flow across the regency are echoed. When a river has been selected, the system will let users choose a sampling point in the river. Finally the system will show data on flow rate, and water quality of the river.

3.3 River pollution control strategy
At the current stage, river pollution control strategy provided by the system adopts results of previous research on Citarum catchment, consultation with several related experts, and government guidelines on river management. Initially, the system shows main principle points of river management. Following that, users may choose a keyword as a guide to flow to sub-principles that consists of management, control, and institution as illustrated in Figure 6. Keywords can be, for example, conservation, public participation, water quality standards, and roles of institutions.

4. Knowledge representation

Knowledge can be classified as facts that tell statically about elements and procedural rules that describe relationships among elements to produce a dynamic action (Rolston, 1988). Representation of the knowledge in a computer program requires predicate logic, argument, variable, and logical inference. Predicates are used to relate elements based on implementation of abstraction as structured in Figure 5. Arguments are associated with

elements within a predicate. The following samples of predicate logic show representation of the knowledge in a computer program.
type_of_industry(Type_of_industry).
location_of_industry(Regency_name).
data_industry(Name_of_industry,BOD,COD,DO,Temp,pH,Cr,Efficiency_of_wastewater_treatment).
industry(Regency,Name_of_industry,Type_of_industry).
name_of_river(Name_of_river).
catchment_area(Name_of_river,Regency_name).
biology(Name_of_river,[biological_parameter]).
organic_chem(Name_of_river,Sampling_point,[concentration_of_organic_ chemistry]).
inorganic_chem(Name_of_river, Sampling_point, [concentration_of_organic_ chemistry]).
area(District, Regency).
area_data(Regency,Area,Population,Number_of_Industry,Agriculture_area,Farm).

Procedural rules relate elements to produce a conclusion based on IF-THEN or condition and action taken. Procedural rules or rules are developed to fulfill the query set for the developed database by utilizing the predicate logic. A sample of visual output is presented in Figure 7. The output is a result of utilizing a rule for finding wastewater data of a certain type of industry at a particular location as follows.

<u>RULE Industry:</u>
WRITE all types of industry
IF type of industry =T_of_I
THEN T_of_I is located in [Regency]
and
IF [Regency] = Reg_N
THEN industry in the Reg_N is [Industry]
and
IF [Industry] = Ind_N
THEN WRITE wastewater characteristics, i.e. BOD, COD, DO, Temperature, pH, Cr, and Efficiency of the wastewater treatment plant.

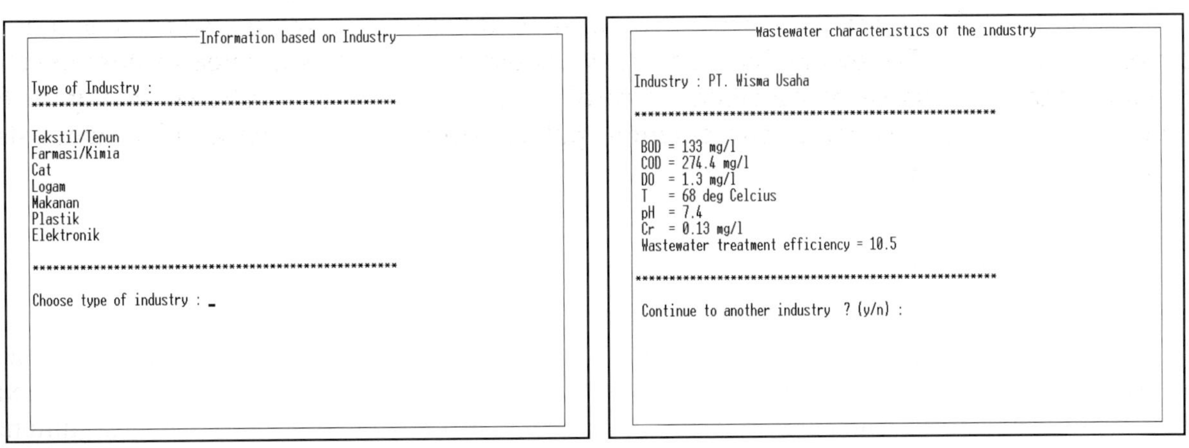

Figure 7. Wastewater characteristics of an industry in a particular regency.

5. Conclusions

The upper part of Citarum receives pollutants from industries located evenly in the catchment area. Thus river pollution control is difficult to implement without a proper structure on water-related data and interpretation from experts. In fact, the size of data on water quantity and water quality is large and it is sparse in several institutions that conduct studies or monitoring projects. In such cases, a system that consists of organized data and knowledge on river pollution control is developed. It is initiated from identification on water related elements, then development on a concept of database entity relationships, setting on query, representation of the data in a computer language, and lastly verification and analysis on the system. Due to specific conditions of the upper part of Citarum catchment area, knowledge about river pollution control is acquired from water-management-experts. Their responses are systematized and analyzed then it is transformed into a query upon the database and rules that should be intact in the system. The developed database consists of information based on industry and river water body, information based on administrative boundary, and river pollution control strategy. It can be operated by a water manager simply by specifying the required data then river water quality can be identified for control and management purposes. Moreover, the principles on river pollution control intact in the system can assist local or regional authorities in order to run daily and long term tasks on river pollution control.

6. References

Amien, L.I., Agus, F., and T. Vadari. (1996). Sistem pembantu pembuat keputusan dalam pengelolaan DAS, Prosiding Analisis Pengelolaan Daerah Aliran Sungai, Pusat Penelitian Tanah dan Agroklimat, pp. 15-29.

Baynon-Davies, Paul. (1991). Expert database systems a gentle introduction, McGraw-Hill Co, London, pp. 108.

Gani, A.H.A., Hatmoko, W., and I.A. Yusuf. (2001). A General overview of decision support system for water resources planning and management in Indonesia, *Jurnal Penelitian dan Pengembangan Pengairan*, 15 (47), pp. 3-12.

Peckham, J., and F. Maryanski. (1988). Semantic Data Models, *ACM Computing Surveys*, 20(3), pp. 153-189.

Rolston, David W. (1988). Principles of artificial intelligence and expert systems development, McGraw-Hill, New York, pp. 31.

Smith, J.M., and D.C.P. Smith. (1977). Database abstractions: aggregation and generalization, *ACM Transaction on Database Systems*, 2(2), pp.105-133.

Sutriati, A., and W. Hatmoko. (2001). Evaluasi basis data kualitas air nasional CIDA sebagai bagian dari sistem basis data sumber daya air nasional, *Jurnal Penelitian dan Pengembangan Pengairan*, **15**(47), pp. 21-28.

Estimation of Downstream Environmental Control Discharges using Artificial Neural Networks

U. Weesakul[1)] and S. Leeratanaruk[2)]

1) Assoc. Prof., Faculty of Engineering, Thammasat University,
Pathumthani, 12120, Thailand
e-mail: wuruya@engr.tu.ac.th
2) Master Student, Joint master degree program: Sirindhorn International Institute of Technology
and Faculty of Engineering, Thammasat University, Pathumthani, 12120, Thailand
e-mail: supree_l@siit.tu.ac.th

Abstract

Environmental problems in estuaries are one of the problems for most rivers. Similar to other rivers, the environmental problems in the lower Chao Phraya river become severe particularly during low flow periods due to less released discharges caused by an increased water demand in the upstream part together with high tidal level in the downstream part resulting in intrusion of salinity into the river. The level of salinity along the river reach has to be maintained at different levels based on the purposes of exploitation of water. Several studies have been conducted for calculation of appropriate amount of discharges to be released from upstream to downstream for environmental control purpose. This study proposed another alternative for estimation of environmental control discharge by using artificial neural networks (ANNs) as a mathematical tool for calculation of minimum discharges released for downstream environmental control in the lower reach of the Chao Phraya river. The results revealed that ANNs were applicable for such a purpose with acceptable accuracy of estimations.

Keywords: ANNs, downstream environmental control discharges, salinity intrusion

1. Introduction

An estuary is a semi-enclosed body of water that is connected between the open sea (salinity ranging from 20 to 35ppt) and river discharge (with salinity ranging from 0 to 0.50ppt). The water in an estuary is called brackish water, which normally has the salinity ranging from 0.50 to 20ppt, (Szuster and Flaherty, 2000). Brackish water is required for the recreation of mangroves as well as living conditions for some animal species (Tingsabadh and Chaturongpalathipat, 2000). Also, the mixture of freshwater and salt water provides a variety of habitats for humans, animals, and plants.

Salinity intrusion in the lower Chao Phraya river during low flow period becomes more and more severe due to a higher demand of water in the upstream part resulting in less released discharge to downstream combining with high tidal level at river mouth, which consequently increases salinity intrusion length to upstream reach (Leewatchanakul, 1988).

Actually, there have been several studies concerning the calculation of salinity intrusion in Chao Phraya River for various aspects. One of these was the study of "Salinity Intrusion in the Chao Phraya and Mae Klong Rivers" (Asian Institute of Technology, 1978). In this study, a tidal dynamic model and salt balance model were developed to investigate salinity intrusion in the lower Chao Phraya estuary. Tidal data at Samut Prakarn and released discharge data from Chao Phraya barrage, Noi river, and Pasak river during January to April 1970 were used to calibrate the tidal dynamic model. The salinity model was calibrated by using salinity conditions during the low flow period i.e. January to April of the years 1970 and 1975. The results of the study revealed that the low flow period i.e. from maximum and mean salinity concentrations, as well as intrusion length under spring tide, were strongly related with released discharge from the upstream reach. These relationships were established in the study so called "operating rules", which were applicable for estimation of released discharge to control salinity intrusion length from the river mouth. This model was further developed in 1980 (Tingsanchali, 1980) for the study of "Salinity Intrusion Problem during Severe Drought in the Chao Phraya River". The model was recalibrated by using data in 1980, which was a drought year. Both spring and neap tides, as well as pumping of raw water from river to water supply treatment plants were also taken into account in the computation and calibration. Results of model simulations revealed that the amount of water pumped from the river for Metropolitan Waterworks Authority had no significant impact on salinity intrusion along the river reach. Another approach to calculate salinity intrusion in the Chao Phraya River is to use a finite element model, which was developed by Leewatchanakul in 1988. It was found from the study that the model could calculate the fresh water discharge to control the most probable salinity intrusion. It could also forecast the salinity content with acceptable accuracy, on a monthly basis with two types of tidal range. However, only monthly discharges were considered in the study, which is not practical for engineering practice.

Deterministic physically based hydrodynamic models as mentioned above are applicable for forecast and Dam operation for minimum discharges, which should be released from upstream for downstream environmental control with satisfactory accuracy. However, the accuracy of computation depends on physically based data, such as river cross-section, man-made interference, which continuously changes with time and needed to be updated causing some difficulties in using such models.

Another new approach, currently and widely used in water resources engineering is a neural network model, as a simple black box model, which is more convenient for engineering practice. There have been several studies using a neural network model for hydrology and hydraulic engineering purposes in Thailand.

An example of such studies is using back propagation neural networks (BPNN) algorithm to forecast discharge and hourly water levels in the lower reach of the Chao Phraya River at Memorial Bridge (48km from river mouth) for one to three hours leading time of forecast (Manusthiparom, 1998). Hourly water levels from 3 stations (Bangsai, Fort Chula, and Memorial Bridge) in year 1995, 1996, and 1998 were used for the MIKE-11 hydro-dynamic model calibration while the available data in year 1991 and 1992 were applied for the MIKE-11 hydro-dynamic model verification. Due to the insufficient data, The MIKE-11 hydro-dynamic model, after calibration and verification, was used to simulate the reference discharges at Bangkok Memorial bridge in training and testing processes of the BPNN for forecasting the discharge at Bangkok Memorial Bridge station. Hourly water level data at Bangkok Memorial Bridge station were used as the target water levels in the BPNN model. Then after training and testing the BPNN model, the ARMA model was used for updating the model output that is the forecasted water levels and discharge at Bangkok Memorial Bridge station. It was found from the study that the combination of MIKE-11 hydro-dynamic model, BPNN model, and ARMA model were applicable for predicting hourly water level and discharges of the Chao Phraya river at Bangkok Memorial bridge with acceptable accuracy. ANNs was also applied for prediction of salinity twelve hours ahead at six potential water abstraction locations along the West Pearl river estuary in Macao (Nguyen Thi Thu Nga, 2002). The study revealed that, with appropriate structures, ANNs was applicable for salinity prediction in an estuary with satisfactory results.

It is therefore interesting to investigate whether ANNs is applicable for calculation of salinity in the Chao Phraya River estuary or not and how high the model performance is. The objective of the study is to propose another alternative for estimation of environmental control discharges by using ANNs as a mathematical tool for calculation of minimum discharges released for downstream environmental control in the lower reach of the Chao Phraya River.

2. Study area

The lower reach of Chao Phraya River, Map of Thailand (Figure 1), mostly influenced by salinity intrusion during low flow period, starts from Bangsai (115km from river mouth) to the Gulf of Thailand (Figure 2). The bed slope of the river reach is approximately 4 to 7cm/km. Both released discharges from Chao Phraya Barrage and tidal range from the gulf of Thailand influence salinity level in this river reach. Generally low flow period in this river reach starts from January until May, with discharges released from Chao Phraya Barrage less than 200m^3/s. The most severe case for the minimum released discharges from Chao Phraya Barrage occurred in 1999 with the value of only 20m^3/s. Normally at Samut Prakarn station average discharge is around 59m^3/s and maximum discharge is about 195m^3/s during low flow period (Table 1). For Khrung Thep Bridge station, average discharge is around 58m^3/s and maximum discharge is 98m^3/s during low flow period (Table 2). Salinity at Samut Prakarn station is as high as 39g/l during low flow period with the average value of 19g/l (Table 1). Maximum salinity at Khrung Thep Bridge station is 22.5g/l during low flow period with the average value of 4.2g/l (Table 2).

Presently, there are only two observed stations [i.e. Samut Prakarn station and Khrung Thep Bridge station] in this area, which collect salinity and water level data. Frequency of data collection is sometime everyday, every 3 days, and every 5 days depending on the situation. Availability of these data is from 1996 to 1999 only. However released discharge data at Chao Phraya Barrage is available for a long time with frequency of everyday data. Tidal level at Fort Chula station was collected every hour with long periods of record.

Table 1. Parameters Characteristics for Samut Prakarn station [during 1996 – 1999]

Parameters	Max	Mean	Min	Unit	Remark
Salinity at Samut Prakarn (S_{SP})	39.05	18.95	2.57	g/l	Observe station
Release discharge 15-days delay duration (Q_{t-15})	195	59.15	20	m^3/s	Discharge from Chao Phraya barrage
Water Level at Pak Nam (WL_{SP})	2.14	1.29	-0.77	m MSL	Observe station
Water Level at Fort Chula (WL_{FC})	2.28	1.39	-0.8	m MSL	Downstream tidal level

Table 2. Parameters Characteristics for Khrung Thep bridge station [during 1996 – 1999]

Parameters	Max	Mean	Min	Unit	Remark
Salinity at Khrung Thep bridge station (S_{KT})	22.46	4.21	0.052	g/l	Observe station
Total release discharge 13-days delay duration (Q_{t-13})	98	57.61	20	m^3/s	Discharge from Chao Phraya barrage
Water Level at Khrung Thep bridge station (WL_{KT})	1.8	0.60	-0.63	m MSL	Observe station
Water Level at Fort Chula with 1hr delay (WL_{FC})	2.1	0.69	-0.9	m MSL	Downstream tidal level

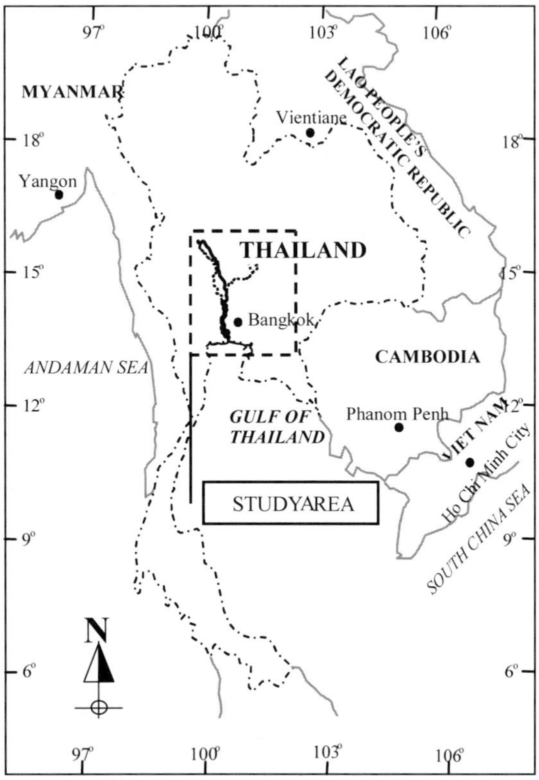

Figure 1. Map of Thailand

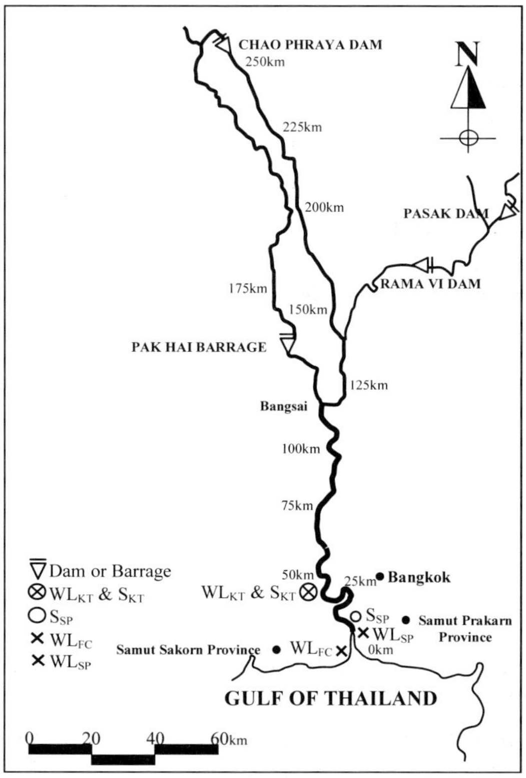

Figure 2. Availability of hydrometric stations

3. Theoretical consideration

A single layer of ANNs was applied as a mathematical tool (Back Propagation Neural Networks Algorithm, BPNN) for estimation of released discharge for control of salinity at Samut Prakarn and Khrung Thep Bridge stations. Artificial neural networks (ANNs) are the information processes which simulate the human's brain solving processes. An advantage of this mathematical tool is that it is able to compact the complicated relationships into the form of direct input-output functions. A fundamental architecture of ANNs is called a single layer network, where there can be a number of input in input layer, a number of hidden nodes in hidden layer, and a result as 1 output in output layer (Freeman and Skapura, 1991).

Despite the apparent success of the BPNN and learning algorithm, there are some aspects, which cannot always guarantee that the algorithm can be universally used. Most of them are the failure of the training algorithm from network paralysis and local minima (Freeman and Skapura, 1991).

Although a single layer net is severely limited in the mappings, ANN can be applied to a multilayer network (more than one hidden layer) and can learn any continuous mapping to an arbitrary accuracy. More than one hidden layer may benefit some applications, but generally a single layer network is sufficient, which is the case that was applied in this study.

4. Model calibration and verification

Variables, which may influence salinity along the lower reach of Chao Phraya River, were investigated by correlation analysis. It was found from the study namely "Analysis of the Low Flow Characteristics for Downstream Environmental Control" (Weesakul and Leeratanaruk, 2003) that during low flow conditions salinity along the lower reach highly depended on discharge released from the Chao Phraya Barrage with the lagging time of 15 days for salinity at Samut Prakarn station and of 13 days for salinity at Khrung Thep Bridge station. Besides upstream released discharges, tidal level at the Gulf of Thailand (Fort Chula Station), as well as their own water level (at the considered station) were also highly related to their salinity.

Therefore, in this study, salinity at the interesting point, their water level and tidal level at the Gulf of Thailand were used as input variables to a single layer network of ANNs to calculate discharges, which should be released from Chao Phraya Barrage. These released discharges were considered as output in output layer of ANNs.

There are many training parameters in an ANNs, which are learning rate, hidden nodes, momentum rate, number of learning cycles, and initial weight. In this study, momentum rate was assigned to be zero, initial weight were set up to be equal for all variables concerned and the same value of initial weight was used for all cases of computation. For the number of learning cycles, 5000 cycles were fixed in this study for all cases. Learning rate was varied from 0.05 to 0.50 (with the value of 0.05, 0.10, 0.25, and 0.50), hidden nodes were varied from 5 to 20 hidden nodes (with value of 5, 10, 15, and 20) resulting in a total of 16 different patterns of calculation for each observed station as shown in Figures 3 to 6.

In order to assess accuracy of computation, root mean square relative error (RMS_{RE}) was adopted in this study, as shown in EQUATION 1, as follow:

$$RMS_{RE} = 100 \times \sqrt{\frac{\sum_{i=1}^{n}\left((Q_{est} - Q_{obs})^2\right)}{\sum_{i=1}^{n}\left((Q_{obs})^2\right)}} \qquad \text{(Equation 1)}$$

Where, Q_{est} is estimated released discharge (m^3/s), Q_{obs} is the observed released discharge (m^3/s), and n is the number of observed data.

Results of computation and error of each case are summarized as follow:

4.1 Samut Prakarn station

The results of computation in the training process revealed that 10 hidden nodes with learing rate of 0.25 provided smallest error (RMS_{RE} = 22.16%, Figure 3) and this was confirmed by the results of testing processes with RMS_{RE} equal to 30.37% (Figure 4)

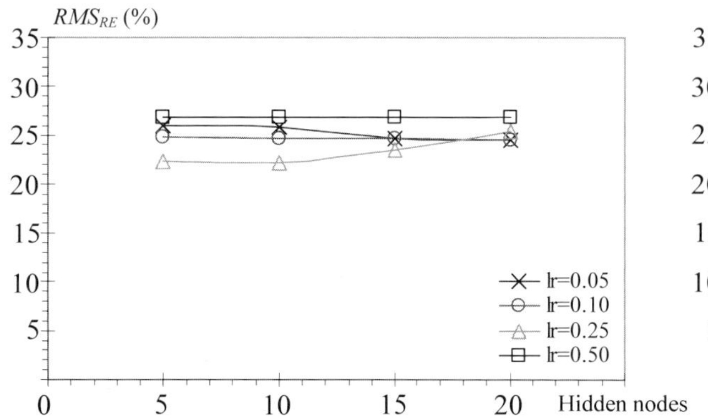

Figure 3. RMS_{RE} of the training processes (Samut Prakarn station)

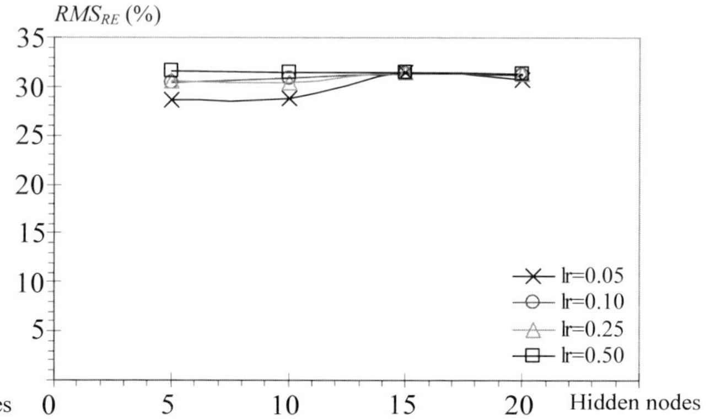

Figure 4. RMS_{RE} of the testing processes (Samut Prakarn station)

4.2 Khrung Thep Bridge station

Figure 5 shows results of computation for the training process at Khrung Thep Bridge station, which reveals that errors of computation are slightly different for almost all patterns except for learning rate of 0.50. Therefore learning rate of 0.25 and hidden node of 10, as previouly selected for Samut Prakarn station, was selected to be used for this case.

The results of computation of this pattern in the testing processes confirmed that this pattern provided the smallest error [RMS_{RE} = 19.96%, Figure 6].

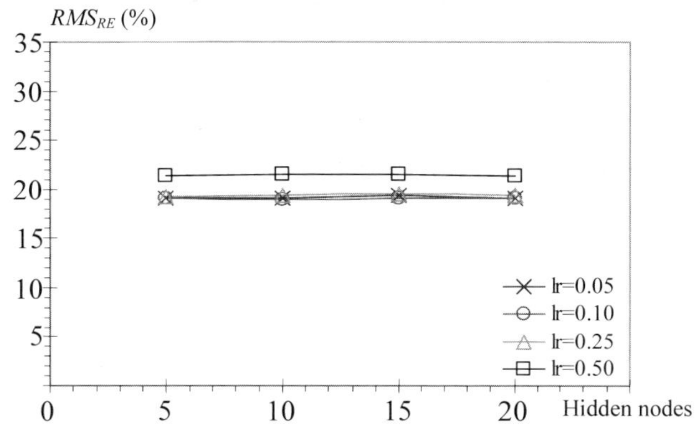

Figure 5. RMS_{RE} of the training processes (Khrung Thep Bridge station)

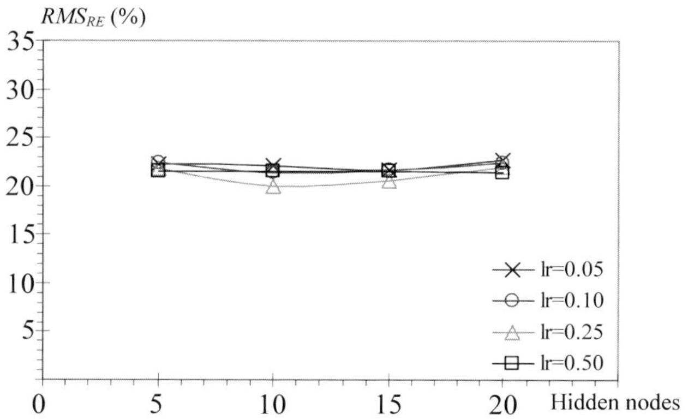

Figure 6. RMS_{RE} of the testing processes (Khrung Thep Bridge station)

Therefore, in the study a single layer network with 3 inputs, 10 hidden nodes, and 0.25 learning rate were used to estimate discharge that should be released from Chao Phraya Barrage as show in Figure 7. For the case of Samut Prakarn station, released discharges was 15 lagging days but was 13 lagging days for Khrung Thep Bridge station.

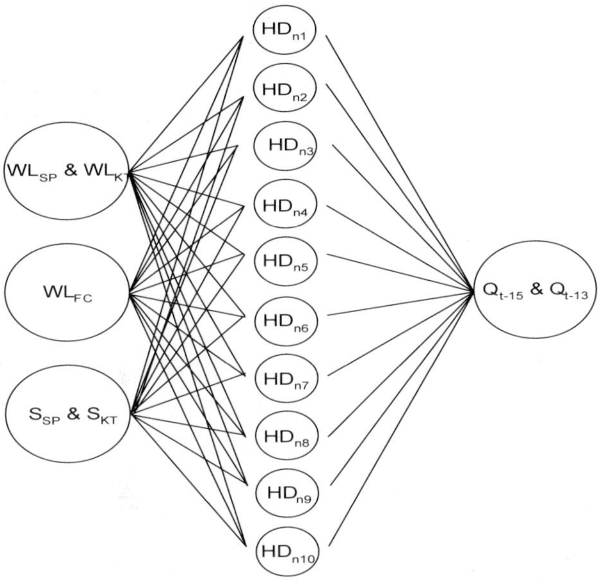

Figure 7. A Single Layer Network with 3-inputs, 10-hidden nodes, and 1-output

5. Conclusions

ANNs was used as a mathematical tool to estimate the amount of upstream released discharge for controlling the appropriate level of salinity in lower reach of Chao Phraya River for environmental control purposes. Analysis of relationships between variables, related to this environmental problem, was conducted between 1996 and 1999 due to availability of data. Such variables are daily salinity and water level at Samut Prakarn station (10km from river mouth), daily salinity data and water level at Khrung Thep Bridge (41.5km from river mouth), daily upstream released discharge at Chao Phraya barrage, hourly tidal level at Fort Chula station (at river mouth). In order to use ANNs to estimate minimum released discharges to maintain required salinity level for downstream environmental control, from Chao Phraya barrage, it is necessary to firstly forecast tidal level at Fort Chula and secondly forecast water level at that required downstream point as input to ANNs and thirdly specify required salinity level as another input to ANNs. Data between 1996 and 1998 was used for model calibration, the results of calculation revealed that ANNs was applicable for this application providing root mean square relative error of 19% at Khrung Thep Bridge and of 22% at Samut Prakarn. Verification of model proposed was conducted using data in 1999, the results of estimation revealed that the model proposed was applicable for estimation of released discharges for salinity control with 20% relative error at Khrung Thep Bridge and 28% relative error at Samut Prakarn.

6. References

Asian Institute of Technology (1978). Salinity intrusion in the Chao Phraya and Mae Klong Rivers, Final report, Electricity Generating Authority of Thailand (EGAT).

Freeman J. A. and Skapura D. M. (1991). Neural networks algorithms, applications, and programming techniques, Addison-Wesley Publishing Company, United State of America.

Leewatchanakul K. (1988). Salinity intrusion in the Chao Phraya River, Master Thesis, Department of Civil Engineering Graduate School, Chulalongkorn University.

Manusthiparom C. (1998). Neural network models for forecasting hourly water level and discharge of the Chao Phraya River at Bangkok Memorial Bridge, Master Thesis, School of Civil Engineering, Asian Institute of Technology.

Nguyen Thi Thu Nga. (2002). A neural network model of the salinity in the west Pearl River estuary, Master Thesis, School of Civil Engineering, Asian Institute of Technology.

Szuster B. W. and Flaherty M. (2000). Inland low-salinity shrimp farming in the central plains region of Thailand, The Chao Phraya delta: historical development, dynamics and challenges of Thailand' rice bowl, http://std.cpc.ku.ac.th/delta/conf/prog_list.htm (p2-i).

Tingsabadh C. and Chaturongpalathipat W. (2000). The costs and benefits from utilizing fresh water for salinity intrusion for tangerine plantation in Bangmod area, Bangkok, The Chao Phraya delta: historical development, dynamics and challenges of Thailand' rice bowl, http://std.cpc.ku.ac.th/delta/conf/prog_list.htm (p3-h).

Tingsanchali T. and Apichatvullop A. (1980). Salinity intrusion problem during severe drought in the Chao Phraya River, Final report, Royal Irrigation Department (RID).

Weesakul U. and Leeratanaruk S. (2003). Analysis of the low flow characteristics for downstream environmental control, *Proceeding of the 1st International Conference on Hydrology and Water Resources in Asian Pacific Region (APHW2003)*. SU-12.

Seasonal Relationship between Flow and Pollutant Concentration in Upper Part of Citarum River

E. W.Irianto[1], A. Yuasa[2], P. Sudjono[3] and B. Machbub[4]

1) Research Institute for Water Resources, Indonesia
Jl.Juanda 193, Bandung, 40153, Indonesia
e-mail: ekowinar@yahoo.com
2) River Basin Research Center, Gifu University
1-1,Yanagido, Gifu, 501-1193, Japan
e-mail: yuasa@green.gifu-u.ac.jp
3) Department of Environmental Engineering, Bandung Institute of Technology
Jl. Ganesha, Bandung, 40132, Indonesia
e-mail: memteq@bdg.centrin.net.id
4) Research Institute for Water Resources, Indonesia
Jl.Juanda 193, Bandung, 40153, Indonesia
e-mail: machbub@bdg.centrin..net.id

Abstract

The relationship between fluctuation of river-flow and pollutants concentration should be assessed, in order to get general phenomena of water quality based on river characteristics and seasons. Unfortunately, analysis of the relationship, between pollutants concentration and river-flow, is rather rare, particularly in Indonesia. This research was held in the upper part of Citarum River. This research uses monthly recorded data between 1990 and 2002, which had been measured at Nanjung Station as an inlet of Saguling Reservoir. The data on river-flow and water quality are plotted to produce graphs, then, the empirical equations of the relationships approximated line are created by trial error method until the correlation coefficient is almost 0.6. It indicates that there is an effect of river-flow on the pollutants concentration such as BOD, COD, Nitrogen and Phosphorous. Those parameters are dominant pollutant parameters that affect Saguling Reservoir. In addition, general phenomena are indicated in this study *i.e* the concentrations of pollutants such as BOD, COD produced from point source reduce during the rainy season because of the dilution effect. Relationship between river-flow and Phosphorous concentrations is also the same with organics phenomenon. However, unique phenomenon is also shown by empirical correlations between the river-flow and Nitrogen *i.e* Nitrogen concentration increases even during the rainy season, mainly in flood conditions. It might be caused by non-point sources originated from the upper part of Citarum Catchment such as heavy-soil erosion in the rainy season.

Keywords: Citarum catchment, empirical equation, pollutant, Saguling reservoir

1. Introduction

Water pollution problems in Indonesia especially in the upper part of Citarum River tend to increase, while efforts on pollution control have not been well programmed mainly by local government. One of the important efforts on water pollution control is water quality monitoring. However, in order to get a water quality model and water quality standards suitable for water pollution control, the results of water quality monitoring should be assessed based on the seasons and the relationships between flow and pollutants concentration.

In addition, river-flow data is also necessary for water quality monitoring because there is a strong relationship between flow and pollutant concentration. Straskaba and Tundisi (1999) state that the type of relationship between flow and concentration of pollutants depends on the type of pollution sources and their intensity. Unfortunately, analysis of the relationships is rather rare, particularly in Indonesia.

Nana and Iskandar (2002) describe that the organics pollutant in the upper part of Citarum Catchment are 77.3 tones/day from domestic, 81.36 tones/day from industries, 25.48 tones/day from agriculture and 14.37 tones/day from livestock. Garno (2002) also estimates that the pollutant loads into Saguling Reservoir are 29.9 million tones/year of organics, 1.36 million tones/year of Nitrogen and 0.21 million tones/year of phosphorous. Those conditions can affect Saguling Reservoir as reported by Brahmana and Achmad (1995). Those studies are very useful to identify the pollutant sources in the river catchment. Nevertheless, those studies have not specifically described the correlation between river-flow and pollutant concentration.

This paper describes the relationship between pollutant parameters, mainly for organic and nutrient parameters, and river-flow in the upper part of Citarum River. The low flow indicates the dry season, whereas the large flow indicates the rainy season.

2. Materials and methods

This research uses monthly record data during 1990 to 2002, which had been measured by the Research Institute for Water Resources. The water sampling is located at Nanjung Station as inlet of Saguling Reservoir as shown in Figure 1. The point is also selected because the pollutant loading originated from the upper part of Citarum Catchment is accumulated into this station.

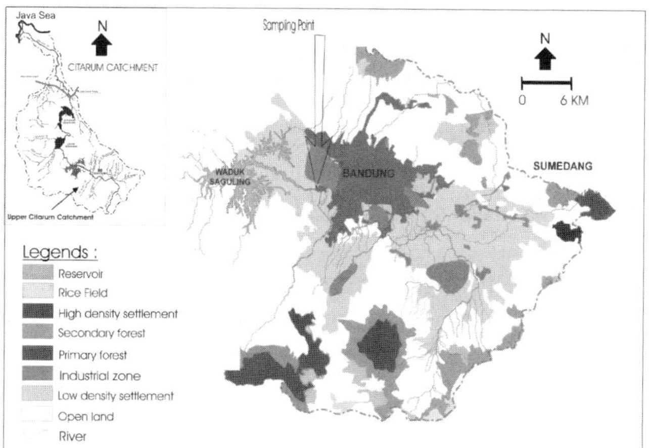

Figure 1. Sampling point in upper Citarum Catchment (Source:Nana and Iskandar, 2002).

The data for both river-flow and pollutant concentration are plotted into graphs, then empirical equations to approximate lines are created. Trial-error method is used to find the empirical equations. The method is stopped if the correlation coefficient between the empirical equation and the real-data reaches around 0.6. After that, approximated line of empirical equations is created to the points representing the relationship between flow and pollutant concentrations. The low flow, until 50 m^3/sec, indicates dry season, whereas the large flow, between 50-150 m^3/sec, indicates rainy season. While, the flow-river of more than 150 m^3/sec indicates rainy season and flood condition.

3. Results and discussion

3.1 Influence of River-Flow Fluctuation on Pollutant Concentration

Organic concentration, as BOD and COD, are general parameters, which are often used to describe organic pollution level. The organics pollution might be caused by activities on domestic, industry, agriculture, and fishery.

Figures 2 and 3 indicate that organics concentrations, as BOD and COD, reduce during the rainy season. This phenomenon might be caused by dilution effect of the river. However, even during the rainy season the organic concentration is still high *i.e* BOD between 10 to 18 mg/l and COD between 20 to 40 mg/l. Those conditions indicate that the natural dilution effect of the river-flow could not reduce the organic pollutant loading.

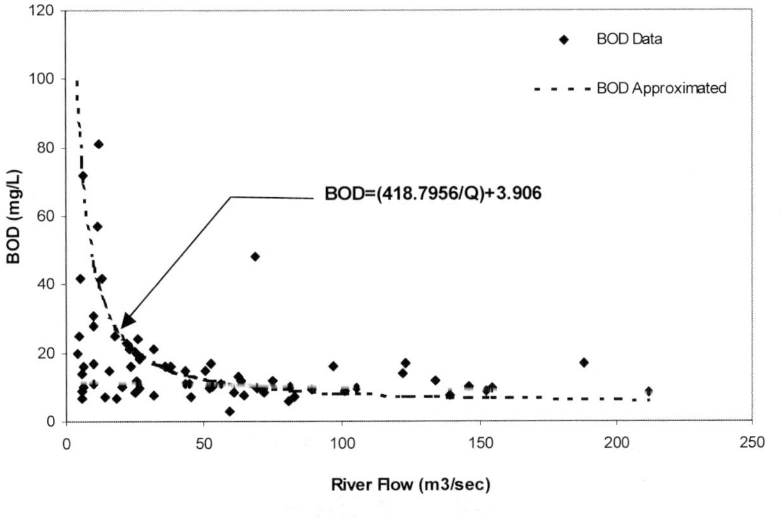

Figure 2. Relationship between river-flow and BOD at Nanjung Station

Dissolved oxygen (DO) is a general parameter that can be used to show the fluctuation of aerobic conditions in the water caused by organic pollutants. However, an interesting phenomenon between river-flow and dissolved oxygen at Nanjung Station is shown in Figure 4. It indicates that the relationship between river-flow and dissolved oxygen tends not to follow a certain equation or a certain line.

However, the figure shows that even in the rainy season the dissolved oxygen tends to be low. The condition might be caused by the high concentrations of pollutants such as organics and nutrients. The changing of Citarum River morphology at Nanjung Station may also affect the low oxygen content even in the rainy season, because the morphology at Nanjung station

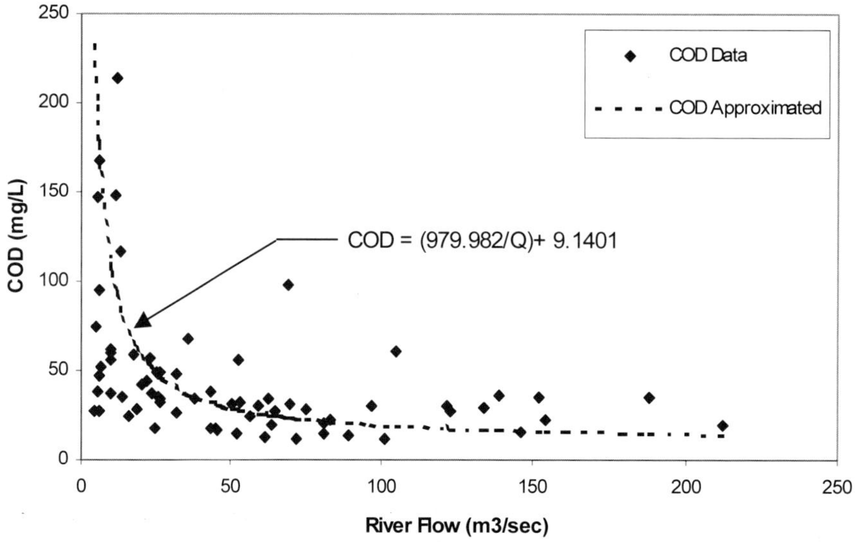

tends to be narrow and deep. Therefore, oxygen in the air may not be absorbed easily.

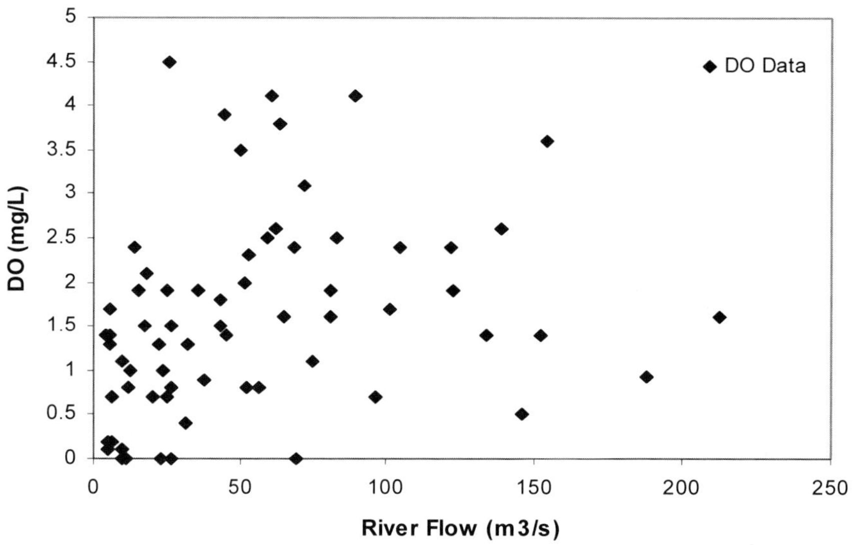

Figure 3. Relationship between river-flow and COD at Nanjung Station

Figure 4. Relationship between river-flow and Dissolved Oxygen at Nanjung Station

Besides organics, this paper also describes nutrient phenomena. Nutrient pollutants are originated by point source *i.e.* domestic and industrial activity, and non-point sources *i.e.* agriculture, fishery and livestock activity.

Nevertheless, the relationship between river-flow and Nitrogen concentration in the upper part of Citarum Catchment has an unique phenomena as shown in Figure 5. There are several interaction stages between river-flow and nitrogen concentration as follows:

a. First stage. In this stage, general phenomena between river-flow and pollutant concentration has occurred until river-flow reaches 100 m^3/sec, that nitrogen concentrations reduce during rainy season. It might be caused by dilution effect.
b. Second stage. The nitrogen concentration relatively tends to plateau at 1.5 mg/l, when the flow between 100 to 150 m^3/sec. The condition could indicate that non-point sources such as activities of agriculture, livestock and fishery are dominant sources polluting the upper part of Citarum River during the rainy season.
c. Third stage. This stage shows that nitrogen concentration tends to increase. The increment of nitrogen concentration might be caused by non point sources mainly heavy soil-erosion in the upper part of Citarum catchment during the rainy season and flood conditions. It indicates that the land erosion process in the upper part of Citarum Catchment might not only bring soil particles to the river, but also contain the nutrients.

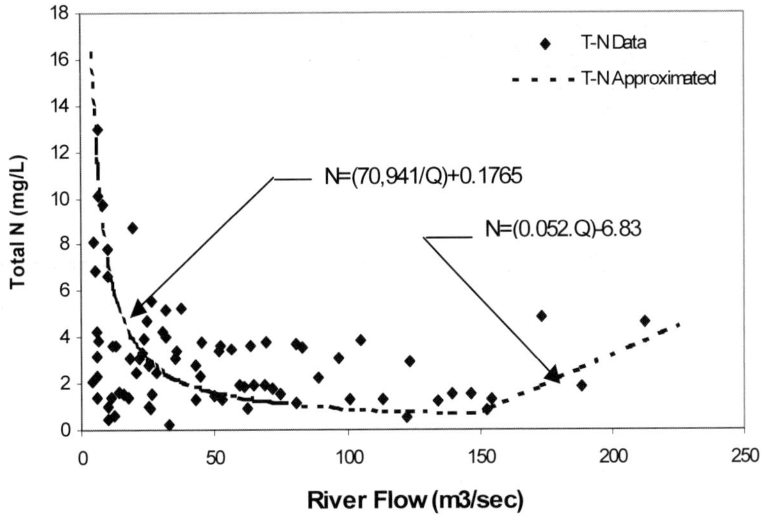

Figure 5. Correlation between River-flow and Total Nitrogen at Nanjung Station

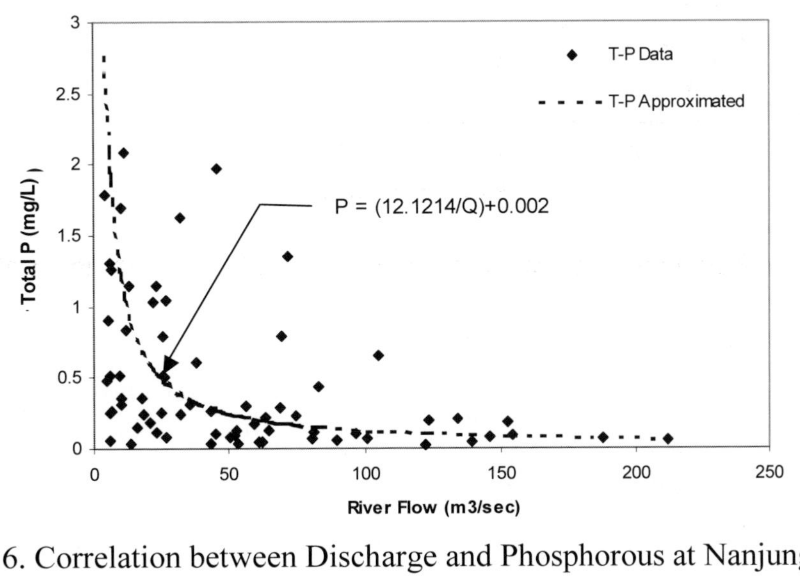

Figure 6. Correlation between Discharge and Phosphorous at Nanjung Station

The other nutrient substance is phosphorous. Phosphorous normally acts as a limiting factor for aquatic life such as phytoplankton in many cases particularly in lakes (Jorgensen, 1998). However, even if the concentration is relatively small, phosphorous with nitrogen can generate an algal bloom in the Saguling Reservoir (Brahmana and Achmad, 1995). With the intention that, the relationship between river-flow and total phosphorous mainly in the upper Citarum must be considered. Similarly, with the organics phenomenon, dilution effect can reduce the phosphorous concentration during the rainy season as shown in Figure 6.

3.2 Empirical Correlation between River Flow and Pollutant Concentration

Dilution effect reduces the organics concentration during the rainy season as shown in Figure 2 and 3. Empirical equations representing the phenomenon are shown in Equation 1 and 2 as follows:

$$BOD = (418.8/Q) + 3.906 \qquad (1)$$

$$COD = (979.98/Q) + 9.14 \qquad (2)$$

While:

BOD = Biological Oxygen Demand (mg/l)

COD = Chemical Oxygen Demand (mg/l)

Q = River flow (m^3/sec)

Equation 1 and 2 show that there is a baseline of organic concentration in Citarum River mainly in the upper part. Those equations also indicate that the baseline concentrations of organics are 3.906 mg/l as BOD and 9.14 mg/l as COD. These values can be used to set up the stream standard in the future.

Equation 1 and 2 also predict that fluctuations of organics loading that flows into Saguling Reservoir is between 18 to 280 tones/day of BOD and 31 to 360 tones/day of COD. Similarly, with those ranges, Nana and Iskandar (2002) estimate, based on the pollutant sources, that BOD loading entering Saguling Reservoir is almost 202 tones BOD/day in 2000, whereas Kurniasih (2002) reported that annual average of COD loading in the upper part of Citarum River is around 280 tones COD/day.

The empirical correlation between river-flow and phosphorous tend to be similar with the organics phenomenon *i.e.* concentrations of total phosphorous reduce during the rainy season. This phenomenon is shown in Equation 3 and Figure 6. Equation 3 shows the empirical correlation as follows:

$$P = (12.121/Q) + 0.002 \qquad (3)$$

While:

P = Phosphorous concentration (mg/l)

Q = River flow (m^3/sec)

Equation 3 also indicates that the baseline concentration of Total Phosphorous in the upper part of Citarum River is almost zero. The correlation line between river-flow and total phosphorous predicts that the fluctuation of phosphorous loading that enters the Saguling Reservoir is between 0.25 to 7.7 tones /day. Brahmana and Achmad (1995) estimate, based on the pollutant sources, that the total phosphorous loading entering Saguling Reservoir is almost 5.8 tones/day.

On the other hand, interactions between river-flow fluctuation and Total Nitrogen concentration have three stages. The first and second stages of the phenomena are shown in equation 4, whereas the empirical equation in the third stage is seen in equation 5.

$$N=(70.941/Q)+0.1765 \quad (4)$$
$$N=(0.052xQ)-6.83 \quad (5)$$

While:

N = Nitrogen concentration (mg/l)
Q = River flow (m^3/sec)

Equation 4 and 5 predict that the fluctuation of total nitrogen loading which enters into Saguling Reservoir is between 0.4 to 85 tones/day. While Brahmana and Achmad (1995) estimate, based on the pollutant sources, that the total nitrogen loading entering Saguling Reservoir is almost 40 tones/day. Those conditions may initiate an euthrophication problem in the Saguling Reservoir (Brahmana and Achmad, 1995).

4. Conclusions

The relationship between river-flow and pollutant concentrations should be assessed in order to get general phenomena of water quality based on river characteristics such as seasons. The relationships are plotted into graphs, and then empirical equations between river-flow and pollutant concentrations such as BOD, COD, Nitrogen, and Phosphor are created by trial-error method. The equations show that the upper part of the Citarum River always contains organic pollutants, such as BOD and COD, even in the rainy season. It means that natural dilution of the Citarum River has little effect on the organic pollutants. Similarly with the organics phenomenon, the concentration of total phosphorous is also affected by the dilution effect.

The empirical correlation between river-flow and Total Nitrogen in the upper part of the Citarum River has an unique phenomenon. First stage, Nitrogen concentrations decrease during the rainy season. Second stage indicates that nitrogen concentration is relatively constant at 1.5 mg/l, when the river-flow is between 100-150 m^3/sec. However, third stage has shown that the nitrogen concentration tends to increase when the river is in a flood condition. Heavy soil-erosion containing nitrogen substances might cause this phenomenon.

The empirical correlation between river-flow and pollutants concentration in the upper part of the Citarum predicts that the fluctuation of organics loading that flows into Saguling Reservoir is between 18-280 tones/day of BOD and 31-360 tones/day of COD. Whereas, the fluctuation loading of nutrients entering into Saguling Reservoir could be predicted between 0.4-85 tones/day of Nitrogen and 0.25 -7.7 tones/day of Phosphorous.

5. References

Brahmana and Achmad. (1995). Euthrophication in three reservoirs at Citarum River, its relation to beneficial uses, In: *Proceeding Workshop on Ecosystem Approach in Lake and Reservoir Management*. IHP-Unesco, Denpasar, pp. 204-205.

Garno,Yudi. (2002). Beban Pencemaran Limbah Perikanan Budidaya dan Eutrofikasi di Perairan Waduk pada DAS Citarum, *Jurnal Teknologi Lingkungan*, 3(2), BPPT, Jakarta, pp. 112-120.

Jorgensen. (1998). A focus on lakes/rivers in environmental education, ILEC Foundation, Shiga, pp. 41.

Kurniasih. (2002). Pengelolaan DAS Citarum Berkelanjutan, *Jurnal Teknologi Lingkungan*, 3(2), BPPT, Jakarta, pp. 82-91.

Nana and Iskandar. (2002). Beban Pencemaran Limbah Industri dan Status Kualitas Air Sungai Citarum, *Jurnal Teknologi Lingkungan*, 3(2), BPPT, Jakarta, pp. 98-106.

Straskaba and Tundisi. (1999). Reservoir water quality management, ILEC Foundation, Vol.9, Shiga, pp. 102-103.

Research Institute for Water Resources. (1999). Data debit di P. Jawa dan Sumatra tahun 1988-1999, Ministry of Public Works Republic of Indonesia , Bandung.

Research Institute for Water Resources. (1997). Data Tahunan Kualitas Air 1990-1996, Ministry of Public Works Republic of Indonesia, Bandung.

GIS Based Watershed Classification in Lao P.D.R.

A. Heinimann

National Centre of Competence in Research (NCCR) North-South
Centre for Development and Environment, University of Berne
Steigerhubelstrasse 3, Berne, CH-3008, Switzerland
e-mail: andreas.heinimann@cde.unibe.ch

Abstract

Every inclined land surface has an inherent potential for soil and water degradation, the seriousness depends on a multitude of parameters such as slope, soil type, geomorphology, rainfall and land use. This potential is checked by the presence of dense natural vegetation cover. In Laos the intensity of commercial logging and agriculture expansion has increased in recent years. This intensified land use leads to reduced vegetation cover, to increased soil erosion, decreasing yield, and finally is likely to influence the hydrological regime. Against this background the Mekong River Commission (MRC) elaborated a spatial explicit Watershed Classification (WSC) for the Lower Mekong Basin. Based on topographic factors derived from a high-resolution Digital Terrain Model, five watershed classes are calculated, giving indication about the sensitivity to resource degradation by soil erosion. An analysis of the WSC in conjunction with forest cover data revealed that more than 37% of Laos can be considered as critical with regard to degradation risk by soil erosion. The WSC allows spatial priority setting for watershed management and generally supports informed decision making on reconnaissance level. In the conclusions the article focuses on general considerations when GIS techniques are used for spatial decision support in a development context.

Keywords: GIS, Laos, soil degradation, watershed classification

1. Introduction

In Lao PDR roughly 80 % of the total area of 236,800 km^2 can be described as mountainous with limited accessibility. It is the Mekong River, which links the remote northern highlands with the floodplains in the South. The Mekong is for the Lao from different views the actual lifeline. The huge potential for hydroelectric production, the potential for irrigation of the fertile floodplains, the important source of fish, the possibility for navigation and the natural political border over a length of 1,500 km makes the Mekong River to be the most dominant natural and economic feature in Laos.

Of its population of 5 million about 85 percent are still engaged in agriculture and just a relatively small portion of the national income is generated in the industrial or the service sector. Consequently the GDP of Laos ranks at the end of the region. With its very high growth of more than 2.9 percent per annum the population of Laos will double within the next 30 years (ADB 2000).

Laos has set up an ambitious programme for hydropower generation mainly in the mountain areas, with a view to exporting most of it for foreign exchange. Currently 65% of the total production is exported to Thailand (ADB 2000). In southern Laos alone, over 30 dam sites have already been identified. Should all these planned projects be realized the hydropower capacity in Lao PDR would increase from the present 203 MW to over 6,200 MW (MAF 2001). The planned hydropower plants will not only create a major source of currency for Laos but will also increase the vulnerability of Laos. The huge investments will on the one hand sharply increase the reliance on the foreign investors and the importing countries and on the other hand the changes in the river regimes and ecosystems could severely affect the rural societies not only in Laos but also in the riparian countries further downstream: Cambodia and Southern Vietnam (Kohler 1998).

Despite significant degradation over the past three decades Lao PDR retains one of the highest proportions of relatively intact forests and one of the richest biodiversities in Southeast Asia. The situation however is not encouraging: Due to the overexploitation and even the extinction of tropical forests in all neighbouring countries except Cambodia the forest areas in Laos are under swelling pressure. The intensity of commercial logging and clearings for agriculture have increased in recent years due to inappropriate land use policies, migration and resettlement, and population growth. In the case of Laos the forest cover is decreasing by 0.5 % per year to around 41 percent in 1997 (FCMP 1997). This intensified land use leads to reduced vegetation cover, to increased soil erosion, decreasing yields and finally is likely to influence the hydrological regime. The rate of the above processes is especially serious in Laos with an average slope of 25.7%.

On the background of these biophysical and socio-economic characteristics of Laos and the associated deterioration of the natural vegetation cover and the related processes, the Mekong River Commission (MRC) initiated the Watershed Classification Project (WSCP). The main aim of the project was to provide a spatially explicit GIS based classification according to the suitability of the major land uses for sound watershed management on a macro level, and to therefore provide a planning tool to facilitate informed decision making in relation to Natural Resource Management (NRM) in Laos and the whole Lower Mekong River Basin (LMB).

The paper at hand aims at giving a brief overview about the mentioned Watershed Classification Project (WSCP), its main products, results and challenges. Furthermore on an aggregated level, the WSCP shall serve as an example to shed some light on critical issues involved when implementing a regional GIS project in a development context.

2. The watershed classification approach

There are as many watershed classifications as there are aims for which the classification is used. A classification for the purpose of hydropower generation will look different from one dealing with forestry and agriculture. The watershed classification which is presented here aims at giving an indication of the sensitivity of watersheds with regard to water resources degradation mainly by soil erosion. On the basis of the topography various parameters crucial to degradation processes such as slope, elevation and landform are linked in a watershed class prediction equation. The potential degradation risk is then calculated for each land unit into five classes: Watershed Class one is the most sensitive and Watershed Class five the least sensitive to potential soil erosion. Watershed Class one and two can be considered as critical and classes three to five as uncritical with regard to soil erosion when cleared of natural vegetation. For each watershed class, general recommendations for sustainable land uses are given:

Watershed Class 1: Protection Forest
Areas with very steep slopes and rugged landforms, commonly uplands and headwater areas. Critical areas for water and soil resources management. Recommended land use: as a rule, these areas should be under permanent forest cover. Other existing land uses based on traditional rights and practices should be considered carefully with regard to their impact on water and soil conservation.

Watershed Class 2: Commercial Forest
Areas with steep slopes, usually at higher elevation. Landforms are in general less susceptible to water and soil degradation than under WSC Class 1. Recommended land use: forest (conservation and production forests), agro-forestry and grazing, if accompanied by strict conservation measures.

Watershed Class 3: Agro-Forestry
Areas with moderate to steep slopes and less erosive landforms. Includes uplands and foot zones of slopes. Wider range of land use tolerable than in WSC Classes 1 and 2 from point of view of water and soil conservation. May be used for commercial forest, grazing and combinations of trees and agricultural crops, if appropriate conservation measures are applied.

Watershed Class 4: Upland Farming
Gently sloping lands. Moderate need for water and soil conservation depending on local conditions. Wide range of land use possible from point of view of water and soil conservation: agriculture and forest.

Watershed Class 5: Lowland Farming
Gently sloping land and flat areas. Suitable for a wide range of land use from point of view of water and soil conservation: paddy rice, other agricultural uses and forest.

It is crucial to note that other land uses are not excluded, provided that adequate care is given to soil conservation. For instance for areas in Class 1, the recommended land use is forest cover, but there are many examples, which clearly show that forest cover is not the only solution for conserving steep watersheds.

3. The watershed classification methodology

The applied methodology is based on a watershed classification originating and used in Thailand by the Royal Forestry Department in the late 1980s (Wooldridge 1986). The method developed for watershed classification in Thailand applied a multivariate statistical analysis for establishing relationships between a number of variables and watershed class numbers. It was based on the best currently available data for the variables of slope, elevation, landform, geology and soil. As there was no soil and geology data available in sufficient quality when the project started in the early 1990s, the variables were reduced to only three, which are slope, landform and elevation. Involving the experts who had developed the method for Thailand, pilot areas were chosen in the Lao P.D.R. and Vietnam, and the parameterisation of the input variables and the coefficients for the equation and their relation to the watershed classes were re-evaluated (MRCS 1997).

WSC equation:

*Watershed Class (WSC) = a – (b * slope) + (c * landform) + (d * elevation)*

Coefficients:

$a = 1.79$ $b = 0.035$ $c = 0.163$ $d = 0.002$

Based on the contours, rivers and spot heights digitized from the 1:50,000 American topographic maps, a Digital Terrain Model (DTM) with a resolution of 50 meters was calculated for the 412 map sheets in Laos (for the entire LMB 1030 sheets have been processed) within a standard Geographic Information System (Arc/Info). In order to overcome various shortcoming of the standard DTM creation algorithms of Arc/Info a sophisticated programme package was developed (NTPHI). Finally the three parameters of elevation, slope and landform were derived from the resulting DTM and fed into the WSC-equation.

4. Results

Based on the DTM the Watershed Classification was prepared in 50 meter resolution for the entire territory of Laos.

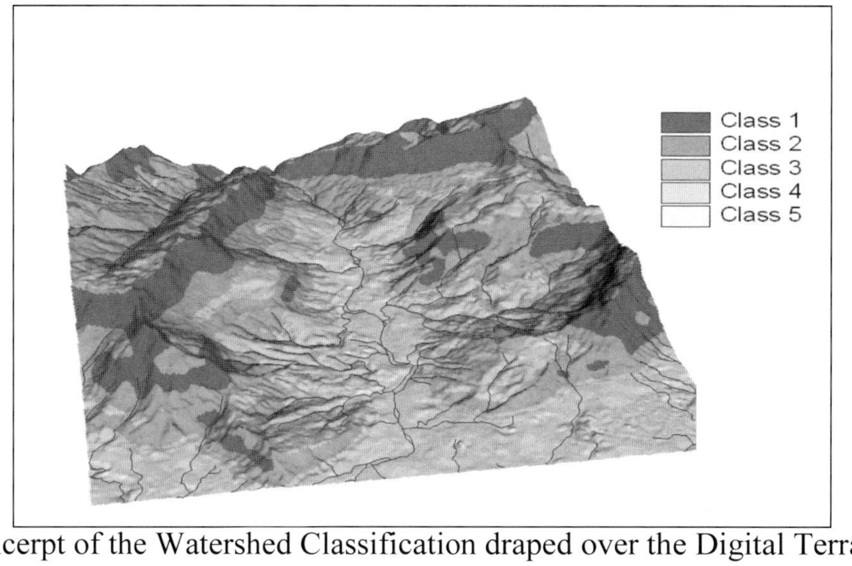

Figure1. Small excerpt of the Watershed Classification draped over the Digital Terrain Model

Almost 60 % (see Table 1) of the country area belongs to Watershed Class one or two, and can therefore be considered as critical concerning potential soil erosion when cleared of its dense natural vegetation cover.

Table 1. Watershed Classes in Laos

WSC Class	*Area in km^2*	*Percentage*
Class 1	48'008	20.3
Class 2	90'973	38.4
Class 3	37'831	16.0
Class 4	31'859	13.5
Class 5	28'130	11.9
Total	*236'800*	*100.0*

To get a clearer picture of the actual area potentially endangered by soil erosion, the Watershed Classification was analyzed in conjunction with the actual forest cover of Lao PDR in 1997. The data has been elaborated by the Forest Cover Monitoring Project (FCMP) of MRC based on visual interpretation of Landsat TM scenes (FCMP 1997). The results show an alarming picture: more the 37 % of the country area falls into Watershed Class one or two, with its steep and rugged landforms, and were not covered by dense forest or bush vegetation in 1997. These areas can be considered as very critical with regard to degradation risk by soil erosion on a macro level and should be subject to closer examination.

Besides the WSC dataset, the Watershed Classification Project (WSCP) elaborated a comprehensive spatial database including national datasets ranging from river network to infrastructure, demography, protected areas and administrative boundaries only to name a few. The potential of this data within a Geographic Information System (GIS) is huge and the resulting analysis can be used for inventorying, monitoring, modelling and project planning on a macro level (national to provincial).

Basically, three potential fields of applications of these databases and related aims can be distinguished:

Mapping: Based on this comprehensive database manifold thematic maps at various scales can be elaborated. The WSCP for example has produced country maps of different themes (e.g. elevation, subcatchments) at a scale of 1:1 Million, and a set of 1:250,00 Watershed Classification maps cover the Lao PDR. Generally speaking the main aim of mapping was to visualize the degradation risk and to therefore create awareness about the existing problems. Furthermore it is obvious that on a large scale the maps can serve as the basis for fieldwork in various domains.

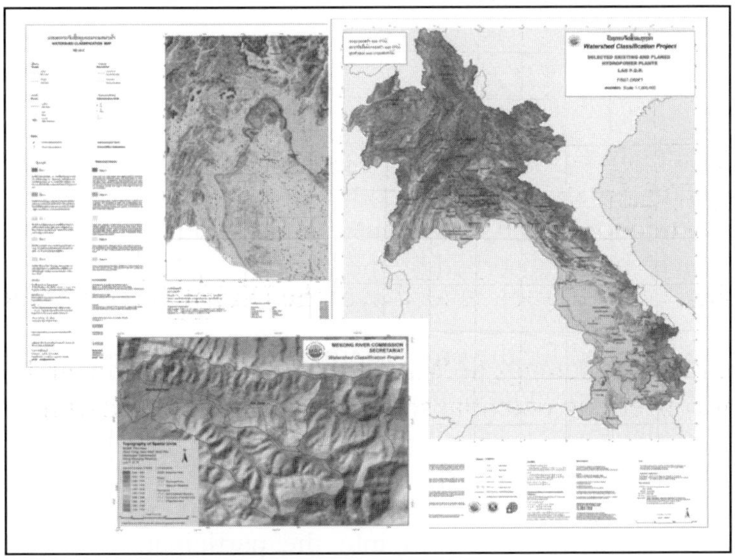

Figure 2. Various thematic maps, from macro to micro level, can be produced based on the WSC database

Analysis: The spatial cross-analysis of the interdisciplinary database has a great potential to assist informed decision making on a national to provincial level in general. As mentioned above, the WSC data in conjunction with forest cover datasets for example allows spatial priority settings for watershed management. By applying GIS analysis functionality, such as network tracing and accumulating as well as location-allocation analysis, problems like equitable water allocation to user sites, water use management during drought situations and planning of new water abstraction permits can be analysed.

Modelling: A raster based GIS in concurrence with an appropriate spatial database opens a wide range of spatial modelling possibilities. Principally modelling always has the aim to either assist in understanding certain phenomena and processes (what is where and why is it there), or to evaluate different scenarios against each other (what if?). An example of simple dam site reconnaissance modelling for the Nam Ngum dam in Laos using the elaborated 50 meter DTM is illustrated in Figure 3.

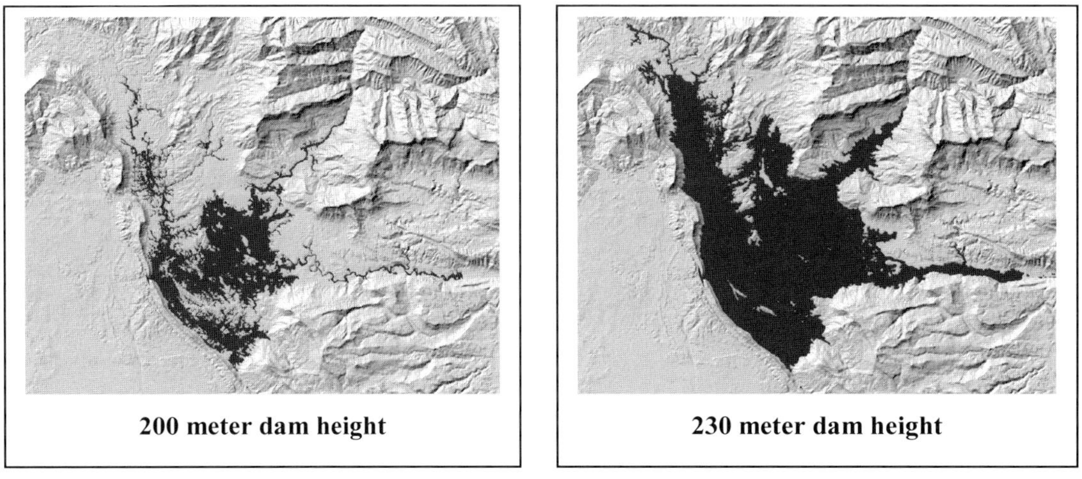

Figure 3. Evaluation of different dam height scenarios and the respective extent of the flooded area for the Nam Ngum reservoir in Lao PDR.

5. Conclusions

It may be concluded that the WSCP data has a great potential for awareness creation among policy makers and the general public for issues relating to sound watershed management. However, awareness creation implies extraction of the main issues at stake and transformation into clear and intelligible messages, which can be absorbed by the targeted audience.

The following principles have to be kept in mind for the proper utilization of the WSCP data:

- The WSCP database is designed for reconnaissance information at the macro-level (basin wide, national, and provincial levels). For micro-level use, the database has to be refined and complemented with other information including field investigations.
- The WSCP data must be put in context with the complex reality of watershed management. The WSCP data should be integrated into the participatory decision making process which is necessary for sound watershed planning and management.

These principles do obviously not only apply to the use of GIS analysis techniques for Watershed Classification, but have a much larger scope. In the following paragraphs some further issues of the application of GIS in a development context shall be discussed. This is especially relevant as the mushrooming of GIS is still ongoing and the potential of GIS is nowadays exploited in an ever increasing number of development projects and programs.

Some reflections on critical issues of GIS in a development context

Since GIS is a "post-industrialized technology" developed of and for the needs of the societies in Europe and North America (Christiansen 1998), a number of critical issues have to be considered when implementing GIS activities in developing or transition countries.

On an organizational level, the introduction of GIS in a particular agency was repeatedly observed to alter power relations in governmental structures. In various countries it was frequently observed that the GIS unit of government departments became prestige objects, and that it entailed an "unhealthy" intra-government competition between different agencies.

Adequate funding and staffing over a reasonable period of time are major factors for a successful implementation. On the one hand maintenance and support must be secured for the rapidly changing hard and software requirements resulting from the short lifespan of computer technologies. On the other hand empowering of local technical staff is perquisite for the sustainability of GIS activities. This can only be achieved by providing know how transfer and continued backstopping through a long-term commitment.

Besides organizational matters, data availability, data exchange and legal issues are of paramount importance for the overall exploitation of the potentials of GIS for sustainable development. Development projects frequently operate in areas, where accurate and up-to-date data are hardly existing or difficult to access. Often it is reported that data sharing and exchange is impeded by power and political motivations within state administrations and civil society organisations. Furthermore the fact that governmental rules and laws concerning data exchange were usually established before the "digital revolution" and that they are therefore not yet adapted to the today's requirements, may be a source of problems. Or as Dunn *et al.* (1997) put it:

'Data sharing may be especially problematic in environments which are not "information – driven" and in which bureaucracy presents particular obstacles. Competitiveness and empire building prevent the free flow of data because common goals are not shared.'

On a technical level data exchange among stakeholders are often further hampered or even made impossible by barely agreed upon data formats and projection standards, leading to unexploited synergies and costly duplication of efforts.

A crucial factor but often neglected in development efforts applying GIS is the close linkage between technical possibilities and thematic focus. Unlike in research, in development programmes thematic knowledge and technical know-how of GIS related activities seldom can be met by one staff member or even within one single unit. On this background, it is no surprise that GIS components tend to become self-dynamic and tend to be detached from the overall respective objectives. This is probably a key, why even technically perfectly shaped GIS are too often not meeting the expectations and their benefit and impact remains little compared to their actual potential (Heinimann et al 2003).

Concluding it has to be noted that better information does not forcibly lead to better decisions, as decisions are often not the result of better spatial information alone, but are guided by factors such as politics, power-relations or personal benefits. Christiansen (1998) concludes 'the less rational the management culture, the less useful an information system will be'.

Nevertheless, experience in Laos and all over the globe have shown that relevant spatial data, keyed to users' needs, well presented and explained, can make a difference and strengthen rationality of decision making at all levels.

6. Acknowledgment

The paper at hand was prepared within the Swiss National Centre of Competence in Research 'North-South' (NCCR N-S) and its Individual Project 1 (IP1) dealing with the methodological framework of syndrome mitigation research. The NCCR N-S is jointly funded by the Swiss National Science Foundation (SNF) and the Swiss Agency for Development and Cooperation (SDC). The author expresses his gratitude to the Mekong River Commission Secretariat (MRCS) in Phnom Penh and the Centre for Development (CDE) of the University of Berne for their kind support and assistance. The author is especially grateful for the invaluable advice and support of Mr. Thomas Breu (CDE) and Dr. Thomas Kohler (CDE).

7. References

Asian Development Bank (ADB). (2000). Environments in transition, Cambodia, Lao PDR, Thailand, Viet Nam, Asian Development Bank (ADB), Manila, Philippines.

Christiansen T. (1998). Geographical Information Systems for regional rural development projects in developing countries, Giessener Geographische Schriften 75, Giessen, Germany.

Dunn C.E., Atkins P.J., Townsend J.G. (1997). GIS for development: a contradiction in terms? Area, Vol. 29, No. 2, pp. 151-159.

Forest Cover Monitoring Assessment and Monitoring Project (FCMP). (1997). Forest cover monitoring project datasets, Mekong River Commission (MRCS), Phnom Penh, Cambodia.

Heinimann A., Breu T., Kohler T. (2003). Geographic Information Systems in mountain development, *Mountain Research and Development,* Vol. 23, No.3 (in press).

Kohler T. (1998). The Mekong Basin: managing the pressures of rapid economic development, In: Mountains of the world: water tower for the 21st century, Commission on Sustainable Development, Mountain Agenda, Haupt , Berne, Switzerland.

Ministry of Agriculture of Lao PDR (MAF). (2001). Concepts and practises of integrated watershed/ river basin management in Lao PDR, Report prepared by the National Capacity Building Project, Vientiane.

Mekong River Commission Secretariat (MRCS), (1997). Mekong watershed classification: the WSC map user guide, Mekong River Commission Secretariat (MRCS), Bangkok.

Mekong River Commission Secretariat (MRCS). (1997). Mekong River Basin diagnostic study, Mekong River Commission Secretariat (MRCS) and UNEP, Bangkok, Thailand.

Woolridge D. (1986). Watershed classification project in Thailand, IUCN (International Union for Conservation of Nature and Natural Resources).

Potential Contamination of Heavy Metals in Fish along Pong and Chi Rivers

B. Sripanidkulchai[1)], W. Sirisangtrakul[2)], A. Priprem[3)], P. Chalorkpunrut[4)] and W. Wirojanagud[5)]

1) Faculty of Pharmaceutical Science, Khon Kaen University, Khon Kaen, 40002, Thailand.
e-mail: bungorn@kku.ac.th
2) Faculty of Pharmaceutical Science, Khon Kaen University, Khon Kaen, 40002, Thailand.
e-mail: jim279th@yahoo.com
3) Faculty of Pharmaceutical Science, Khon Kaen University, Khon Kaen, 40002, Thailand.
e-mail: aroonsri@kku.ac.th
4) Department of Fisheries, Faculty of Agriculture, Khon Kaen University, Khon Kaen, 40002, Thailand.
e-mail: pracha@kku.ac.th
5) Department of Environmental Engineering, Faculty of Engineering, Khon Kaen University, 40002, Thailand.
e-mail: wanpen@kku.ac.th

Abstract

Levels of heavy metals including Cd, Co, Cr, Cu, Ni, Pb and Zn in fish from 8 predetermined sampling sites along the Pong and Chi Rivers were investigated. Fish were randomly collected from each site and categorized as herbivorous and carnivorous. The sites, marked A, B, C, D, E, F, G, H, located downstream between Ubolratana reservoir and Mahasarakham weir covering a river length of about 180 km. The livers of the fish were immediately separated, processed and analyzed by atomic absorption method. The hepatic levels of Cd, Co, Cr, Cu, Ni, Pb and Zn of the herbivorous fish were 0.0874-3.2477, 6.7-84.98, 3.68-43.2, 60.3-1,620.6, 2.6-41.5, 0-6.32 and 86.8-5923 µg/g dried weight; those of carnivorous being 766-1.7976, 8.51-47.69, 2.68-34.7, 135-807.6, 1.39-15.8, 0.5-13.86 and 6603.3-3,106 µg/g dried weight, respectively. The overall results suggest that site B exhibited the highest levels of the heavy metals accumulated in the fish livers, and, that the herbivorous fish showed higher levels of metal contaminants than the carnivorous. The highest hepatic cadmium content of herbivorous fish was found at site H, whereas those for carnivorous fish were at site B. Both herbivorous and carnivorous fish from site B also contained the highest hepatic Pb levels. Investigation of heavy metals contained in fish should be one of the key indicators that reflect the severity of heavy metal contamination in natural water resources.

Keywords: carnivorous, fish tissue, heavy metal, herbivorous, Pong River

1. Introduction

The Pong River is an important water resource of the Northeast Region of Thailand, particularly for Khon Kaen Province. The Pong River runs downstream towards the Chi River, which runs through Mahasarakham Province and merges with the Moon River and then subsequently flows towards Cambodia. Any toxic moieties that affect living creatures in the Pong River could thus generate significant incidences of pollution outbreaks. There are some industries located along part of the Pong River. Incidences of pollution of the Pong River by accidental releases of chemical residues from the industries located along the river were reported in 1986, 1993, and 1994 (Khon Kaen University, 1995). Since then, a national alert has called for rehabilitation of the environmental quality of the Pong River. Poor water quality periodically kills many fish and other living creatures along the river. In 1995, this phenomenon occurred in the Pong River, particularly in the parts of the river where the water quality was classified as class 3-4 as per the Ubolratana Reservoir's water quality report, Chod lagoon, Nong Wai irrigation weir. Chod lagoon, in particular, was classified as class 4. Most of the rehabilitation efforts mainly focused on information based on the quality of water. However, monitoring and thorough investigation of the living creatures in the ecosystem of the Pong River, such as fish, shells fish, plants, plankton, etc., could constitute vital information for environmental management of the river in the future.

All heavy metals are potentially harmful to most organisms at some level of exposure and absorption. Certain heavy metals such as Cd, Co, Cr, Cu, Ni, Pb and Zn are known to be related to the industry. The pollution of fresh water fish with heavy metals is an issue both from a hygienic and from an eco-toxicological point of view. Industries and communities situated on or nearby the riverbanks are usually the major causes of the waste generation.

Recent reports using levels of heavy metals in fish to indicate and assess the contamination of heavy metals in water were found to be very useful for environmental monitoring (Yilmaz, 2003, Szefer et al., 2003). There is little or no information regarding heavy metal accumulation in fish habitats along the Pong and Chi Rivers, particularly with respect to the distribution of heavy metals. The distribution of heavy metals in the fish could result in bio-transference and bio-magnification of the hazardous heavy metals to humans, because of our culture of fish consumption (Mansour and Sidky, 2003). It is also suspected that the water might be receiving some heavy metal(s) from various sources, not only from the industrial wastes but also from households and animal farming nearby the river. The liver of fish is one of the organs, which is most susceptible to storage of heavy metals and therefore commonly used in analysis (Yilmaz, 2003). The study of heavy metals such as Cd in the living organs usually emphasizes on the analysis of the content in the liver and/or some other related tissues (Sirisangtrakul, 2000). In our study, the liver of the fish living in the predetermined area of the Pong and Chi Rivers were the primary object intended for investigation. Muscles and some adipose tissues might be of particular interest because of the use as human food. Gills are the target organs for assessing metal accumulation that reflects the concentration of metals in water where the fish species live.

In this study, we have measured the levels of Cd, Co, Cr, Cu, Ni, Pb and Zn by weight in fish from the Pong River, in order to determine the heavy metal accumulation in the fish at different sites and to investigate the distribution of the heavy metals in various fish organs.

2. Study sites

The Pong River originates from Phoo-Kra-Dueng Mountain and the Chi River sub-watersheds. It flows through Amphur Phoo-Kra-Dueng in Loei province to Ubolratana reservoir and Nong Wai weir, as shown in Figure 1. The river has many tributaries. The downstream watershed area comprises the Pong River and its tributaries from Ubolratana reservoir downstream to Mahasarakham weir on the Chi River and a short river section of the Chi River from Ban Tha Pra to its confluence with the Pong River. The down stream watershed area is divided into various sub-sections: Ubolratana reservoir to Nong Wai irrigation weir, Nong Wai irrigation weir to Ban Na Piang; Ban Na Piang to Mahasarakham weir (site H in Figure 1) and Ban Ta Pra downstream to the Chi and the Pong River confluence.

Major tributaries which were predetermined to be the sites of sampling in this study, included Huay Sai (site A as shown in Figure1), Huay Kum Mum, Huay Chod and Chod lagoon (site B in Figure 1) and Huay Sue Ten and Sue Ten lagoon (site C in Figure 1). Nong Wai irrigation weir (#2 in Figure 1) is located about 34 km from Ubolratana dam (#1 in Figure 1). The community near Nong Wai weir, namely Nong Bua Noi (site D in Figure1.) was one of the predetermined sites of sampling in this study.

Downstream from Nong Wai irrigation weir (#2), the Pong River flows to join the Chi River. Between Nong Wai irrigation weir and 150 km downstream to Mahasarakham weir (site H), two lagoons were selected for sampling, i.e. Tung Taew lagoon (site E) and Bua Noi lagoon (site G). The sampling station at Ta Kra Serm (site F) was located between site E and G, and the last site of study was at Mahasarakham weir (site H).

In summary the industries situated within the area of study comprise a pulp and paper industry and an electrical generating plant near site B; and some foods and drinks production plants near sites E, F and G.

There are numerous communities along the river section studied, and the land use is dominated by agriculture mainly paddy or rice fields. The most common period of rice planting starts in April or May, which is when the fields and soils are prepared for cultivation. The use of fertilizers is increased from then on until the harvesting time in November and December.

Potential pollution affecting the Pong and Chi Rivers could be detected as discharges from various sources such as industries, animal farms, affected soils and land used for agricultural activities, waterweeds and community (domestic) wastes. Some examples of the industries along the rivers include distillery, sugar mill, tapioca mill, and a paper and pulp mill. The Pong and Chi Rivers support large commercial and recreational fishing communities. Decrease in fish populations has been observed with a decline in water quality.

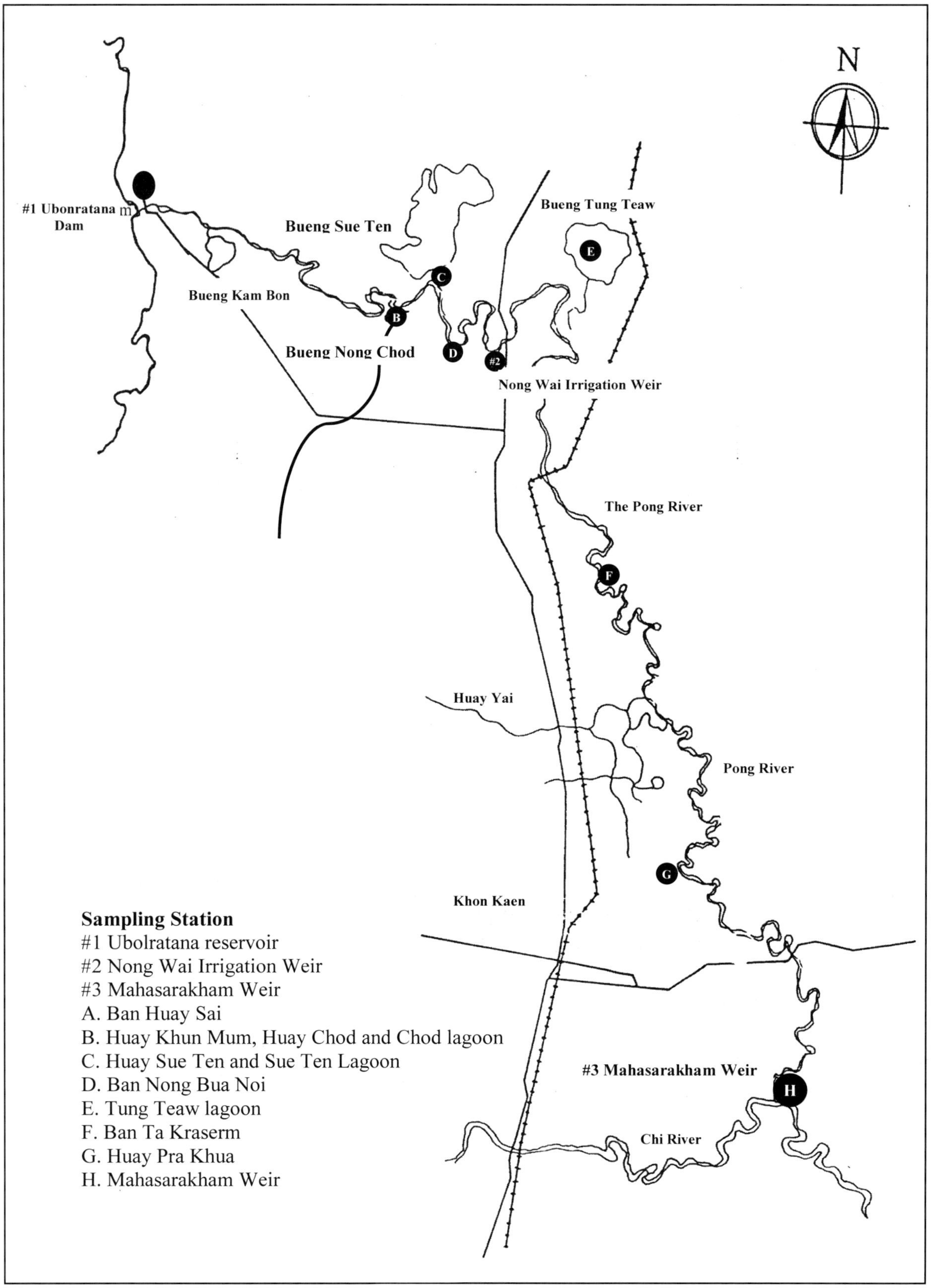

Figure 1. Sampling Stations along the Pong River

3. Materials and methods

3.1 Samples collection

A survey prior to sample collection was performed to determine the sites and the duration of sampling. The sampling was carried out during the dry season between April and November of 2002 when the water level of the river would be lowest and, thus, the highest concentration of most of the contaminants and pollutants would be expected. November is the end of the cultivation and the beginning of harvesting season, and this month would usually be the time when the land would be peaked with fertilizers and other agricultural substances. The sites for sample collection were chosen due to high concentrations of pollutants as observed in previous studies (Khon Kaen University, 1995) and shown in Figure 1.

Two categories of fish, herbivorous and carnivorous, were randomly sampled from 8 predetermined sites of the Pong River. The fish samples were collected using a beach seine from a boat. Mature full-sized individuals of each fish species, were categorized by a fishery scientist to minimize variations in metal concentration due to size of the organisms. The fish samples were then collected and immediately transferred to the laboratory in sealed plastic bags for sample analysis.

3.2 Preparation and analysis of samples

The whole liver of each fish was separated, weighed and subjected to pretreatment and analysis, as follows:

Drying:	Each sample was dry-heated at 70°C until a constant weight was obtained.
Digestion:	The dried sample was soaked with concentrated nitric acid until completion and hydrogen peroxide was added to bleach the digested sample. The dried sample was dispersed in de-ionized water and separated using a paper filter.
Analysis:	The sample was analyzed by using flame atomic absorption. The method was validated prior to use. The metal levels were calculated in micrograms per gram of wet weight and dried weight.

4. Results and discussions

The random sampling of fish in habitats along the Pong and Chi Rivers at site A, B, C, D, E, F, G and H resulted in 3 species of herbivorous fish, namely *Trichogaster trichopterus, Henicorhynchus siamensis, Osteochilus hasselti* and two species of carnivorous fish, namely *Hampala dispar, Channa striata*, as shown in Figure 2.

2.1 Herbivorous fish

a) *Trichogaster trichopterus*

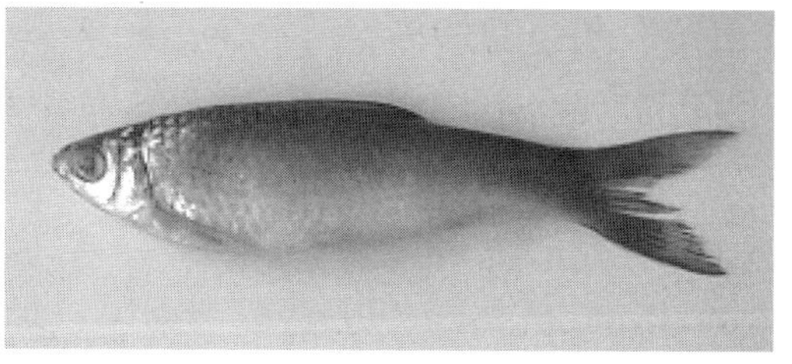

b) *Henicorhynchus siamensis*

c) *Osteochilus hasselti*

2.2 Carnivorous fish

a) *Hampala dispar*

b) *Channa striata*

Figure 2. Illustration of fish sampled from sites A-H categorized into 2.1 herbivorous and 2.2 carnivorous fish.

Mean levels of heavy metals in the liver of herbivorous fish in habitats in the Pong and Chi Rivers collected from sites A-H is shown in Table 1. The range of the hepatic levels of Cd, Co, Cr, Cu, Ni, Pb and Zn of herbivorous fish samples were 0.026-0.964, 1.88-14.78, 1.24-14.7, 14.94-281.8, 0.77-7.22, 0.15-1.1, 21.5-1030 µg/g wet weight; 0.0874-3.2477, 6.7-84.98, 3.68-43.2, 60.3-1,620.6, 2.6-41.5, 0-6.32 and 86.8-5923 µg/g dried weight, respectively. It was observed that the majority of heavy metals; Co (14.78 µg/g wet weight or 84.98 µg/g of dried weight), Cu (281.2 µg/g wet weight or 1,620.6 µg/g of dried weight), Ni (7.22 µg/g wet weight or 41.5 µg/g of dried weight), Pb (1.1 µg/g wet weight or 6.32 µg/g of dried weight) and Zn (1030 µg/g wet weight or 5932 µg/g of dried weight), were observed in the samples from site B. Those from site F and H were Cr (11.1 µg/g wet weight or 43.20 µg/g of dried weight) and Cd (0.964 µg/g wet weight or 3.248 µg/g of dried weight), respectively. For site A and D, the samples of fish liver were too small to assure the quality of the analysis, because the size of the fish and also the liver of the fish were smaller when compared with the carnivorous, causing difficulties and insufficiencies in the separation of an adequate amount of liver needed for the analysis.

It is noticed that the livers of herbivorous fish collected from sites B, C, E, F and H exceeded the permissible tolerable weekly intakes of Cd and Pb of 0.043 and 0.033 µg/g wet weight (FAO/WHO, 1989). These tolerable limits are usually applied for the muscles of the fish we intake for food (the results of which are being investigated), however, this application was intended to suggest the safety of fish consumption.

Table 1. Levels of heavy metals in the liver of herbivorous fish sampled from sites A-H

site	*Cd*		*Co*		*Cr*		*Cu*		*Ni*		*Pb*		*Zn*	
	a*	b**	a*	b**	a*	b**	a*	b**	a*	b**	a*	b**	a*	b**
A	-	-	-	-	-	-	-	-	-	-	-	-	-	-
B	0.046	0.265	14.78	84.98	-	-	281.8	1620.6	7.22	41.5	1.1	6.32	1030	5923
C	0.032	0.013	1.88	7.600	7.53	30.4	14.94	60.3	2.01	8.1	0.15	0.6	21.5	86.8
D	-	-	-	-	-	-	-	-	-	-	-	-	-	-
E	0.026	0.087	14.15	47.86	4.67	15.8	119.5	404.3	6.32	21.4	0.85	2.88	229	775
F	0.799	3.110	8.79	34.2	11.1	43.2	149.7	582.5	3.85	15	0	0	0	160
G	0.035	0.103	3.57	10.55	1.24	3.68	96.32	284.5	5.79	17.1	0	0	594	1755
H	0.964	3.248	1.99	6.7	4.36	14.7	22.07	74.4	0.77	2.6	0.24	0.8	45.3	153

* values expressed as average data of duplicate or triplicate samples in the unit of µg/g of **wet** weight of fish.
values expressed as average data of duplicate or triplicate samples in the unit of µg/g of **dried weight of fish.

Mean levels of heavy metals in the livers of carnivorous fish sampled from the Pong and Chi rivers are shown in Table 2. The range of the hepatic levels of Cd, Co, Cr, Cu, Ni, Pb and Zn of herbivorous fish samples were 0.06-0.413, 1.86-9.63, 0.56-11.21, 45.8-177.8, 0.47-4.04, 0.13-3.18 and 131.6-683.9 µg/g wet weight or 0.177-1.798, 8.57-47.69, 2.68-34.7, 135-807.6, 1.39-15.8, 0.5-13.86 and 603.3-3,106 µg/g dried weight, respectively. The highest levels of the heavy metals detected from the liver of the carnivorous fish were as follows from site A - Cu (177.8 µg/g wet weight or 807.6 µg/g of dried weight) and Zn (683.9 µg/g wet weight or 3106 µg/g of dried weight); from site B - Cd (0.413 µg/g wet weight or 1.798 µg/g of dried weight) and Pb (3.18 µg/g wet weight or 13.86 µg/g of dried weight); from site F - Ni (4.04 µg/g wet weight or 15.80 µg/g of dried weight); from site G - Cr (11.21 µg/g wet weight or

34.70 μg/g of dried weight); and from site H - Co (9.63 μg/g wet weight or 47.69 μg/g of dried weight).

As the levels of Cd and Pb from the liver of the herbivorous fish were over the FAO/WHO limits, one could anticipate that the carnivorous fish would also contain high levels of Cd and Pb. This was the case in this present study. The livers of carnivorous fish collected from all sites exceeded the permissible tolerable weekly intakes of Cd and Pb of 0.043, 0.033 μg/g wet weight, respectively. These tolerable limits are usually applied for the muscles of the fish we intake for food (the results of which are being investigated), however, this application was intended to suggest the safety of fish consumption.

Table 2. Levels* of heavy metals in the liver of carnivorous fish sampling from the Pong river

site	*Cd*		*Co*		*Cr*		*Cu*		*Ni*		*Pb*		*Zn*	
	a*	b**	a*	b**	a*	b**	a*	b**	a*	b**	a*	b**	a*	b**
A	0.192	0.709	3.120	14.160	8.440	31.200	177.8	807.6	2.48	11.27	1.8	8.18	683.9	3106
B	0.413	1.798	6.700	29.170	5.780	21.200	95.2	414.2	3.08	13.42	3.18	13.86	248.1	1079
C	0.060	0.177	3.170	9.340	1.160	3.420	45.8	135	0.47	1.39	0.52	1.53	251.2	739.8
D	0.023	0.767	1.860	8.570	8.390	31.100	37.65	172.6	3.18	11.8	0.13	0.5	131.6	603.3
E	0.054	0.260	2.770	13.250	0.560	2.680	60.44	289.3	1.06	5.1	0.32	1.54	175.4	839.3
F	0.389	1.523	7.290	29.230	4.110	16.100	75.66	303.4	4.04	15.8	1.02	4.09	206.6	828.4
G	0.299	1.274	5.620	23.970	11.210	34.700	88.09	375.5	1.78	5.5	0.81	3.44	243	1036
H	0.052	0.256	9.63	47.69	0.9	4.44	67.5	334.3	0.32	1.59	0.32	1.59	192.2	951.9

* values expressed as average data of duplicate or triplicate samples in the unit of μg/g of wet weight of fish.
**values expressed as average data of duplicate or triplicate samples in the unit of μg/g of dried weight of fish.

Heavy metals in fish were observed in most of the investigated sites (sites A-H) along the rivers. Some of the heavy metals studied are essential trace elements required for human function. The toxic substances, which should be monitored closely are Cd and Pb. Generally, herbivorous fish are more susceptible to the pollutants than carnivorous because of the food chain. In our case, although the number of fish population was not the objective of our study, the types of heavy metals found in the samples from random collection of fish population could indicate that these mobile living organisms were affected by pollution in the environment.

The causes of the heavy metal contamination of the rivers could not be deduced from the results of this study. However, when comparing all sampling sites and keeping in mind that the periods of sampling were when the highest possibilities of observation of heavy metal contamination would be expected; the comparison of sites provided some clues. A potential contamination in the fish at site B was indicated by the high level of Cd and Pb detected in the liver, regardless of fish species, herbivorous or carnivorous. The levels of these toxic heavy metals in the fish livers at sites F, G and H were observed to be high and should also be monitored. Incidentally, some industries are located around these sites but whether the industries were the major cause needs further thorough investigation.

The intensive evaporation of water in the dry season and the end of the harvesting season gives a peak of fertilizer contamination in the rivers, and the levels of major components are thought to reach the highest concentrations in this period of the year. Fish could intake some of the pollutants and could be affected. It would be very interesting to study the results obtained from water, sediment and living organisms collected from these sites in order to conclude the causes of the contamination. The overall results from this study reveal the potential contamination of heavy metals in fish habitat in the Pong and Chi River (site A-H), which had never been previously attempted.

5. Conclusions

Whatever the sources and causes of metal contamination may be, the results presented in this study indicate that the potential heavy metal contamination of the Pong and Chi Rivers may have been accumulated in the food chain of its habitat. To identify the source of such contamination, it is important to further study and elaborate on the source of the pollution, the habitat and behavior of the fish. It might be interesting to compare the results obtained from various tissues and from various species of fish with known life style and life cycle. This would show the accumulation and distribution pattern of pollutants. Furthermore, since fish are mobile and diverse, the overall data which would help us to explain the heavy metal contamination should at least consist of data on heavy metals in water, sediment, fish, plankton, shells and waterweeds (Yap et al., 2002, Mansour and Sidky, 2003).

6. Acknowledgements

The National Research Council of Thailand is acknowledged for the financial support, which enabled the implementation of this project.

7. References

FAO/WHO (1989). Evaluation of certain food additives and the contaminants mercury, lead and cadmium, WHO technical report Ssries No. 505.

Khon Kaen University. (1995). Study project for preparation of the action plan for the Pong river water quality rehabilitation, Khon Kaen province.

Mansour S.A. and Sidky M.M. (2003). Ecotoxicological studies. 6. The first comparative study between Lake Qarun and Wadi El-Rayan wetland (Egypt) with respect to contamination of their major components, *Food Chemistry,* 82, pp. 181-189.

Yilmaz A.B. (2003). Levels of heavy metals (Fe, Cu, Ni, Cr, Pb and Zn) in tissue of *Mugil cephalus* and *Trachurus mediterraneus* from Iskenderun Bay, Turkey, *Environmental Research*, 92, pp. 277-281.

Sirisangtragul W. (2000). Effect of cadmium and selenium on hepatic xenobiotic metabolizing and antioxidant enzyme systems in hamster, M. Sc. thesis in Medical Biochemistry, Graduate School, Khon Kaen University.

Szefer P., Domagala-Wieloszewska M., Warzocha J. (2003). Distribution and relationships of mercury, lead, cadmium, copper and zinc in perch (*Perca fluviatilis*) from Pomeranian Bay and Szczecin Lagoon, southern Baltic, *Food Chemistry*, 81, pp. 73-83.

Yap C.K., Ismail A., Tan S.G., Omar H. (2002). Correlation between speciation of Cd, Cu, Pb and Zn in sediment and their concentrations in total soft tissues of green-lipped mussel *Perna viridis* from the west coast of Penninsular Malaysia, *Environmental International*, 28, pp. 117-126.

Environmental Issues and Constraints in Lake Management: a Case Study from India

P. Mishra[1)], P. K. Mohanty[2)], H. Takada[3)] and T. Sugimoto[4)]

1) Ocean Research Institute, The University of Tokyo
1-15-1, Minamidai, Nakano Ku, Tokyo, 164-8639, Japan
e-mail: babuli@ori.u-tokyo.ac.jp
2) Department of Marine Sciences, Berhampur University
Bhanja Bihar, Berhampur, 760 007, India
e-mail: pkm@bu.ori.nic.in
3) Faculty of Agriculture, Tokyo University of Agriculture and Technology
3-5-8, Fuchu, Tokyo, 183-8509, Japan
e-mail: shige@cc.tuat.ac.jp
4) Ocean Research Institute, The University of Tokyo
1-15-1, Minamidai, Nakano Ku, Tokyo, 164-8639, Japan
e-mail: sugimoto@ori.u-tokyo.ac.jp

Abstract

Chilka Lake (19° 28'- 19° 54'N; 85° 05'- 85° 38'E) spreading over 1100 km^2 along the east coast of India is the largest brackish water body in tropical Asia. Since the mid-eighties, the lake is allegedly dying and on the verge of environmental catastrophe. Natural forcing through heavy siltation, eutrophication and biological sedimentation has transformed a major portion of the lake into swamp, threatening its existence. Excessive human activities such as fishing, aquaculture, agriculture, recreation and tourism have caused severe biodiversity loss and pollution. Depletion of fisheries and resources has triggered several socio-economical problems in the last few years, thus, the lake has received a lot of national and international support and funding for the conservation of its ecosystem. Since 2000, the government has initiated many developmental projects in an attempt to restore the lake from further degradation, however many issues remain untouched and unachieved. This paper makes a comprehensive assessment of the integrative environmental factors that are essential to lake functioning and suggests some of the key points that need to be incorporated in the ongoing management plan for the lake.

Keywords: biodiversity loss, Chilka, environmental issues, India, lake managemnt

1. Introduction

Coastal lagoons are one of the most productive, extensively exploited habitats and in many instances, natural forces multiplied with human factors throw them into a critical condition and irreversible state. Chilka lake in India attests to this very fact and is a good example to study the impact of man on nature. The lake supports about 150,000 livelihoods and contributes a large proportion to the local economy through tourism and export of fisheries. In the last two decades, natural factors have caused rapid environmental degradation affecting the lake water quality and biodiversity. In order to conserve it, the lake was listed as a "Ramsar site" by the Government of India in 1981 and since then it has received a lot of attention through various national and international projects and fundings. In 1992, the state Government set up a body- Chilka Development Authority (CDA) presently responsible for the management of the lake. Recently, it is being realized that despite the decade-long existence of goals and strategies for sustainable management of the lake resource and biodiversity conservation, most efforts remain in the early stages or unachieved. This paper discusses the major ecological issues and points out some of the lake management constraints and strategies that need to be incorporated in the present ongoing restoration efforts for the lake.

2. Study area and data source

Located in the eastern Indian seaboard, facing the Bay of Bengal, Chilka lake (19° 28'- 19° 54'N; 85° 05'- 85° 38'E) is the largest partially enclosed brackish water body in tropical Asia (Figure 1). With a length of 65 km and a maximum width of 20 km, the lagoon is quite shallow with an average depth ranging seasonally from 1.5 to 1.7 m. The seasonal water spread area ranges from 906 to 1165 km^2. Earlier the lagoon was separated by a sand bar of 60 km length and connected through a long (35 km) narrow (150m) intricate channel. In order to improve the circulation pattern, a new inlet was constructed on 23rd September 2000 by excavating the barrier spit at the the village Sipakuda, the nearest approachable point from the lake to the Bay of Bengal (Figure 1). Agricultural fields and human settlements surround the lake on all sides except in the southern, which is predominantly occupied with hills. On the seaside of the lake, there are few islands occupied with human settlements and their mode of communication with the main land is mostly by the small-mechanized boats or catamaran.The information used in this paper is culled from some of our past and recent field investigations and reports (ISRO, 1998; Mohanty et al., 2001; KWEF, 2002).

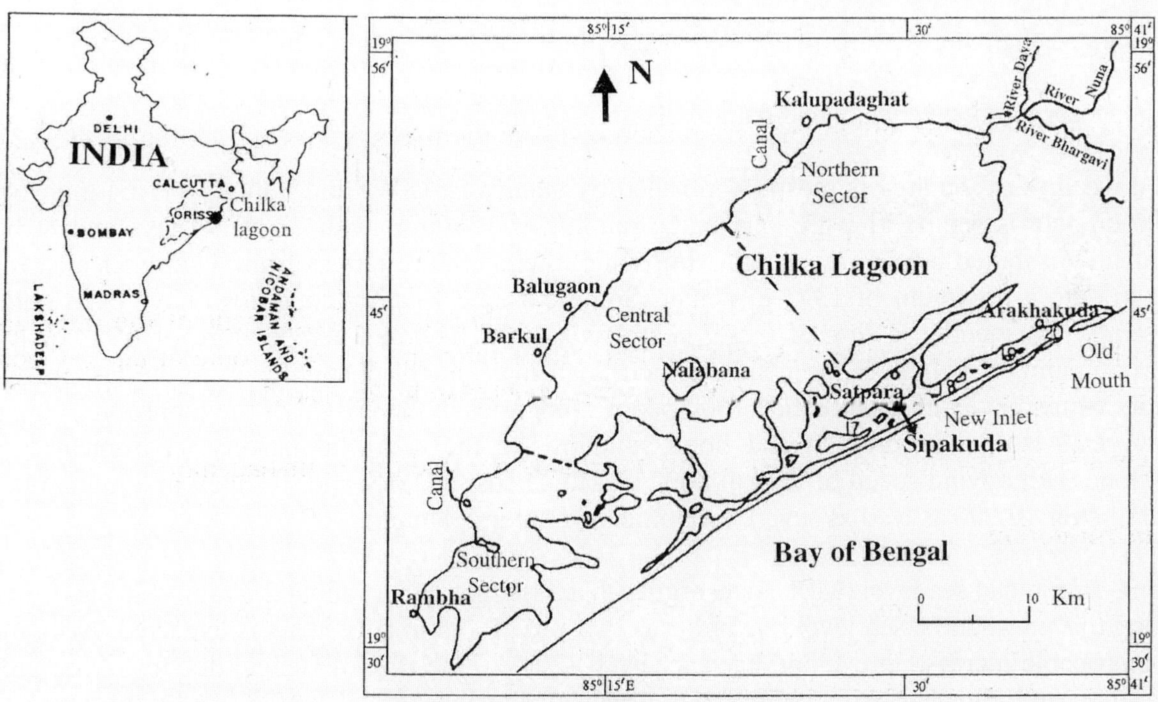

Figure 1. Study area.

3. Results and discussion

The primary influencing factors in the lagoon functioning are physiography and climate, the hydrology, the salinity regime, the sedimentation pattern, and importantly the population and their resource utilization behavior.

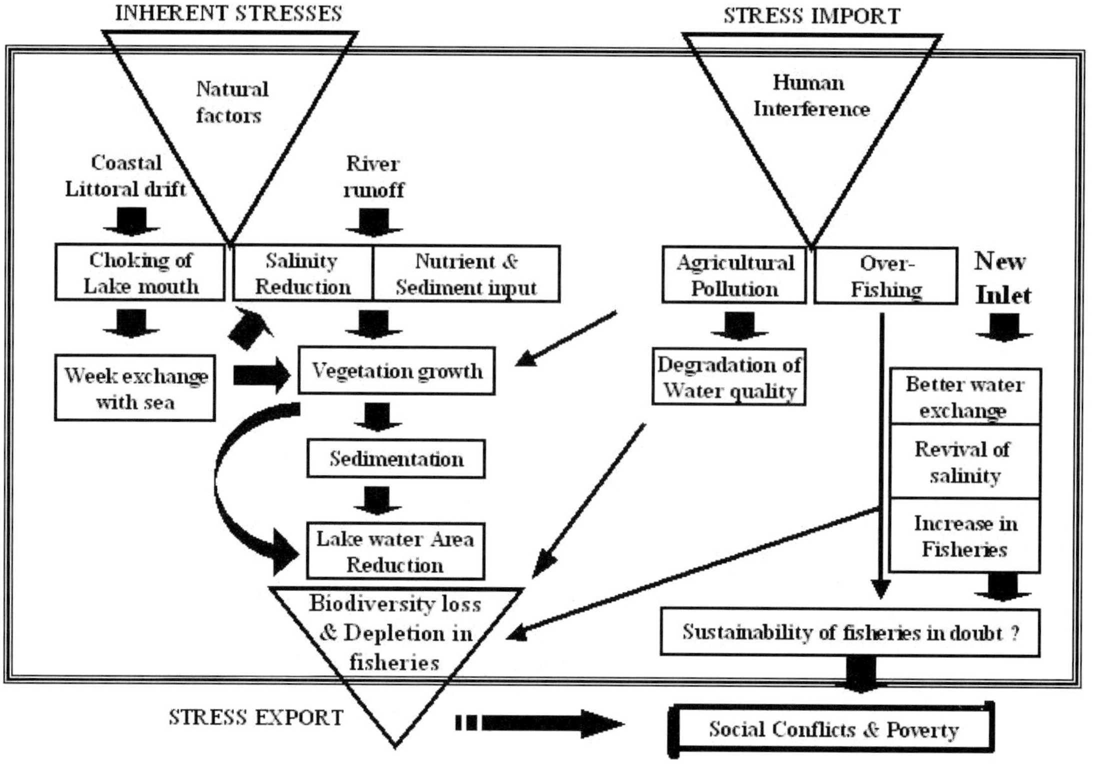

Figure 2. Schematic chart showing the various components affecting Chilka's environment.

3.1 Effect of climate change and natural disasters on the lake

The weather of the area is mostly controlled by southwest (SW) and northeast (NE) monsoons. The climate is semi-arid with an average rainfall of 1240mm with 72 rainy days and 80% of which contributed by SW monsoon. Atmospheric temperature ranges from a summer maximum of 40°C to a minimum of 13°C in winter. Some extreme weather conditions have been noticed recently. To mention a few of these, heat-wave spells having a temperature of 45°C or more during summer months in the neighborhood is an intermittent phenomenon in the last four/five years which have killed more than 1000 people. Vagaries in the arrival of the monsoon in the last few years is causing drought and flood situations in the area alternatively. A tropical super-cyclone with a wind speed of 260km hour^{-1} hit this coast in 1999 and resulted in a human death toll of over 10,000 as well as severe economic and natural damage.

Long-term mean sea level (MSL) data argues that a rising trend of about 2.5 mm year^{-1} is ocurring along this coast and a sea level rise of between 15 to 38 cm in fifty years is expected. Also, there is a probable increase in cyclone intensity by 10-20% in the Bay of Bengal due to future global warming and climate change which may affect mostly the coastal low-income rural population (Aggarwal and Lal, 2000). Associated with tropical cyclones, storm surges could be amplified and pose a greater threat in inundating larger coastal areas. The present coastal erosion rate is quite high and lowers the beach elevation by 3-4m during the SW monsoon months (Mishra et al., 2001). Thus, the present sand bar of 7 to 8m high and 300-500m wide that separates the lake from the Bay of Bengal is highly at stake and quite trivial in protecting its brackish characteristics and biodiversity from a consequent sea level rise and associated storm and surge events.

3.2 Hydrology and sedimentation process in the lake

About 52 rivers, streams and rivulets of a total catchment area of 3210 km^2 are debouching to the lake. Their flow pattern coincides with a high discharge during peak flood seasons of the SW monsoons and heavy rainfall during tropical cyclones and low during rest of the year. The fresh water flow contributed by the distributaries of the major Indian River Mahanadi accounts for about 61% (850 m^3 s^{-1}) of the total, whereas, 39% (536 m^3s^{-1}) comes from the other small streams, rivers and rivulets. During the monsoon months 0.365 x 10^6 tons of sediment enters into the lagoon, out of which 75% is contributed by the major distributary rivers and 25% by different small rivers and streams. Besides that, about 1.5 x 10^6 m^3 yr^{-1} of sand moves as littoral drift from south to north during March to October along the coast and choke and/or initiate northward shifting of the inlet (Chandramahan and Nayak, 1998).

The lagoon and the rivers experience heavy flooding during the monsoon months. Associated with strong wind, the sea level is elevated and affects immediate drainage of flood water to the sea resulting in frequent inundation of the lowland areas in the lake's upper regions. So, the lake receives a large amount of silt load and fresh water inflow at one end and no outflow, this obviously turns it to a "sink" slowly increasing its shallowness and creating a chaotic environment condition.

3.3 Lake water quality, pollution and eutophication problems

The fresh water and sediment input to the lake at its northeastern end and restricted outflow establishes a remarkable bathymetry and hydrographic gradient. Apparently, the northern sector is the shallowest (20-170cm) and fresher (0.4-2.0 psu) region, the southern sector is comparatively deeper (120-260cm) and saline (7.3-11.7 psu) whereas the central sector has a moderate (37-205cm) depth and salinity (7.4-9.2 psu) structure and the channel region is the deepest (>300 cm) and most saline (29-33 psu) region. Hydrographic data reveals that on average a high saline condition prevails during the dry season (May and June) when there is less river runoff and high evaporation, lowest salinity in flood seasons and an intermediate regime during winter months.The progressive decrease in the salinity from an average 22 psu to 3 or 4 psu in the last forty years and the dominance of seasonal mode of variations over the interannual mode is a serious threat to the lake water quality and ecology (Pal and Mohanty, 2002). An increase of nearly 10 to 15 psu in salinity has been observed after the construction of the new inlet (KWEF, 2002).

The pollution aspects have not been studied properly yet. The major source is presumed to be fertilizers and pesticide residues related to phosphorous and nitrate compounds from the adjoining croplands. In the southern region, a chloro-alkali factory uses mercury in its production and for a good period of time was releasing some of its effluents into the neghiboring environment and mercury contamination could be suspected through leaching effects. Nearly 400 mechanized boats and 5000 non-mechanized boats in the lake and various types of paints and chemicals may be affected by hydrocarbon compounds and under investigation. Sediment samples recently analyzed for ten target sterols shows coprostanol, a human fecal pollution indicator, in the lake ranges from 0.0045 μg/g-dry to 27.5 μg/g-dry and are relatively low in all stations except at one station that is located within three kilometers from the human settlement. It has an extremely high value(27.5 μg/g-dry). The sterol profile at another location nearer to the migratory birds wintering ground is highly affected by fecal material of bird origin. Thus, the coastal innershore region of the lake closer to human settlements are affected by domestic or fecal inputs and pollution.

Luxuriant vegetation growth in the northern sector and along the border area of the lake occurrs because of shallowness, sediment texture, low salinity, availability of high nutrients brought by fresh water flow from the land. Various species of aquatic macrophytes having a wide range of saliny tolerance have been noticed in recent years. Our analysis of Indian Remote Sensing Satellite data reveals that the total vegetation coverage in the lake from 1993 to 2002 varied from 210 to 508 km^2 with an alarming increment at a rate of 21.31 km^2 $year^{-1}$ noticed for the period 1993-1995 (ISRO, 1998). The macrophyte biomass (wet-weight) ranging in between 5 to 6 kg m^{-2} indicates that the vegetation infestation is at an alarming stage. Though, the presence of thick vegetation supports a number of fauna and regulates the nutrient load of the lake, however, they also initiate other problems i.e., entrapment of the sediment and decay of macrophytes contributing to the biological sedimentation processes, ultimately settling to form new land masses. After the excavation of new in-let, weed growth has ceased appreciably due to high salinity conditions.

3.4 Biodiversity issues

In terms of biodiversity, the lake is extremely rich in flora and fauna and earmarked as a "hotspot" in the federal white paper. The major flora of the lake constitutes 97 species of phytoplanktons, 22 species of algae and about 150 species of vascular plants.The macrophytes can be classified under algae, pteridophytes and angiosperms. Nearly 800 species of fauna mainly zooplanktons, and 37 species of amphibians and reptiles, 217 species of fishes, and 156 species of birds have been recorded. Benthic fauna in association with macrophytes play a major role in influencing the biological sedimentation rate (Mohanty *et al*., 2001).

The composition of fish species in the lake is a mixture of brackish, catadramous and anadarmous fish and a major percentage of the commercial fishes are of migratory origin. Recently, replacement of brackish water fish species by fresh water species and a decline in the total number fish species has been reported. Two varieties of export oriented prawn *i.e*., *Paneaus monodon* and *Paneaus indicus* and two major commercial important crab species *i.e.*, *Scylla serrata* and *Neptunus pelagicus* were on the verge of extinction because of over-exploitation. The total fish landing for the lake during 1985-86 was about 8900 metric tons and declined to 1300 tons in 1994-95 and almost remained stagnant hovering at 1700 tons until 1999. The reasons attributed to decline in fisheries are high fishing pressure, poaching of juveniles for aquaculture, natural blockage of the lake inlet resulting in an hindrance in migration of economic species, decrease in lake total water spread area and water quality. After the opening of a new mouth in 2000, the fisheries increased to three times (about 4900 tons) that of 1999, however, far from the catch of two decades back. The occurance of an endangered Irrawaddy Dolphin species *Orcaella brevirostris* has declined drastically because of a decrease in food supply (fish), siltation and habitat loss. A full phase plan is underway to conserve the species.

In the middle of the lake, a swampy island of 16 km^2 hosts a number of migratory birds in winter (November to February) from the Caspian Sea, Lake Baikal and parts of Russia and Siberia. An estimation shows that about 160 species of birds of a total population of between 500,000 and 700,00 used to visit the lake in the early part of nineties. In recent years, there is a tremendous decrease in their total numbers and species composition. In 1996, only 85 species and in 2000, 68 species were sighted in the lake and their total numbers were less than 100,000 in 2002. The reasons for their decline are mainly the loss of habitat, shortage of forage and illegal poaching. Presently, the area is protected as a bird sanctuary.

3.5 Human dimension

A large population (about 12500 active fishermen famillies) depends directly on fishing or related trades. Most of them are poor, large, do not possess any personal assets and have an annual income of about 400-500 US$. There are no other opportunities to develop any other business or trade. Earlier studies show that until the 1970s, the fishermen population and fish production in the lake grew at the same rate. The advent of commercial aquaculture and the monetary benefits involved with it and intermittent crop failures attracted many non-traditional fishermen and farmers to adopt fishing increasing the dependency on the lake. In between 1980 and 1990, there is nearly a population growth of 4 percent per year whereas there was a 50% decrease in the total fish landings by the end of 1990 showing over-exploitation is the only reason for the biodiversity loss. In India, the caste decides the occupation and the fishermen community belongs to the lowest caste in the caste hierarchy system. Though, legally they have a right over the lagoon water and fishing in the lake, in the last few years, considering the money involved in aquaculture and fishery, people from higher castes have entered into prawn farming and the fishing business. Thus, forceful encroachments into the lake water by non-traditional caste, group clashes ensued in which a few people lost their lives every year. Thus, poverty and lack of alternatives are working as a vicious circle *i.e.*, poverty triggers over-fishing and abuse of the lake environment ultimately decreasing the lake resources and the local economy and further allowing the poverty to widen.

3.6 Future needs and management plan

The factors affecting the lagoon processes as discussed, disparate as they are, illustrate that major biodiversity loss of the lake lies with natural pressure compounded with human activities, the resource use conflict as well as threats and constraints for sustainable management and developments. Notwithstanding these negative signs, options are still open to restore the lake. Any restoring program for the lakes needs to be viewed from two angles - i.e., long-term plans to conserve the lake resources and the other is to attain a sustainability in the fishery which can bring some economic comfort to the local population. Figure 3 shows an integrated management plan for the lake.

To comply with the natural or inherent problems of the lagoon, many projects are undertaken and are also in the pipeline. The recently excavated new inlet has improved the salinity, sedimentation and circulation pattern of the lake and is regarded as a major accomplishment in restoring the lake, however, the ancillary effects are many. The cost-profit factor in maintaining the inlet and its long-term sustainability could possibly pose as a major threat for the management in the future. In June-2003, heavy erosion was reported on both sides of the newly excavated mouth and it has threatened some of the neighbouring villages and the stability of the sand bar is under question? Regular sand dredging of the lake inlet might trigger nutrient alternation leading to undesirable phytoplankton blooming. Unrestricted exchange of water between the lagoon and the sea might turn the brackish water into a more saline condition and may induce complete collapse of its indigenous biologically important species.

Our past working experience with the lagoon shows that, we may require a controlled salinity system to maintain its brackish characteristics and keep the biodiversity intact, hence, construction of a barrage system with gates at the present lake mouth can be a suitable option, however, this needs a model study and investigation. Managing the watershed area and diverting some of the major river's water and sediment directly to the sea through small canals should also be included. The other problems that need constant addressing are improving lake fisheries and ecologically based tourism and pollution. No doubt, the post effect of the new-mouth has affected positively on the fisheries due to the return of some marine species of fish to the lake, but, attaining sustainability in the fishery is the prime requirement. Recently taking the advantage of the revival of the salinity structure, the lake use pattern or the fishing behavior of the local people has changed tremendously. The usage of fixed nets as a trap with wooden and/or bamboo barricades (locally known as *gheries*) for shrimp culture has mushroomed all over the lake and the peripheral area and creeks of the lake is taken over by unauthorized fishing activities. These *gheries* are preventing the free flow of sediment from the lake into the sea, instead accumulating more vigorously in the lake shore and channel areas. The existing fishing methods are quite unsustainable and also affect the auto-recruitment of biological species from and into the lagoon. Usage of improper mesh-size nets and other unscientific nets in the lake and also at the lake mouth cause the decline and obstruct the migration of economic species.

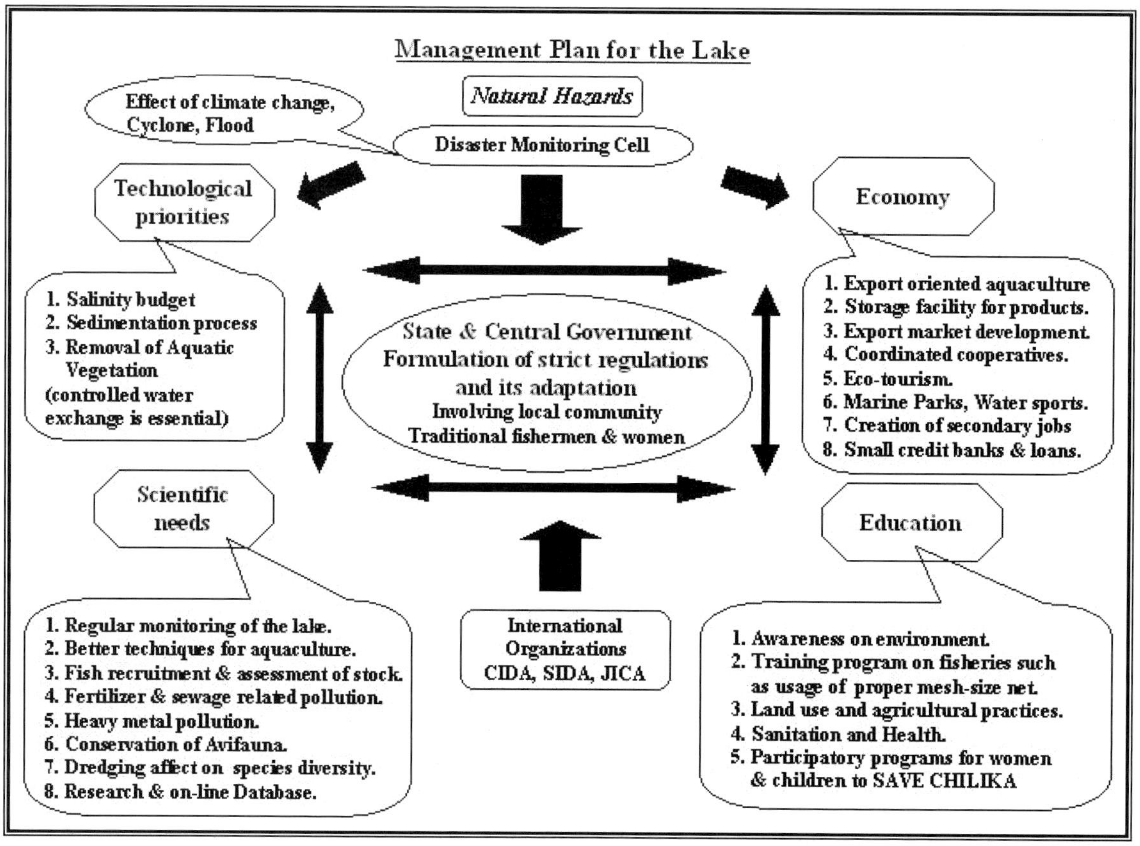

Figure 3. A schematic view of the management plan.

Framing of rules and regulations on fishing and their strict enforcement is also a prime requirement. The rules need to focus on the usage of proper mesh size, restriction of fishing periods, the provision of fishing and non-fishing areas taking different communities views into consideration. Along the lake artificial fish breeding centers, better technology for fish recruitment, retention and stock assessment need to be properly adapted. The concept of eco-tourism with proper care needs should be developed to creat jobs for the local population. Regular scientific monitoring, creation of an online database and environmental education to the local population are highly essential for the better management and resource utlilization of the lake.

4. Conclusions

The major environmental degradation in Chilka is partly due to the inherent characteristics that arise through natural processes complicated by over-exploitation of its resources exporting stress to the ambient social network. This has been reflected through social conflicts, crimes and poverty that might multiply in the future. Thus, efforts to conserve or restore the lake biodiversity urgently demand the improvement in the socio-economic condition of the area and this can be achieved only through proper village and/or community based planning and coordination. The recently constructed new-inlet shows a positive effect on the lake fisheries temporarily, however, associated problems demand constant monitoring. Importantly, the population growth and sustainability of lake resources in relation to future climate change or natural calamities are some of the trivial questions which need to be included in the present management policies. Finally, as the lake is firmly entrenched into the life style of the people culturally and spiritually, the opinion of the local people and local knowledge should be given priority in deciding the policies for resource utilization and biodiversity management.

5. Acknowledgement

We kindly acknowledge the financial support of KWEF, Tokyo, Japan.

6. References

Aggarwal D. and Lal M. (2000). Vulnerability of Indian coastline to sea level rise, Proceedings of APN/SURVAS/LOICZ joint conference on Coastal conference on coastal impacts of climate change and adaptation in the Asia-Pacific region, Kobe, Japan, pp. 65-73.

Chandramohan P. and Nayak B. U. (1994). A study of the improvement of the Chilka lake tidal inlet, East coast of India, *J. Coast. Res.*, 10, pp. 909-918.

Kurita Water and Environment Foundation (2002). Monitoring of water quality conditions and weed cover in Chilka lagoon through field survey and Indian remote sensing satellite (IRS) data, Final report, Tokyo, pp. 20.

Mishra P., Mohanty P.K., Murty A.S.N. and Sugimoto T. (2001). Beach profile studies along Goplapur coast, India. *J. Coast. Res.*, Special Issue 34, pp. 164-171.

Mohanty P. K., Pal S.R. and Mishra P. (2001). Monitoring ecological condition of a coastal lagoon using IRS data: a case study in Chilka, East coast of India, *J. Coast. Res.*, Special Issue 34, pp. 459-469.

Pal S. R. and Mohanty P.K. (2002). Use of IRS-1B data for change detection in water quality and vegetation of Chilka lagoon-east coast of India, *Int. J. Remote Sens.*, 23, pp. 1027-1042.

Indian Space Research Organization (1998). Genesis, extent and future of biological edimentation in Chilka lagoon, Orissa: an attempt with IRS-1B, RESPOND project report, Bangalore, India, pp. 42.

Farmer Managed Irrigation Systems of Nepal: Balancing Water Uses and Environment Conservation for Sustaining Livelihood

G. P. Shivakoti[1)] and S. G. Shrestha[2)]

1) Associate Professor, School of Environment, Resources and Development, Asian Institute of Technology
PO Box 4, Khlong Luang, Pathumthani 12120, Thailand
e-mail: ganesh@ait.ac.th

2) Doctoral Student, School of Environment, Resources and Development, Asian Institute of Technology
PO Box 4, Khlong Luang, Pathumthani 12120, Mail Box 4, Thailand
e-mail: shiddis@yahoo.com

Abstract

In Nepal, Farmer Managed Irrigation Systems (FMISs) dominate the number and the irrigated area. They were built and managed by farmers themselves for centuries. The main objective of this paper was to analyze the factors that made these FMISs function well for so long. FMIS communities have indigenous but ultra-modern knowledge of layout, construction, management, ecology and hydrology. They have built irrigation structures in exceptionally difficult and varying geophysical environments without the aid of modern equipment. FMISs were constructed, operated and maintained for centuries with least damage and threat to environment and humanity. This study has found that the perpetuation of accumulated experience, skill and the knowledge have enabled them balance water use, conserve natural ecosystems, generate employment and feed rapidly growing populations, build social capital for collective action and maintain communal integrity, make best use of local resources and indigenous technology, and keep several FMISs alive. Centuries old live FMISs testify it. An in-depth study on FMISs doctrine and fine-tuning them to respond to the current challenges will benefit not only the FMISs but also the entire humanity.

Keywords: downstream, farmer managed irrigation systems (FMISs), livelihoods, upstream

1. Introduction

Nepal falls between the monsoon-region stretching from Japan to India (MOWR and JICA, 2000). Hence, it is believed that irrigation systems were developed to grow rice since ancient times. FMISs in Terai (plain land) of Nepal were developed even before the sixth century (Shah and Singh, 2001). FMISs cover about 70 percent of the total irrigated area in the country (Shukla, et al., 1997, Pradhan, 2003) and over 89 percent of the irrigated area in hills and mountains (Parajuli, et al., 2001). There are more than 18 thousand FMISs in the country (Pradhan, 2003). They have been developed and managed by farmers themselves. Historically their explicit goal has mainly been to ensure food supply for household consumption through supplemental irrigation to the monsoon paddy. But rapid population growth and increased demand for cash to meet other livelihood needs has induced FMISs adopt various measures to grow more food and earn cash. It has sustained livelihood for so long. However, the process is slow to address growing complexities and challenges. On the other hand modernization has brought several negative impacts on the environment. Therefore, the concern now is to look into the possibilities of linkage between modernization and traditional FMIS technology to effectively address current challenges of food and cash requirement with least or no damage to the environment.

The main objective of this paper was to analyze factors that kept thousands of FMISs functioning well and address challenges ahead. Hence, this study has focused on the analysis of complex relationships that sustained FMISs so long and made them environment friendly.

2. Materials and methods

Although indigenous irrigation development technologies look simple, they are the result of trials done for centuries. The indigenous technologies that we see today are the outcome of accumulated experiences, knowledge and skills. These were carried out over generations without any written documents. This paper focuses on the technologies adopted in FMISs and the factors that kept them moving for centuries with least damage to the environment and at the same time coping with challenges ahead. This study has analyzed five livelihood assets in relation to various equities, food production, employment generation, environment conservation and water use balance among various uses and users. Five livelihood assets or capitals selected were human, natural, physical, financial and social. Economically active family members with different ages and genders constitute the human capital. Similarly, water and land resources constitute the natural capital, irrigation infrastructure, road and agro-processing units constitute the physical capital, on-, off- and non-farm incomes constitute the financial capital, and socioeconomic and water allocation equities, literacy rate, access to credit and innovations constitute the social capital. Livelihood as defined by Chambers and Conway (1997) comprises capabilities, assets and activities required for means of living.

The information for this paper was mainly derived from the census survey of downstream and upstream FMIS households of Chitwan, Nepal conducted from May-December 2002. Surtani FMIS was selected to represent a downstream or lowland system whereas Pampa represents an upstream or upland system.

The frequencies, means, proportions and indices of different livelihood capitals and their variables were computed to analyse factors that kept FMISs functioning for so long and at the same time balanced water use for various uses and users and conserved ecosystems with the least damage to the environment. Lowland and upland systems were compared to analyze upstream-downstream relationships.

3. Results and discussion

There are as many as nine FMISs within a stretch of about 15 km in Pampa stream. All FMISs opted for indigenous weirs despite frequent damage by flash floods. Huge man, money, material and time resources are invested in the maintenance of such weirs. The upstream Pampa FMIS alone mobilized 530 labor days in 2001 and 1440 in 2002 just to rehabilitate indigenous headwork damaged by 3-4 flash floods within two months of the monsoon season in both years. Close network and trusts among users and an exceptionally high man-irrigated land ratio of 13 in both systems have made such labor intensive and arduous task continue for centuries through community participation. In addition Water Users' Association (WUA) had to use its Rs. 67,000 (1 USD = Rs. 78) cash saving and about Rs. 100,000 equivalent gabion wire grant provided by the Department of Irrigation, Nepal in the later year.

Despite this all FMISs preferred for temporary diversion to keep downstream systems alive. Generally WUA appoints person/s as a messenger to communicate about meeting, decision and labor mobilization among community members. Communications during natural disasters are exceptionally quick and effective. Remuneration for messengers comes from the contribution from all members of the irrigation community.

The community people have several years' accumulated knowledge and experiences about bed flow and downstream recharge through seepage and percolation from farm fields. Hence, there is at least the same or increased volume of stream flow after a few meters of diversion structure. These indigenous weirs are simple run-of-the-stream type made of locally available stone and brushwood. The species used for diversion are Baddhairo (*Legestroemia parviflora*), Banmara (*Eupatorium odoratum*), Dhursillo (*Solanum erianthum*) Lampate (*Aesculus indica*) and Sindoore (*Melotus philipinsis*). In due course few systems have replaced local materials by gabion wire but never with concrete structures. The goal of a traditional weir has always been an optimization of social benefit but not the maximization of water acquisition at source.

Development and adoption of such indigenous weirs have enabled nine FMISs to survive and irrigate more than 750 ha of their land at various locations. The accumulated knowledge and experiences and adoption of indigenous technology have been functional in balancing water use among communities and creating harmony and integrity within and between them. It is an excellent example of social capital formation being perpetuated for years. Local communities themselves have crafted and maintained such virtues. Paddy fields that have been mainly irrigated through FMISs were in fact water reservoirs to control flood and stabilize downstream flow. FMISs irrigation water has also been used for washing, bathing, livestock feeding, construction and in some instances drinking too.

Over the time FMISs have developed strong communal ownership on water resources and it has been strengthened by usufruct right as has been defined in the Water Resource Act and Regulation of Nepal (MOLJPA/HMGN, 2001). Litigation on water acquisition at source between Pampa and Chipletee FMISs reached up to the Supreme Court. This is a measure of the degree of community ownership on water resource.

In spite of these FMISs are often criticized for overuse of irrigation water especially for paddy that is 6.3 lps (liters per second) per ha against the standard 1.5-3.0 lps/ha (Halcrow *et al.*, 1997, Parajuli *et al.*, 2001). But it has been useful in recharging downstream flow and controlling floods as described above. Use of shrubs and branches in headwork construction has also been perceived damaging the environment. PRA with the community people revealed the fact that such use enhances coppicing as being pruned. Hence, such use is not necessarily damaging and is least harmful as compared to free grazing of livestock, forest clearing for cultivation and gathering wood, fuelwood and herbs from the forest. Downstream flow resulting from the adoption of indigenous headwork has preserved aquatic species particularly the fish. Hence, some households were able to continue fishing in downstream Pampa.

Crop diversification was high in both systems and both were growing as many as 14 different types of crops in one crop year. Their choice of particular crop was dictated by water adequacy and farm households' access to support services specially the market and market information. Hence, upstream farm households were found growing water-economizing crops during dry seasons and none grew paddy in the spring season. On the other hand, more than 93 percent downstream (a water adequate system) farm households were growing spring paddy. It is the reflection of higher natural capital index (0.91) particularly higher water index (0.95) as indicated in livelihood asset pentagon (Figure 1).

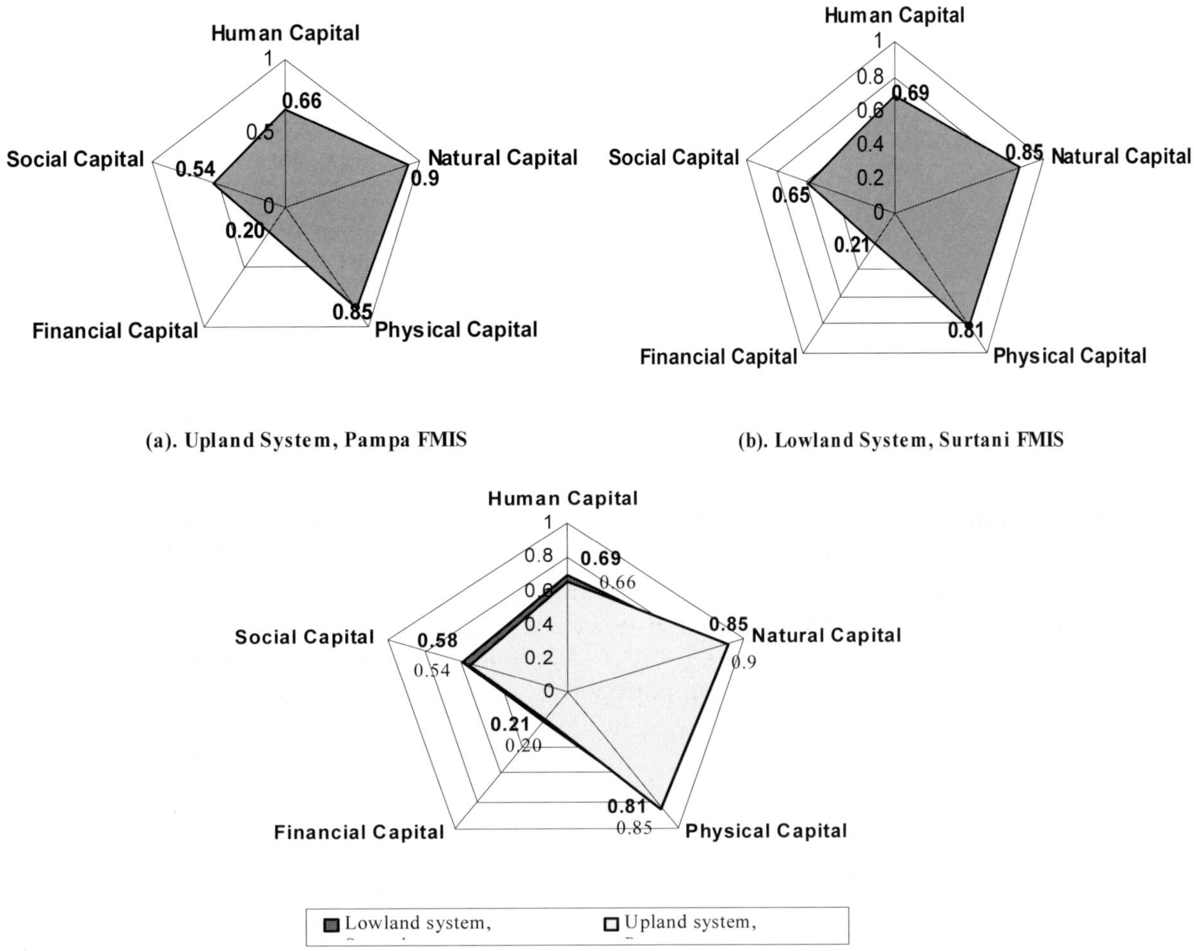

(a). Upland System, Pampa FMIS

(b). Lowland System, Surtani FMIS

(c). Lowland and Upland FMISs

Figure 1. Livelihood capitals and their variables.

Cropping intensities in head, middle and tail portions were computed at 288, 284 and 284 percent for downstream and 265, 274 and 264 percent for upstream FMISs. The differences in CIs in head, middle and tail locations of irrigation systems were not significant in both FMISs ($p=0.05$). It implies that irrigation water was equitably distributed in all the locations of the irrigation systems. Relatively upstream irrigation system faces water scarcity during winter and spring seasons mainly due to temporary diversion constructed from gabion wire and brushwood and observably higher seepage and percolation losses from cultivated lands owing to its location in an elevated peninsula. Some portions of cultivated lands have sandy soil as well. Hence, Water Users' Association (WUA) in upstream FMIS has made an additional day of irrigation arrangement for the third group farms located in the middle part of the system.

Equitable and judicious allocation of irrigation water in all reaches of irrigation systems has created trust towards the system and among users. This has integrated people and ensured their participation to keep system functioning for centuries. Non-significant differences in cropping intensities and yields in head, middle and tail portions of both FMISs as obtained from analysis of variance (ANOVA) indicates water allocation equity. Democratically constituted WUA dynamism has made this possible. WUAs are formed through a democratic process. Problems and issues are discussed openly and decisions are generally made by consensus. All decisions and accounts are transparent. This has enhanced trust, mutual benefits and a strong network between and among members making all accountable for collective action.

Fairly equitable land distribution among farm households in both systems has kept the community intact. Land distribution among large, medium, small and land-less (cultivating on rented-in lands) in lowland farm households was 3, 43, 50 and 4 percent with an average farm size of 3.75, 1.04, 0.25 and 0.11 ha respectively. Such distribution was 1, 47, 52 and 0 percent with an average farm size of 2.00, 0.82 and 0.28 in lowland system. The proportion of large and land-less was very low as compared to small and medium. Difference between small and medium was also very low.

Despite traditional irrigation infrastructure development technology both systems produced surplus cereal, which is their staple food. Average annual household cash earning from the trading of such surplus was Rs. 47,841 in downstream and Rs. 34,592 in upstream systems. Downstream systems produced net consumable (after deducting seed, post harvest losses and recovery percentage) 157.65, 16.10 and 8.85 tons of cereal, pulse, and oilseed against annual requirement of 84.32, 9.29 and 3.37 ton respectively (Table 1). Similarly upstream systems produced net consumable 213.9, 5.74 and 17.05 tons against annual requirement of 107.32, 11.82 and 4.29 tons of above crop categories respectively. The food requirement was computed based on PRA on farm households' food habit and daily requirement of 2550 calories for an average person.

Table 1. Average annual food production, market and energy value of productions from one ha of land and food production from the entire system of downstream (Surtani) and upstream (Pampa) systems.

Food items	Downstream system (n = 76)				Upstream system (n =110)			
	Food production, ton	Market value of production, Rs.	Energy value of production, 10^6 Kcal	System food production (35.3 ha), ton*	Food production, ton	Market value of production, Rs.	Energy value of production, 10^6 Kcal	System food production (52.5), ton*
Cereal	7.95	62378	2740	279.93	6.30	50376	2163	300.46
Pulse	0.54	12062	186	18.96	0.11	4235	36	8.03
Vegetable & potato	0.24	883	11	11.42	0.55	3256	50	22.60
Oilseed	0.01	28	6	15.08	0.01	105	2	2.06
Total	8.74	75351	2943	325.39	7.97	57973	2251	333.15

* Based on average yield rate and corresponding cropped area.

Both systems were employing large numbers of human and animal labor (Table 2). Family labor contributed more than 70 percent of total human labor employed in both systems for all categories of labor disaggregated by gender and age viz. adult male and female and child labor. The rest was contributed by hired labor. Female, male and child labor stands in their hierarchical order of employment in both systems. Only downstream systems used contract labor for sowing/transplanting and harvesting. Good road networks in downstream systems facilitated seasonal flow of contract labor from neighboring districts. Tractors and threshers were found being used for land preparation and threshing. Use of machines has substituted as much as 103 (27%) labor in downstream and 88 (21%) in upstream systems. This has provided a very good opportunity for off- and non-farm employment to farm family members.

Overall labor employment (including labor equivalent of contract work and machine costs) per year in one ha of land was 388 in downstream and 395 in upstream systems. It meant upstream generated more employment than downstream systems.

Table 2. Employment in one hectare of land in downstream (Surtani) and upstream (Pampa) systems.

Labor types	Downstream system (n = 76), No./ha/year	Upstream system (n = 110), No./ha/year
1. Male:		
Total	56	86
Hired	13 (23)	26 (30)
Family	43 (77)	60 (70)
2. Female:		
Total	175	219
Hired	44 (25)	43 (20)
Family	131 (75)	176 (80)
3. Child:		
Total	13	2
Hired	1 (8)	0.7 (35)
Family	12 (92)	1.3 (65)
4. Total human	244	307
5. Animal*:		
Total	1	8
Hired	0.8 (80)	4 (50)
Family	0.2 (20)	4 (50)
6. LE	144	88
7. LS	103	88
8. Overall LE	388	395

** Bullock pair with a human labor.*
Figures in parentheses are percent of total.
LE = Labor equivalent of contract work and machinery costs, LS = Labor substitution by machine.

Both irrigation systems are vulnerable to financial, social and human capital shocks owing to lower access to or endowments of these capitals in the study FMISs. Higher food security was associated with higher natural and physical resource endowments as indicated in asset pentagon. Similarly, moderate to slightly high human capital index in the asset pentagon was related to higher levels of employment.

4. Conclusions

The study shows that several FMISs virtues have made them sustainable since time immemorial with minimum damage to natural environment and at the same time meeting growing demands for food and employment. Some important virtues were accumulated - experiences, knowledge and skills on ecology, hydrology and layout, construction, operation and maintenance of irrigation systems. In addition, democratic and transparent decision making, socioeconomic and water allocation equities and strong communication mechanisms were some of the foundations for social capital building. These have enhanced the development of trust, mutual benefits and networking among uses and users. These virtues have enabled FMISs amass collective action at all times and levels and to be sustained for centuries.

5. Acknowledgement

The study was financed by the World Bank assisted by His Majesty's Government of Nepal, Agricultural Research and Extension Project and International Water Management Institute, Sri Lanka which are duly acknowledged.

6. References

Carney, D. (1998). Sustainable rural livelihoods: what contribution can we make?, Paper presented at the Department for International Development's Natural Resources Advisors' Conference, July, United Kingdom.

Halcrow, GEOCE and East Consult (1997). System performance and impact evaluation of selected ILC sub-project, Department of Irrigation of His Majesty's Government of Nepal, Kathmandu, Nepal.

MOLJPA/HMGN (Ministry of Law, Justice and Parliament Affairs of His Majesty's Government of Nepal) (2001). Water Resources Act 2049 (1992), Water resources regulation 2050 (1993) (written in Nepali) and irrigation regulation 2056 (2000), MOLJPA/HMGN, Kathmandu, Nepal.

MOWR (Ministry of Water Resources of His Majesty Government of Nepal) and JICA (2000). Sinchai Hatepustica (Handbook of Irrigation) (translated in Nepali), MOWR and JICA, Nepal.

Parajuli, U., Sharma, C., Panta D., Gautam S. R. and Yadav R. P. (2001). Hill irrigation: the question of rehabilitation, Research report No. 46, Winrock International, Kathmandu, Nepal.

Pradhan, P. (2003). Towards self-management of irrigation system: experiences from Nepal, In: *Participatory Management of Irrigation Systems, Water Utilization Techniques and Hydrology,* A.W. Abdelhadi, T. Hata and A. Mishra (Eds), Proceedings of International Workshop. Kobe University, Japan, pp. 25-41.

Shah, S. G. and Singh, G. N. (2001). Irrigation development in Nepal: investment, efficiency and institution, Research report series No. 47, Winrock International, Kathmandu, Nepal.

Shukla, A. K. (2002). Learning from farmers in large and complex participatory systems, In: *Improving Irrigation Governance and Management in Nepal,* G. P. Shivakoti and E. Ostrom (Eds), Institute for Contemporary Studies, Okland, California, pp. 179-203.

Shukla, A. K. and Sharma, K.R. (1997). Participatory irrigation management in Nepal: a monograph on evolution, processes and performance, Research and Development Branch, Irrigation Management Division, Department of Irrigation of His Majesty Government of Nepal.

Study on the Reasonable Utilization of Transboundary Water Resources and Environmental Protection in Southwest China

H. Daming[1)], Z. Wenjuan[2)] and G. Junhua[3)]

The Asian International Rivers Centre of Yunnan University, Kunming, Yunnan 650091, P.R.China
1) hedaming@public.km.yn.cn
2)wjzhao0772@sina.com
3) gjunhua@ynmail.com

Abstract

In recent years, transboundary water resources of international rivers have attracted global attention. Whereas in China, the study is at the initial stage. Southwest China is the area where high concentration of international rivers and headwaters of major Asian international rivers can be found. It enjoys abundant transboundary water resources and its water system evolution exerts widespread impact. Hence, the reasonable utilization of transboundary water resources and ecological protection will be of great importance for the southwest region, for China as well as the Asian continent. Integrated multi-disciplinary study on the reasonable allocation and utilization, sustainable maintenance and ecological conservation of transboundary water resources in Southwest China, undoubtedly, will offer scientific grounds for the formulation of transboundary water resources security strategies, promoting regional economic cooperation with Southeast and South Asian countries, reducing transboundary conflicts, protecting biodiversity and propelling socio-economic and ecological development in Southwest China as well.

Keywords: international rivers, Southwest China, water resources

1. Introduction

Since the 1990s, much attention has been attached to the importance of international rivers, whose transboundary water resources has become the focal issue of international study. The United Nations and World Bank have repeatedly warned the world of global water shortage, water conflict and even water war in the 21st century. In 1999, the first international symposium on Cooperative Development and Coordinated Management of International Rivers was held in China. In 2000, the build-up of Database of Transboundary Fresh Water Resources Conflict was accomplished with the support of the United Nations University. On the "International Water Day" in 2001, Annam, the secretary general of the UN pointed out that, "the intense scramble for fresh water would possibly be the roots for future conflicts or even wars." In 2002, Oregon State University, Yunnan University and another 8 universities in the world jointly formed the University Partnership of Transboundary Waters.

Currently, there are 261 international rivers shared by different countries, involving 45.3% of continental mass and 40% of the world's population, whose transboundary water resources account for about 50% of the world's total runoff volume and 60% of the world's fresh water supply (Wolf, 2001).

The UN's *Comprehensive Assessment Report on World Water Resources* points out, "Water issue will seriously restrict global economic and social development, and probably result in conflicts between different countries. Therefore, studies on national strategy on water resources and relevant scientific problems in the 21st century shall be one of the key issues different governments encountered at the turn of the century"(Liu and He, 1996; Liu and Chen, 2001).

China ranks the top globally in the number of international rivers and quantity of transboundary sharing water resources, with catchment area accounting for 1/3 of the nation's total territory and transboundary water resources involving 18 riparian countries and approximately 3 billion population. After the implementation of reform and opening-up, China has given priority to international rivers. The State Department has jointly organized 23 coordinating groups for cooperative development of Tumen and Lancang River together with various ministries and departments.

West China possesses plenty of international rivers. Of all the 15 major international rivers in China, 11 are in the west. The southwest provinces of Tibet and Yunnan, in particular, have the highest concentration of international rivers, and contain the headwaters of major Asian international rivers, thus widely regarded as the "water tower" of Asia. Hence, in China's Western Development strategy, special attention should be attached to the sustainable development and transboundary ecological security of international rivers. And the equitable allocation, reasonable utilization and sustainable maintenance of transboundary water resources, will not only be of great importance to the sustainable development of western China, but also be of strategic significance to the insurance of regional food security, human health and ecological function for China, south and southeast Asian countries.

2. Strategic position of transboundary water resources in Southwest China

The total mean annual runoff volume of China's international rivers amounts to $7{,}674\times10^8m^3$, accounting for 28.3% of the nation's total. Due to the fact that China possesses the upstream of most international rivers, its natural outflow is far greater than its inflow, with annual volume of $7{,}320\times10^8m^3$ (accounting for 27% of the nation's total annual natural runoff and 95% of transboundary water resources respectively) flowing into the neighboring countries. China's total inflow is about $170\times10^8m^3$, accounting for less than 1% of its total runoff and 2.3% of its outflow. These transboundary water resources are not only very important for China, especially western China's sustainable development, but also strategically significant for other 17 riparian countries' socio-economic development and ecological security. Therefore, much emphasis should be placed on the reasonable utilization, protection and management of those transboundary water resources, so as to guarantee regional water and ecological security (He and Gou, 2000).

The equitable and reasonable utilization of transboundary water resources in southwest China mainly focus on "Southwest Rivers", including Yarlung Zangbo-Bulamput River, Irrawaddy River, Nujiang-Salween River, Lancang-Mekong River and Yuan-Red River. With remarkably abundant water resources, southwest rivers are regarded as the most concentrated and potential water reserve in China (see Figure 1). The annual mean runoff yielding in southwest region is about $5{,}853\times10^8m^3$ (equal to 21.6% of the total runoff in China) and its average runoff depth amounts to 687.5mm. The annual flow shows great regional differences, for instance, the annual runoff depth of Irrawaddy River and rivers in southern Tibet reach 1,483mm and 1,253mm respectively, which are the highest in southwest China; the bordering area between southwest Tibet and India, in particular, amounts to 4,000mm and ranks the top in China; whereas in western Tibet, the figure is below 50mm.

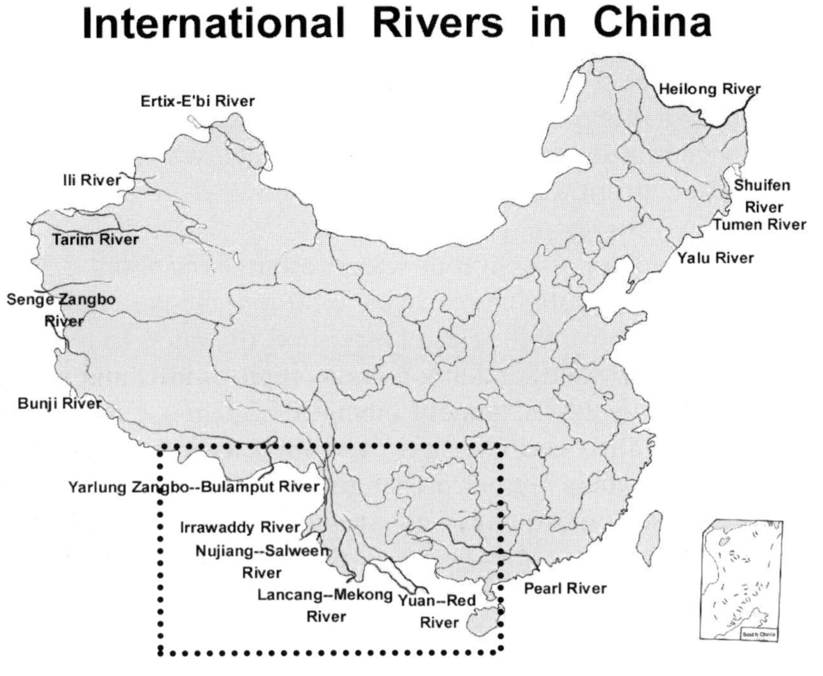

Figure 1. Location of "Southwest Rivers"

Southwest rivers are located in the plateau, mountainous and remote area, where backward economy, low productivity and poor development of water resources are witnessed. Statistics show that in 1998, the actual utilization of surface water by various national economic departments in Yunnan Province is 6.512 billion m^3, with an average utilization rate of 3.97%, the highest being the Red River basin (5.56%) and the lowest the Irrawaddy River (2.62%). While water resources' utilization rate is rather low, the general shortage rate is very high. According to Yunnan Province Water Resources Bureau's forecast, even by the year 2010 when the dependable output reaches 75%, surface water available in the international rivers basin will be $81.16\times10^8 m^3$, with a mere utilization rate of 5.96%. While the quantity of water demanded is $99.02\times10^8 m^3$, the shortage quantity is $17.86\times10^8 m^3$, with a shortage rate of 18.0%; by the year 2020, with a dependable output of 75%, surface water available will be $100.49\times10^8 m^3$, utilization rate 7.39%, water demanded $118.20\times10^8 m^3$, water shortage $17.71\times10^8 m^3$ and the shortage rate is still 15.0%.With water conservancy works almost all concentrated on the outflow system, Tibet's water utilization rate has long been very low through history, while demand increases year after year along with the economic development. In 1999, the total water supplied amounted to $25.84\times10^8 m^3$, increased by 56.7% as compared to that of 1998, among which surface water constituted 94.4%, with a mere utilization rate of 0.57% (Tibet Water Conservancy Bureau, 1998, 1999). In the past 50 years after the establishment of new China, despite Tibet's rapid development of water resources utilization, its water development is far below the national average level, with an overall utilization rate less than 1.0%.

3. International challenges for utilization and conservation

Of all the continents, Asia represents the lowest per capita use of available water resources. With rapid population growth and socio-economic development, the decrease of available water resources will become more serious, which highlights the international water issue. With a stream area representing sixty-five percent of its continental mass, Asia ranks the top globally in the number of international rivers and quantity of transboundary sharing water resources. In the last 50 years, due to population growth, the per capita use of available water in south and middle Asia dropped by 70%, in north Asia 60% and southeast Asia 55%. All of these regions are concentrated areas of international rivers, whose transboundary water issues deserve unprecedented concern worldwide.

Currently, China's per capita possession of water resources is about 2,300m^3. It is predicted that by the year 2030 when population reaches 1.6 billion, the per capita possession will drop to 1,760m^3, which means China will be a country short of water. In terms of per capita water resources, among all the provinces and autonomous regions in China, the first five are in the west (*Reference Information*, 2000), namely Tibet (206,892m^3), Qinghai (13,978m^3), Yunnan (5,953m^3), Xinjiang (5,774m^3) and Guangxi (4,415m^3). Being the upstream or headwaters, these provinces and autonomous regions are strategic resource bases for China's water safety and sustainable development, meanwhile key areas for ecological security.

There are 17 riparian countries bordering on China, most of which are short of water (see Table 1) and are determined as countries potential of water crisis by UN's Economic and Social Council except Laos and Burma. Afghanistan, India, Iran and Pakistan, in particular, are regarded as countries of high possibility of water crisis. Meanwhile they are downstream countries closely related to the development and utilization of transboundary water resources in western China.

Table 1. Relationship of Water and Population between China and Riparian Countries

Nations	Per capita regenerative fresh water in 1998 m^3 / person·year	Population growth / million persons		
		1998	2025	2050
Afghanistan**	2354	23.364	45.262	61.373
Bengal *	10940	124.043	179.980	218.188
Bhutan *	49557	1.917	3.646	5.184
Cambodia *	8195	10.751	16.990	21.394
China *	2231	1,225.091	1,480.430	1,516.664
India **	1896	975.772	1,330.201	1,532.674
Iran **	1755	73.057	128.251	170.269
Kazakhstan	4484	16.854	20.047	22.260
Korea **	2887	23.206	30.046	32.873
Laos	50392	5.358	10.202	13.889
Mongolia *	9375	2.624	4.052	4.986
Burma	22719	4.7625	67.643	80.896
Nepal *	7338	23.168	40.554	53.621
Pakistan **	1678	147.811	268.904	357.353
Tadzhikistan	11171	6.161	9.747	12.366
Thailand *	1845	59.612	69.089	72.969
Russia	29115	147.231	131.395	114.318
Vietnam *	4872	77.896	110.107	129.763

Source: World Resources, 1998~1999, P. 244~305.

* Nations potential of water crisis determined by UN's Economic and Social Council in 1996;

**Nations of high possibility of water crisis.

The table reflects that the outside environment for international rivers water resources in western China is not so admirable, and it is potential for intense competitive utilization in the near future. Furthermore, since the downstream countries have developed water resources to a much higher extent, they are sensitive to what has happened in the upstream and show great concern over upriver development and water quality changes.

After the admission into WTO and under the call of globalization, China has not yet accumulated enough research results, management schemes and countermeasures in line with international practices and laws, so as to respond to the competitive utilization and coordinated management of transboundary resources, as well as cope with the issue of transboundary conflicts (Zhu, 1993).

4. Major issues of sustainable utilization and regenerative maintenance

4.1 Serious environmental impacts of human activities in mountainous area

Most of the Southwest Rivers originated in the high mountains characterized by topographic unconformity and ecological fragility, with a lot of environmentally distressed population. Since the 1950s, human activities such as large scale logging, slope farming, pasture and wetland exploitation and the construction of highways and hydropower plants, have resulted in serious environmental deterioration, obvious hydrology uncertainty change and increasing disappearance of biodiversity. With the worsening of mountainous disasters, i.e. torrential flood, landslide, mudflow, drought and low temperature, the livelihood and production system including farmland, towns, transportation, irrigation works, hydropower plants and human settlements in mountainous areas are seriously endangered. Take Hengduanshan mountainous area in northwest Yunnan as an example, in 1998, there were still 495,000 people (equal to 16% of its total population) who did not have enough food and clothing, among whom 194,400 were environmentally distressed due to loss of livelihood; 79% of the population depended on firewood for energy, which resulted in the annual disappearance of 129,300 hectare forest. The annual surface soil loss amounted to 615t/km^2, close to that of the Loess Plateau; the annual mean erosion modulus of slope reclamation above 25° reached 1,500 t/km^2; while over 60% of the townships were imperiled by mudflow.

During the late stage of Manwan's construction, in the reservoir area within Manwan town, Jingdong county, slide and collapse ruined 319.10mu (equal to 21.27 hectare) cropland and forest, imperiled three villages and forced 161 people from 32 households to move out, with an extra settlement cost of 600,000 RMB. In July, 2000, torrential floods made 953 people from 222 households displaced, and 510 houses with 2,426 rooms were destroyed or damaged.

4.2 Ecosystem changed quickly by the interruption of infrastructure

From the present to the near future, a series of large scale actions carried out in western China, including the construction of expressways, railways and cascade hydro-stations, as well as the exploitation of mineral, biological and tourism resources, will put eco-environment and socio-economy under continuous disturbance and speed up their evolution.

Those actions will dramatically alter the original resource, economic and environmental patterns in the region, which brings to light the relevant issues of environmental deterioration, biodiversity disappearance, major projects' safety and benefit, regional economic cooperation and transboundary ecological maintenance. Therefore, how to balance development and protection has become an important and urgent issue in the region.

4.3 Worsening of soil erosion as a result of ecological degradation

The distribution of Southwest Rivers' sedimentation generally presents a degressive trend from east to west, the highest is Red River basin, the lowest the Qinghai-Tibet Plateau. For most of the rivers, greater sediment concentration is usually found in the middle and lower reaches. Comparison between the series results of 30 (1953-1983) and 39 (1953-1991) years downstream of Manwan dam-site shows that, the variation of long-term annual runoff is relatively slight, merely increased by 0.81%; while the sediment concentration increased by 36.6%, and annual sediment runoff by 37.3%.

With less cropland, cultivation activities and human disturbance, the original vegetation in Tibet is properly protected, which best explains why rivers in Tibet show relatively lower sediment concentration. Weathering is the main cause of sedimentation, meanwhile collapse, landslide and mudflow along the banks also carry a lot of silt into the river. The long-term average sediment concentration of Yarlung-Zangbo's mainstreams, Nugesha, Yangcun and Nuxia read 0.72kg/m^3 (1961-1995), 0.42kg/m^3 and 0.26kg/m^3 (1956-1995) respectively. Among the tributaries, Nianchu River is the highest (with an annual mean sedimentation of 1.36kg/m^3), and the rest are below 1.0kg/m^3 (Lhasa River 0.11kg/m^3 and Niyang River below 0.10kg/m^3).

Yunnan is a province imperiled by serious soil erosion. At present, over 40% of the land is endangered by soil erosion, and the annual sediment run out of the province and the border amount to 3×10^8t. The long-term average sediment concentration of Red River basin is 0.36-4.93kg/m^3, which ranks the top among Yunnan's major rivers.

For Lancang River, in the upstream of Changdu, the sediment concentration is rather low due to slight peak-valley difference and human disturbance. From Changdu to Gongguoqiao, the landform is characterized by high mountains and deep valleys, with sparse cropland. Most of the solid sediment is bedload rather than suspended load, and the annual mean suspended sediment is 0.73kg/m^3. In this reach of the river, tributaries represent greater sediment concentration than the mainstreams. Pijiang basin, in particular, faces increasing problems of soil erosion. In the reach from Gongguoqiao to Nuozadu, deep valley and unconfirmed topography are found. Steep slope, barren soil, low vegetation coverage and serious erosion make this region the most severely erosion-stricken area, with Weiyuanjiang as a typical example.

For Nujiang, the reach above Huitongqiao mainly runs through Qinghai-Tibet Plateau and Hengduanshan mountainous area, and the sedimentation is relatively low. The annual mean sediment concentration of Daojieba (a hydro-station on the mainstream) from 1965 to 1979 is 0.38kg/m^3. With population growth, slope reclamation, deforestation and worsening of soil erosion, the figure increases to 0.43kg/m^3 from 1965 to 1985. Below Huitongqiao, most of the sedimentation comes from Kuke River and Nanting River. The annual mean sediment concentration of Kejie station on Kuke River and Laolao station on Nanting River read 3.17kg/m^3 and 2.23kg/m^3 respectively. Substantive sediment brings about silt piling-up, riverbed instability and worsening of flood.

For Irrawaddy River, sediment concentration is relatively low. The flow in the upper and middle reaches of Dayingjiang, one of the major tributaries, is mainly contributed by Binglang River and Nandi River. Almost the same in catchment area, Binglang River contributes 70% of the flow, while Nandi River 75% of the sediment, which corresponds to the local saying of "Binglang's water and Nandi's sand". Lianghe and Yingjiang reaches of Nandi River are well-known for soil loss. Although the annual mean sediment concentration of Xialaxian station downstream is merely 0.79 kg/m^3, the silt mainly concentrates around farmland and villages, potential of great damage. The annual mean sediment concentration of Tenglongqiao station in the upstream of Ruili River is 0.33 kg/m^3, in recent years, however, the figure is increasing owing to deforestation and slope reclamation in the upper stream.

4.4 Increasing hazard of ecological degradation on large scale projects

In southwest China, intense geologic formation results in frequent occurrence of mountainous disasters. The increasing regional ecological degradation brings light to issues concerning the ecological security and benefits of major projects. Facing greater ecological pressure, it is imperative to find effective ways to lessen ecological deterioration's adverse impact on major projects for seeking of better economic benefits.
Take Manwan station, the first cascade dam on Lancang River as an example. On January 7, 1989, a landslide occurred during the excavation of Shangzuo dam bottom. The excavation was suspended and the whole project was delayed for one year, with a total loss of 1 billion RMB. In 1990, a mudflow struck the sand yard of Manwan station which caused 14 deaths and 33 injuries. After the construction of the dam, improper settlement resulted in the worsening of soil loss. Owing to the rapid concentration of sediment, the silting area further extends to the upstream and leads to annual increasing loss of beddings and bays.

5. Conclusions

5.1 China is the headwaters of major international rivers in Asia, the reasonable utilization and conservation of its abundant transboundary water resources will greatly affect water utilization and ecological security in China and riparian countries, influencing regional cooperation and development.

5.2 The annual mean runoff of 5 international rivers in southwest China amounts to $5{,}850\times10^8$ m^3, accounting for 21.7% of the nation's total and 80% of its outflow. Southwest China is the region where "Water Tower" of Asia locates. Currently, the water utilization rate is only 1%, whereas it is the most concentrated and potential water reserve in China, deserving special attention as strategic resource for China, south Asia and southeast Asia in the 21st century.

5.3 In the past, due to the lack of transportation access and scientific research, some regions are extremely poor in scientific studies, and even have some data gaps. Recent studies show that water systems in southwest China present distinct uncertainty changes. In Yunnan Province, for example, contrast analysis of two runoff series of 1956-1979 and 1956-1992 show that: for rivers influenced by southwest monsoon and supplied by melting water, such as Irrawaddy River, Nujiang and Lancang River, the long-term average runoff volume shows an upward trend. The more west-biased the location, the bigger the amplification, as the increment rate of Irrawaddy River is 4 times of that of Lancang River; whereas for rivers influenced by the southeast monsoon and without melting water supply, such as Red River and Pearl River, the long-term average runoff volume shows a downward trend. The more east-biased the location, the smaller the amplification, which can be reflected by the following decreasing sequence: Lisiangjiang (4.7%)<Yuan River (5.1%)< Panlongjiang (5.4%)< Nanpanjiang (5.9%)< Beipanjiang (7.6%). Those uncertainty changes are closely related to local human activities and global climate changes, but no relevant research is being carried out currently.

5.4 In the last decade, highway construction in mountainous areas, cascade hydro-station development and channel improvements resulted in increasing changes of water systems, rapid disappearance of biodiversity and worsening of soil loss. For instance, in Yunnan Province, serious soil loss would occur annually in its 52 rivers or reaches as a result of bank scouring. Currently, there are 38 rivers and 41 reaches that need urgent harnessing, with a total length of 260.56km and estimated investment of 494,338,000 RMB. The research and planning for the eco-environmental protection of international rivers in this region must be strengthened.

5.5 Located in the plateau and mountainous areas, southwest rivers lack hydrological and water quality monitoring networks and the corresponding standards are inconsistent with international ones. Hydro-stations on Tibet's international rivers are very scarce and some rivers even have none. More stations on major rivers should be set up for better assessment of transboundary water resources and its changes in the region, so as to provide reliable scientific data for reasonable formulation of international water allocation schemes and basin development planning.

6. Acknowledgement

The research is supported by the Ninth and Tenth Five-Year Plan national key science and technology project of China (No. 2000-K01-02-01, No. 2002BA901A22) and key project of Yunnan Natural Science Foundation (No. 2001Dooo2Z).

7. References

Aron T. Wolf (2001). Transboundary waters: sharing benefits, lessons learned, Thematic Background Paper, International Conference on Freshwater, Bonn, pp.1-2.

Changming Liu, Xiwu He (1996). China's water issues in 21st century (in Chinese), Science Press, Beijing, pp. 58-72.

Changming Liu, Zhikai Chen ed. (2001). Current status assessment and development trend analysis of water resources in China (in Chinese), China Water Resources and Hydropower Press, Beijing, pp.1-2.

Daming He, Changming Liu & Zhifeng Yang (1999). Study for the sustainability of international rivers in China (in Chinese), *Acta Geographica Sinica*, 54(Supplement), pp. 2-9.

Daming He, Junhua Gou (2000). Study for international rivers and its reasonable utilization in China, In: *International Symposium on Water Resource Development and Environmental Problems*, Xinjiang University ed., Urumchi, pp. 2-9.

Dexiang Zhu (1993). Significance and development of international rivers Studies (in Chinese), *Geography Studies*, 12(4), pp. 85-93.

Heng Liu ed.(1998). Guidance for water and sustainable development: principles and policies (in Chinese), Asian-Pacific Economic and Social Council, pp.7-8.

Nanjing hydrology and water resource institute, water resource institute of China water conservancy and hydropower academy (1998). *China's Water Supply and Demand in the 21st Century* (in Chinese), Hydropower Press, Beijing, pp. 18-21.

The United Nations (1997). The law of Non-navigational Uses of International Watercourses, In: *Work Report of the 46th Meeting of International Law Committee*, New York, pp.152-255.

Tibet Water Conservancy Bureau (1999). Tibet water resources bulletin (in Chinese), Tibet Water Conservancy Bureau, Lhasa, pp. 6-9.

Water Vapor Estimates from Forests and Croplands to Study the Hydrological Balance and Generation of Ecosystem Services: a Case Study of National Capital Region, India

V. Joseph[1)] and J.S. Pandey[2)]**

1) School of Environmental Studies, Cochin University of Science and Technology
Kochi-682 016, Kerala, India
e-mail:valsammajoseph@rediffmail.com
2) National Environmental Engineering Research Institute, Nagpur, 440020, India
e-mail: jaispandey@yahoo.co.in
** *Corresponding author*

Abstract

The critical biophysical water challenge in future will be whether the available freshwater is sufficient to support the food production as well as to generate ecosystem services, because of the pressure on ecosystems due to an increasing population and extensive land use changes. A recent interest in the fresh water assessments is to distinguish between green water (water vapor) and blue water (liquid water) for the integrated management of ecosystems. In the present study, we have attempted a generalized estimation of the mean water vapor from different biomes viz. forests and croplands of the National Capital Region (NCR) of India. The water vapor estimation was done based on previous studies on NCR and following the methodology of Rockstrom *et al.* (1999). From the values of mean water vapor estimates from forests and croplands for each zone within the NCR, we have derived a ratio between the vapor estimates from forests and croplands and termed it as local ratio (LF/LC). The ratio of the same for the total NCR is also calculated and termed as regional ratio (RF/RC). The results showed that when comparing with the global ratio (GF/GC) which we derived from Rockstrom *et al.* (1999), LF/LC and RF/RC can be used as a new tool for assessing the impact of land use pattern on the hydrological balance and ecosystem service generation in the region.

Keywords: blue water, ecosystem services, freshwater, green water, hydrological balance, water vapor estimates

1. Introduction

The major impacts of climate change in India would be on the surface and ground water hydrology and the agriculture of the country. An overview of India's water resources reveals that it is much below the natural capacity and technical feasibility (Bobba *et al.*, 1997). The interregional variability in climatic change induced by green house warming together with interregional heterogeneities due to differences in spatial factors makes the estimation of water requirements a complicated process. When considering the increasing population and demand for water, it becomes essential to know with sufficient accuracy the future availability of water. The complex interactions between chemical, biological and hydrological processes occurring over any spatially extended region and most especially in river basins involve all kinds of possible dynamics operating over a wide range of spatial and temporal scales (Rodriguez, 2000).

Precipitation over land is seen as the overall freshwater availability and is portioned in to green water (evaporating vertically) and blue water (flowing more or less horizontally) (Falkenmark and Lundqvist, 1998). FAO (1997) estimated that 54% of accessible blue water is already appropriated for direct and indirect human use, while 26% of the available green water is already appropriated for agriculture and forestry. Generally, it is only the liquid water or blue water that is perceived as freshwater resource for socioeconomic development, whereas, the reason to be concerned over future liquid water use originates from rain fed land use and there is a critical trade off between flows of water vapor for food production and for other welfare supporting ecosystem services.

Vast quantities of green water evaporate in to the atmosphere when stomata open to absorb CO_2, the water loss being proportionate to the evaporative demand of the atmosphere. Green water consumption of 4000 m^3 per ton of grain produced and sometimes more is not expected in sub tropical regions, whereas, arid regions typically suffer from green water scarcity which makes crop production dependent on the addition of blue water through irrigation (Falkenmark and Lundqvist, 1998). The liquid water or blue water part of water flow consist of surface runoff and the ground water recharge, whereas the water vapor or green water is the return flow of water to the atmosphere as evapotranspiration (ET) which includes transpiration by vegetation and evaporation from soil, lakes and water intercepted by canopy surfaces (Rockstrom *et al*., 1999). Thus, the evapotranspiartion is considered as the work of the whole ecosystem including the resilience it needs for securing the generation of ecosystem services in the long run. Vapor flow accounts for 64% of the available water resources in the global hydrological cycle while the realistically accessible blue water accounts for 11% (Rockstrom, 2001)

The major portion or up to 80% of the required food production will come from irrigated agriculture (FAO, 1997). Postel *et al.* (1996) had appropriated the renewable fresh water for the increasing population demand on a global scale. The World Bank/ United Nations Development Programme, estimated that there is considerable scope for expanding the acreage under irrigation by about 110 million ha, most of which might be in Asia (World Bank/ UNDP, 1990). Competition for water is increasing in highly irrigated countries like India. The projected water requirements for India by 2050 for the production of food grains to feed ever-increasing population are very high (Agarwal, 1999). The average share of irrigated area in gross cropped area is expected to be 41-45% by 2010 and 59-76% by 2050. The average yield of food grains in the rain fed area is expected to be 1.10 tonnes/ha by 2010 and 1.50 tonnes/ha by 2050, while the yield from irrigated areas is excepted to be 3.0 tonnes/ha in in 2010 and 4.0 tonnes/ha in 2050. When rain fed production becomes less, it requires providing blue water through irrigation and this redirection of water will reduce the water for human consumption as well as water for ecosystem service generation. Therefore, it becomes imperative to estimate the water vapor flow both regionally and locally, as the critical issue of how to feed a growing population through agricultural food production cannot be tackled in isolation from the freshwater dependent generation of ecosystem services in the surrounding landscape. In the present study, we have attempted a generalized estimation of the mean water vapor from different biomes viz. forests and croplands of the National Capital Region (NCR) of India following Rockstrom *et al.* (1999) and further derived the ratio of the estimate from forests and croplands of each zone and the NCR region as a tool to study the hydrological balance of the region.

2. The study area

The Nation Capital Region is an environmentally critical region covering and area of 30242 km^2 (Figure 1A) and includes sub-regions from the Union Territory of Delhi and from the states of Haryana, Rajasthan and Uttar Pradesh (Figure 1 B). The regions included in the present study were Union Territory of Delhi, Panipat, Rohtak, Sonipat Gurgaon, Faridabad, Rewari (Haryana); Meerut, Gaziabad, Bulandshahr (Uttar Pradesh) and Alwar (Rajasthan). The change in land use characteristics, particularly the conversion of agricultural land for non-agricultural uses is alarmingly at a larger scale over the years. The major factors influencing the land use characteristics in the region are the rapid increase of economic and industrial activities, the pressure of population and extensive cultivation (NCRPB, 1998). The total forest cover of the region is only 2.1% of the total area.

The NCR region in general, is a part of the well-integrated drainage system of Ganga and Yamuna. These rivers meet the bulk of the water requirements of the region. The other important rivers are the Hindon, the Kali and the Sahibi. Rivers, canals and lakes are the major sources of water supply. Wide flood plains and high banks along with clay and silt deposits are common features in the course of the Ganga and Yamuna. .Ground water resource is mainly controlled by geology and precipitation in the area. About 80% of the annual precipitation occurs during the monsoon months. The general water table ranges between 6-15 m below the ground level. The aquifer systems go up to the depth of 230 m and most of the bore wells are dug into these systems and their recharge is mainly through the rivers and precipitation. In summer months these tube wells dry as the water table sinks (NEERI, 1996).

3. Methodology

In the present study, the quantification of the regional and local water vapor flow was based on the first bottom-up estimation of Rockstrom *et al.* (1999) for quantifying continental water vapor flows. For the quantification of Water Vapor Flow (km^3/yr) from forests, the area (km^2) was multiplied by Annual Mean Actual Evapotranspiration (km/yr) for the forest type. The water vapor flow from croplands was calculated by multiplying the annual crop production (in Mega grams per year (Mg/yr) 1 Mg =1 tonne) by the water use efficiency (km^3/Mg) of the respective crop type.

The ratio of water vapor estimates from world's croplands to forestlands was calculated based on the estimates by Rockstrom *et al.* (1999) and termed as global ratio. This ratio (GF/GC) derived by dividing the total water vapor estimates from world's forests (40,000 km3/yr) by that for croplands (5400 km^3/yr) was taken as a standard for comparison. From the values of mean water vapor estimates from forests and croplands for each zone within the NCR, we have derived a ratio between the vapor estimates from forests and croplands and termed it as local ratio (LF/LC).

The ratio of the same for the total NCR is also calculated and termed as regional ratio (RF/RC).

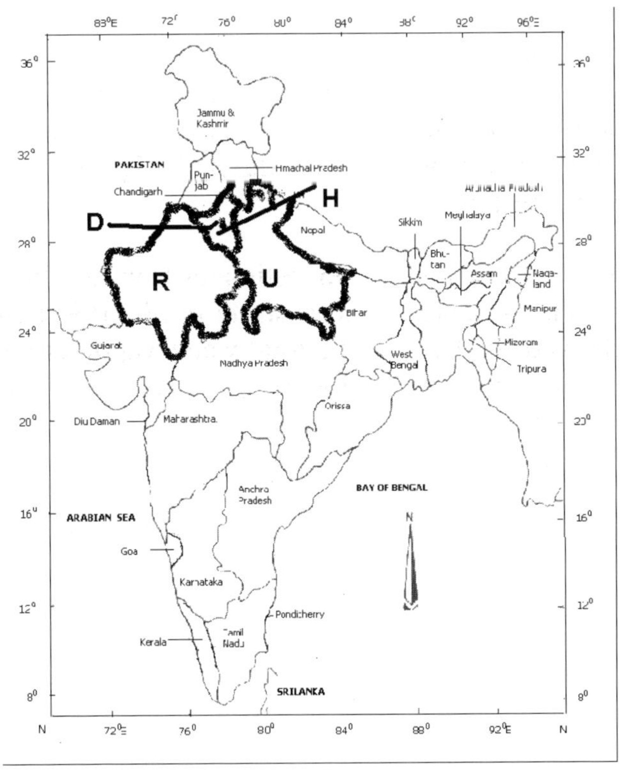

Figure 1. **A:** Map of NCR region showing different regions (NEERI, 1996); **B:** Map of India showing the location of different states constituting the NCR region, and the Union Territory of Delhi (D -Delhi, R -Rajasthan, U- Uttar Pradesh , H- Haryana)

Study Area: The National Capital Region

The results showed that taking the global ratio (GF/GC) as a standard, LF/LC and RF/RC can be used as a new tool for assessing the impact of land use pattern on the hydrological balance and ecosystem service generation in the region.

The data on the areas of forests of NCR region was collected from earlier studies conducted by the (National Environmental Engineering Research Institute) NEERI (NEERI 1993, NEERI 1994 and NEERI, 1996). The data for the mean annual production during the period 1989 to 1991 were collected for different cropland types including cereals, fibers, cotton, oil crops, roots and tubers, spices, sugarcane and vegetables (CMIE, 1996). The crop types under cereals included rice, jowar, bajra, maize, ragi, wheat and barley. Gram and arhar are the main pulses in the region. The oil crops included groundnut, sesamum, rapeseed and mustard, linseed, castor, sunflower, niger seed and safflower. Potato and sweet potato were classified under roots and tubers. The fibers included jute and mesta. The major crops included under vegetables were coriander, chilies, garlic and onion whereas the spices included ginger and turmeric. In the present study, the water vapor estimates from the croplands were carried out based on the data available for each crop type. The data for crop production in the zone Bawal (Rajasthan) was not available and is not included in the study. If we take into consideration the data for more crop types which were not available, the water vapor estimates will increase. The other ecosystem types like wetlands and grasslands in the region were also not taken into consideration.

4. Results and discussion

The water vapor estimates from croplands of NCR is 6.95 km^3/yr. The water vapor estimate was maximum for Gurgaon and Meerut whereas it was minimum for Delhi (Figure 2). The water vapor estimate from forests is 2.619 km^3/yr, the values being lowest for Delhi and highest for Rohtak (Figure 2). The forests in the region are dry deciduous type and the mean actual evapotranspiration was taken as 0.000792 km/yr following Rockstrom *et al.* (1999). The low estimates of water vapor from forests compared to croplands in the NCR region points to redirection of liquid water for irrigation and also the low area of forests in the region (Table 1). This redirection will continue to deplete the rivers and cause down stream ecological impacts and subsequently impact the generation of ecosystem services which depends on the water transpired by vegetation and on evaporation. Annual evapotranspiration is generally greater for forested than non forested catchments, the catchments having values in between that observed for fully forested and fully cleared catchments (Zang *et al.*, 2001). The plant available water capacity is primarily responsible for greater evapotraspiration from forests than from pastures and crops. The trees usually maintain a relatively constant evapotranspiration rate over time even while soil moisture is less, due to the ability of the trees to access soil moisture from depth when compared to the shallow rooted plants.

The ratio of water vapor estimates from croplands to forests was calculated for each sub-region in the NCR region (Figure 3). This ratio termed local ratio (LF/LC value) is used as a tool to predict the hydrological balance of the region and the sub-regions. The LF/LC values for almost all places in the region are very low and ranged from 2.22 to 0.06. The regional ratio RF/RC is 0.377. The low values of LF/LC and RF/RC show a hydrological imbalance in the region when compared with the global ratio of 7.5. The more the ratio approaches the global ratio, the more water vapor will be available for the generation of ecosystem services in the region.

Table1. Land Use Pattern (%) in the National Capital Region (NEERI, 1994)

Zone within the Region	Fallow Land	Forest Land	Pasture and Other Uses	Land Use other than Agriculture	Culturable Wasteland	Unculturable Wasteland	Net Sown Area
Panipat	0.4	2.9	3.1	3.5	0.8	0.5	88.8
Rohtak	2	1.9	-	7.4	3.4	0.9	84.4
Sonipat	2.9	5.9	1.5	6.4	0.3	3.1	79.9
Meerut	3.6	2	0.2	12.4	1.2	1.4	79.2
Gaziabad	3.6	1	-	14	2.9	3.6	79.2
Delhi	8.7	1.1	1.4		7.1	48.9	32.8
Gurgaon	12.5	5.3	0.7	12.6	0.7	2.6	65.9
Faridabad	2.7	1	0.6	16.1	0.1	4.1	75.4
Bulandshahr	4	1.9	0.8	9.2	2.8	3.5	77.8
Rewari	1.5	2.8	-	8.7	0.7	3.4	82.9
Alwar	-	2.3	3.3	4.2	1.7	13.5	75

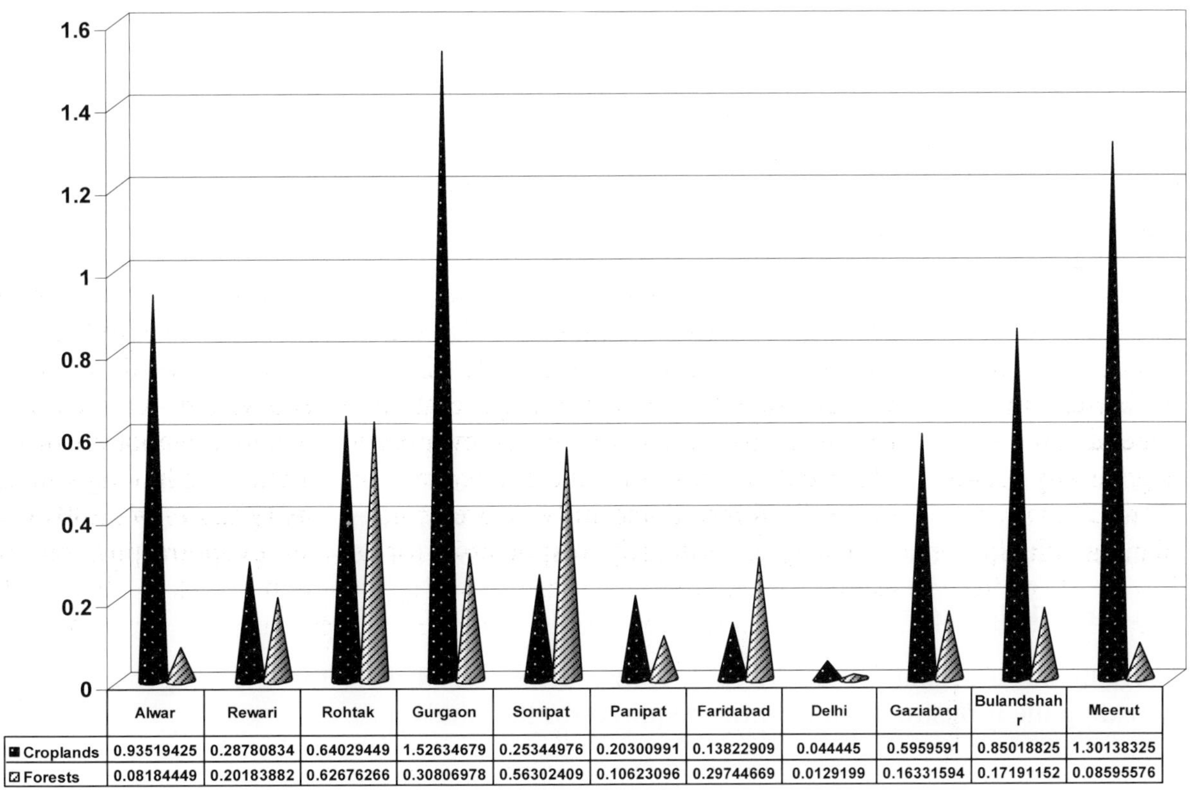

Figure 2. Mean Water Vapor Estimate (km^3/yr) from Forests and Croplands of NCR.

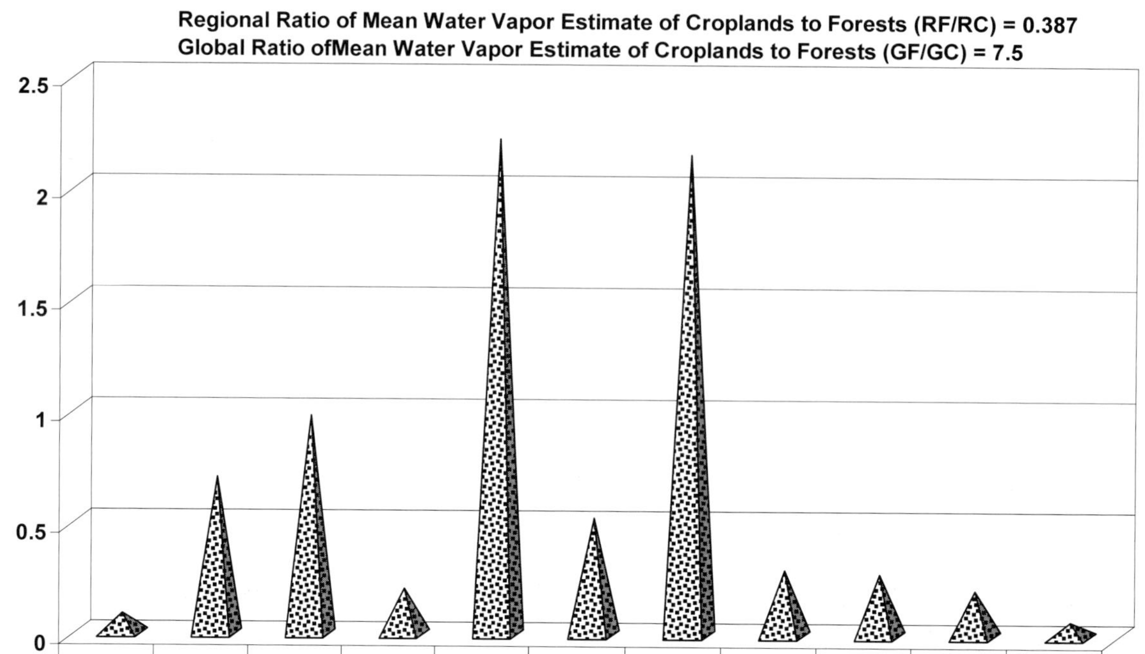

	Alwar	Rewari	Rohtak	Gurgaon	Sonipat	Panipat	Faridabad	Delhi	Gaziabad	Bulandsh ahr	Meerut
LF/LC	0.087516	0.7012959	0.9788662	0.2018347	2.2214426	0.5232797	2.1518386	0.290694	0.2740388	0.2022041	0.0660495

Figure 3. Local Ratio of Mean Water Vapor Estimates of Croplands to Forests (LF/LC) of NCR.

5. Conclusion

It is apparent from the results of the study that there is a hydrological imbalance in the different zones studied in the NCR region. Thus, the quantification of the trade offs of water vapor between forests and croplands can be an important tool in estimating the impact of land use pattern on hydrological balance of a region. Without an understanding of how hydrology and biotic processes interact to sustain ecosystem functioning and the provision of ecosystem services, it will be very difficult to place either an ecological or social value on alternative water management scenarios (Strange *et al.*, 1999). It is, therefore, imperative to further study the dynamics incorporating factors like climate, soil, vegetation characteristics, and the population in the region. It is also required to develop models incorporating important ecological characteristics for readily predicting the long-term consequences of land use change on the hydrological balance and the generation of ecosystem services of the region.

6. Acknowledgements

The authors are thankful to Dr. R.N. Singh, Director, NEERI for encouragement and for providing the infrastructural facilities required for this study.

7. References

Agarwal A., Narain S., Srabani S. (1999). The citizen's fifth report, Part – II, Statistical database, *Centre for Science and Environment*, New Delhi.

Bobba A.G., Singh V.P., Bengtsson L. (1997). Sustainable development of water resources in India, *Environ. Manag* 21 (3), pp. 367-393.

CMIE (1996). India's agricultural sector, A compendium of statistics, July 1996, *Center for Monitoring Indian Economy* Pvt. Ltd., Mumbai, pp. A-62 –A-329.

Falkenmark M., Lundquvist J. (1998). Towards water security: political determination and human adaptation crucial, *Water Resources Journal*, September, pp. 12-28.

FAO (Food and Agricultural Organization of the United Nations). (1997). Food production: the critical role of water, Technical background document7, prepared for the World Food Summit, 13-17 November 1996, Rome.

NEERI. (1993). NEERI 93-annual report of National Environmental Engineering Research Institute, Nagpur, India, pp. 101-113.

NEERI (1994). NEERI 94-annual report of National Environmental Engineering Research Institute, Nagpur, India, pp. 96-101.

NEERI (1996). Carrying capacity based developmental planning in National Capital Region Report, National Environmental Engineering Research Institute, Nagpur, 440 020, India.

National Capital Region Planning Board. (1998). Regional plan 2001, National Capital Region, Ministry of Urban Development, Government of India, New Delhi.

Postel S.L., Daily G.C., Ehrlich P.R. (1996). Human appropriation of renewable water, *Science* 271, pp. 785-788.

Rockstrom J., Gordon L., Floke C., Falkenmark M., Engwall M. (1999). Linkages among water vapor flows, food production and terrestrial ecosystem services, *Conservation Ecology*, 3 (2), pp. 7-9.

Rockstrom J. (2001). Green water security for food makers of tomorrow: windows of opportunity in drought-prone savannahs, *Water Sci.Tech.*, 43(4), pp. 71-78.

Rodriguez-Iturbe, I. (2000). Ecohydrology: a hydrologic perspective of climate-soil-vegetation dynamics, *Water Resour. Res.*, 36 (1), pp. 3-9.

Strange E.M., Fausch K.D., Covich, A.P. (1999). Sustaining ecosystem services in human – dominated watersheds: biohydrology and ecosystem processes in the South Platte River Basin, *Environ. Manag.*, 24(1), pp. 39-54.

Zhang L., Dawes W.R., Walker G.R. (2001). Response of mean annual evapotranspiration at vegetation changes at catchment scale, *Water Resour. Res.*, 37(3), pp. 701-708.

World Bank/UNDP. (1990). A proposal for an internationally supported programme to enhance research in irrigation and drainage technology in developing countries, Vol.2. Washington D.C.

Impact of Pollution Sources on the Pong River Water Quality

P. Mungkarndee[1)] and W. Wirojanagud[2)]

1) Department of Environmental Engineering
Faculty of Engineering, Khon Kaen, 40002 University
e-mail: pinthita@kku.ac.th
2) Department of Environmental Engineering
Faculty of Engineering, Khon Kaen, 40002 University
e-mail: wanpen@kku.ac.th

Abstract

The Pong River is an important water source for the upper northeastern provinces of Thailand, especially Khon Kaen Province that uses this river as its sole water supply source. Land use activities along the river, which relate to pollution sources, have an impact on the river water quality. The Pong River reach that is interesting to study is divided into two sections. The first section is the downstream of Ubonratana Dam downstream to the Nong Wai irrigation weir of which the river length is about 34 km. The second section is downstream of Nong Wai irrigation weir to the Mahasarkham Dam of which the length is about 180 km. The major industries situated along the first river section are a pulp and paper mill and a thermal power plant. While sugar, distillery, paper board and tapioca mills are located along the second river section. In addition, there is intensive floating basket aquaculture along the first river section. The irrigation area is mostly along the area of the second river section. Therefore, pollution sources consist at both point of sources including industries, communities and aquaculture, and non-point sources comprising runoff from the irrigation area. Seventeen sampling stations were designated according to the pollution sources to investigate the water quality in terms of general physical and chemical characteristics as well as heavy metals and mutagens. The water quality at these sampling stations were generally similar and classified as moderate water quality. The exceptions were some sampling stations located downstream of the pulp and paper mill and in the part of the river that receives effluent discharge from the municipal wastewater treatment plant which showed higher levels of conductivity, color, COD, and Hg than other stations as well as mutagens. Therefore, the different pollution sources have potential impact on water quality. It is necessary to monitor the water quality of the Pong River by emphasizing some specific parameters including mutagens and heavy metals as these substances could affect bio-accumulations and bio-concentrations in the food chain and thus become a risk to human health.

1. Introduction

The Pong River is the main water source of Khon Kaen province in the Northeastern Region of Thailand. The river water is used for community water supply, industry, agriculture and aquaculture. At the same time, the river is a recipient of wastewater from communities and industries and runoff from agriculture. Water pollution would occur if the pollution load exceeds the capacity of the river. Water pollution of the Pong River has occurred frequently. Pollution has been observed since 1986 and became very serious in 1992 when molasses from a storage tank of a sugar factory leaked into the river and contaminated the water downstream from the factory to the Chi and Mun Rivers. The environmental agencies of the Government have realized these problems and have seriously monitored the industrial effluents discharged into the river. However in 1993-1994 and also in 1997, 1998, and 1999 up-to now, the phenomena of fish death, especially in the hot and dry seasons, has occurred.

In this study, besides the general physical and chemical water quality parameters, heavy metals, phenolic compounds and mutagens were analyzed to investigate the concentration levels of these parameters and their potential impact on the aquatic life and the food chain.

1.1 Study Area and Scope

The Pong River originates at Phu Kradueng in Loei Province and flows through Phu Kradueng District, Ubonrattana District and Muang District of Khon Kaen Province and then joins with the Chi River and flows through Mahasarakham Province. Ubonrattana Dam was built across the Pong River in Ubonrattana District to generate electricity and to serve for irrigation in the area. The Pong River is divided into 2 sections. The first section starts at Ubonratana Dam and runs 34 km downstream to Nong Wai irrigation weir. The second section runs from Nong Wai irrigation weir 180 km downstream to Mahasarakam Dam. Starting from Ubonrattana Dam, 17 sampling stations were selected covering both river sections. The sampling stations located in the Pong River itself and the tributaries like lagoons or streams were designated according to pollution sources. The sampling stations started near Ubonratana Dam downstream to Mahasarakham Weir where the land use is likely to be different. The land use along the first river section from Ubonratana Dam to Nong Wai irrigation weir comprises a large pulp and paper mill, a thermal power plant, some rainfed agriculture areas, and villages. At this river section, there is intensive floating basket aquaculture production. The land use along the second river section consists of villages and big communities (e.g. municipalities), a large irrigation area, and large industries including sugar production, distilleries, paper board, and tapioca mills. These various land uses are generally considered as potential pollution sources that might cause water pollution problems in the river.

Water quality monitoring was conducted 3 times during 2002 (January, April and November) to represent the condition in all three seasons. According to the location of pollution sources, the river was divided into 6 sub-sections each with a set of sampling stations as described in Table 1 and shown on the map in Figure.1.

Table1. Sampling Stations

Sub-Section	Description	Sampling Station
1	Downstream of Ubonrattana Dam	Sampling station 1 at Ban Huai Sai (river) Sampling station 2 before the Chot lagoon (river)
2	Chot stream and lagoon (Pulp and paper industry)	Sampling station 3 in the Chot stream Sampling station 4 before passing through project green of the industry Sampling station 5 after project green area Sampling station 6 at the Chot lagoon outlet
3	Aquaculture Area	Sampling station 7 at Ban Huai Sue Ten (river) Sampling station 8 at Ban Nong Bua Noi (river)
4	Before flow through industrial area, farm and communities	Sampling Station 9 in front of Nong Wai Weir Sampling Station 10 inTung Tieu lagoon Sampling Station 11 before sugar factory (river)
5	At industrial area and villages	Sampling Station 12 at Ban Ta-kraserm (river) Sampling Station 13 at the outlet of Yai stream (river) Sampling Station 14 at Ban Napieng (river) Sampling Station 15 in the Phra Krue streaam
6	After passing through industrial area and villages	Sampling Station 16 at Ban Ta Hin (river) Sampling Station 17 in front of Mahasarakham Weir

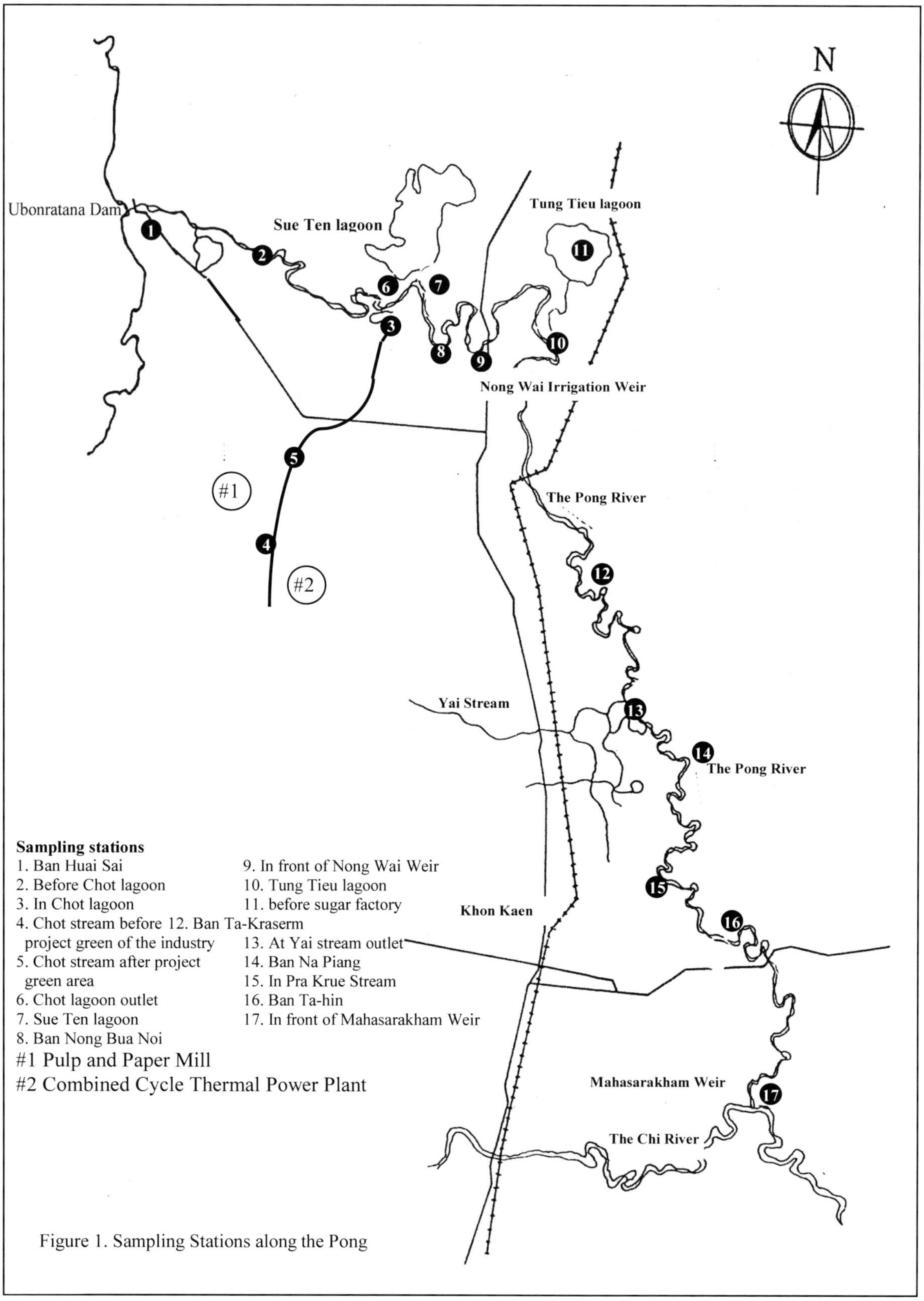

Figure 1. Sampling Stations along the Pong

All the collected water samples were analyzed for the following water quality parameters:

- Physical parameters: turbidity, color and conductivity
- Chemical parameters: pH, BOD, COD, TKN, NH_3-N, PO_4-P, Cl^-
- Heavy metals: Hg, Pb, Cd, Cr, Cu, Zn
- Mutagens: Test of possibility by Ames Test
- Toxic substances: Phenolic compounds, organochlorine pesticides

1.2 Pollution Sources

There are two main pollution sources of the Pong River which are point sources from industries (sampling station 3-6, 12-13) and non-point sources from villages, aquaculture, agricultural area and animal farms (sampling stations 1-2, 7-9, 16-17).

Point sources: Types of land use of the Pong River watershed from Ubonratana Dam to Mahasarakham Dam are shown in Table 2. The pulp and paper industry and thermal power industry are situated between sections 1 and 4 of the study area. The distillery, sugar, paper-board and furniture, ice production, sand and gravel and tapioca industries are between section 4 and 6 (Fig.1).

Non- point sources: The non-point source of pollution is theoretically defined as the runoff from the agriculture area. However, there are some pollution point sources that are not equipped with a treatment system, which would discharge wastewater as non-point sources as well. These pollution sources include domestic wastewater discharge from the villages, animal farms, aquaculture farms and solid waste duping sites, from which the loading and type of pollutants are difficult to determine.

Table 2 presents pollution sources of industry, community and agriculture.

Table 2. Land use in the Pong watershed downstream from Ubonratana Dam to Mahasarakham Weir

Type of Land use	Description	BOD loading kg/d	Volume of wastewater (m^3/d)	Treatment system	Effluent management
1. Industrial area	*Production Rate*				
- Pulp and Paper Mill	240,000 ton/yr	125	18,000	Activated sludge and Land treatment	40% Reuse 60% land and discharge to the Pong
- Combined Cycle Thermal Power Plant	300 megawatts	-	-	Cooling System	Recycle
- Sugar Mill	7,460,600 ton/yr	-	240	Oxidation Pond	Reuse and maintaining zero effluent discharge
- Card-board and furniture	35,000 ton/yr	-	120	Oxidation Pond	Maintaining zero effluent discharge
- Brewery product	46,800 m^3/yr	-	100	Septic tank for domestic waste	Discharge to the Pong River
- Tapioca Mill	21,900 ton/yr	-	1,200	Oxidation Pond and Land Treatment	Discharge to Huai Yai and the Pong River
- Ice production industry	14,706 ton/yr	-	-	No Wastewater	-
- Sand and gravel product	9,636,000 m^3/yr	-	160	Sedimentation	Discharge to the Pong River
- Household industries; Thai rice noodle production, meat processing (Chinese sausages and dried pork) and brick production	-	-	-	-	- -
2. Significant community area	*Population (persons)*				
- Ubolratana Sanitary District	6,105	77	-	-	Discharge to paddy fields
- Nam Pong Sanitary District	11,885	150	-	-	Discharge to the old Pong River
- Ubolratana District	2,612	8	-	-	25% of its total discharge to the Pong river
- Wang Chai Sanitary District	11,019	138	-	-	Discharge to Huai Chan lagoon
- Amphoe Nam Pong	13,557	21.4	-	-	25% of its total discharge to the Pong river
- Khon Kaen Municipality	129,401	1,804	-	Aerated lagoon	-
- Villages located along the river banks include Ban Na Piang, Ban Huai Sun, Ban Yang Yong, Ban Bueng Nium and Ban Pak Pueay	-	-	-	-	-
3. Agricultural area	324,251 rai	-	-	-	-
4. Aquaculture	-	-	-	-	Discharge to the Pong River

Source: Research on Environmental Management of the Pong River, Khon Kaen University, 2001.
Study project for prepsration of the action plan for the Pong River water quality rehabilitation, Khon Kaen Province, Khon Kaen University, 1995.

2. Results and discussion

Physical Analysis: The analysis of turbidity and color of most of the sampling stations were found to be acceptable with values of 5-13 NTU and 20-35 units, respectively (Figure 2 and Figure 3). However at station 13 to 17, the turbidity and color were slightly higher with the values of 26-90 NTU and 53-240 units, respectively. These stations received the effluent discharges from the tapioca industry (station 13), sand and gravel industry as well as the village (Ban Huai Phra Krue, station 15) where there are rice noodle household industries. Phra Krue stream also receives the effluent discharge from the municipal sewage treatment plant. The conductivity of the river was in the range of 140-200 µs/cm, except at stations 3 and 5 which receive seepage from the land treatment (called Project Green) of the pulp and paper industry and station 15 which receives domestic and industrial wastewater, the conductivity values were at 316, 2,046 and 316 µs/cm, respectively (Figure 4).

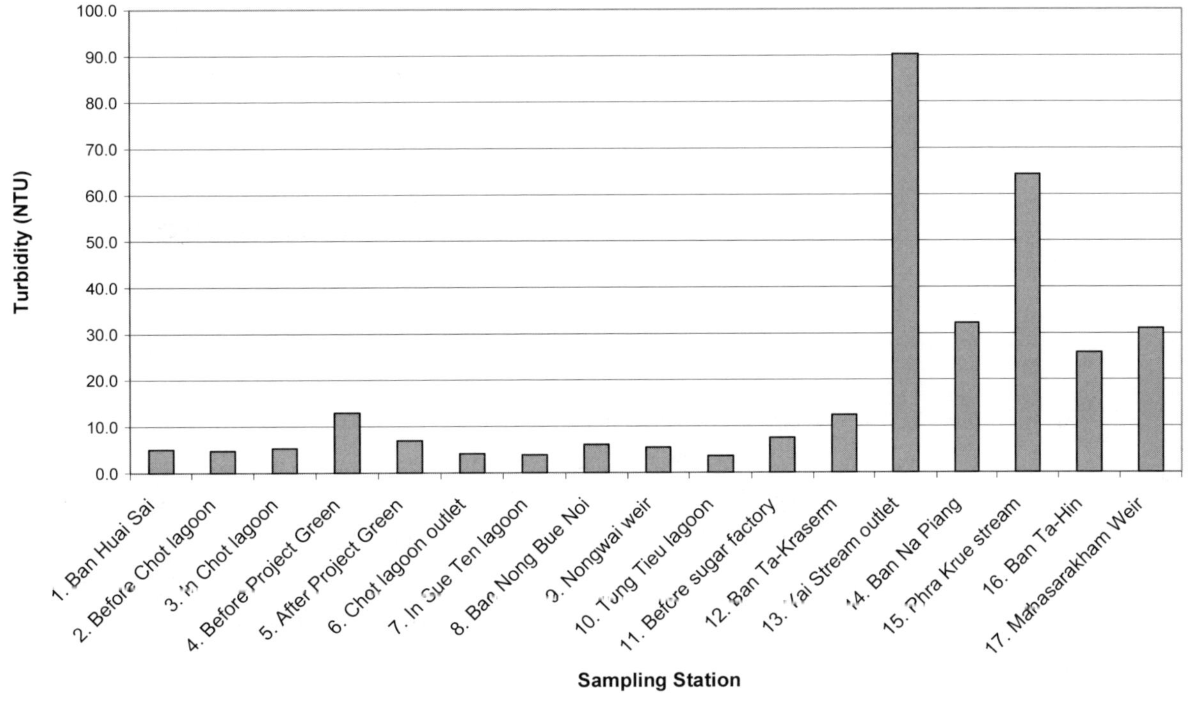

Figure 2. Results of turbidity analyses at 17 sampling stations in the Pong River from Ubonrattana Dam to Mahasarakham Weir

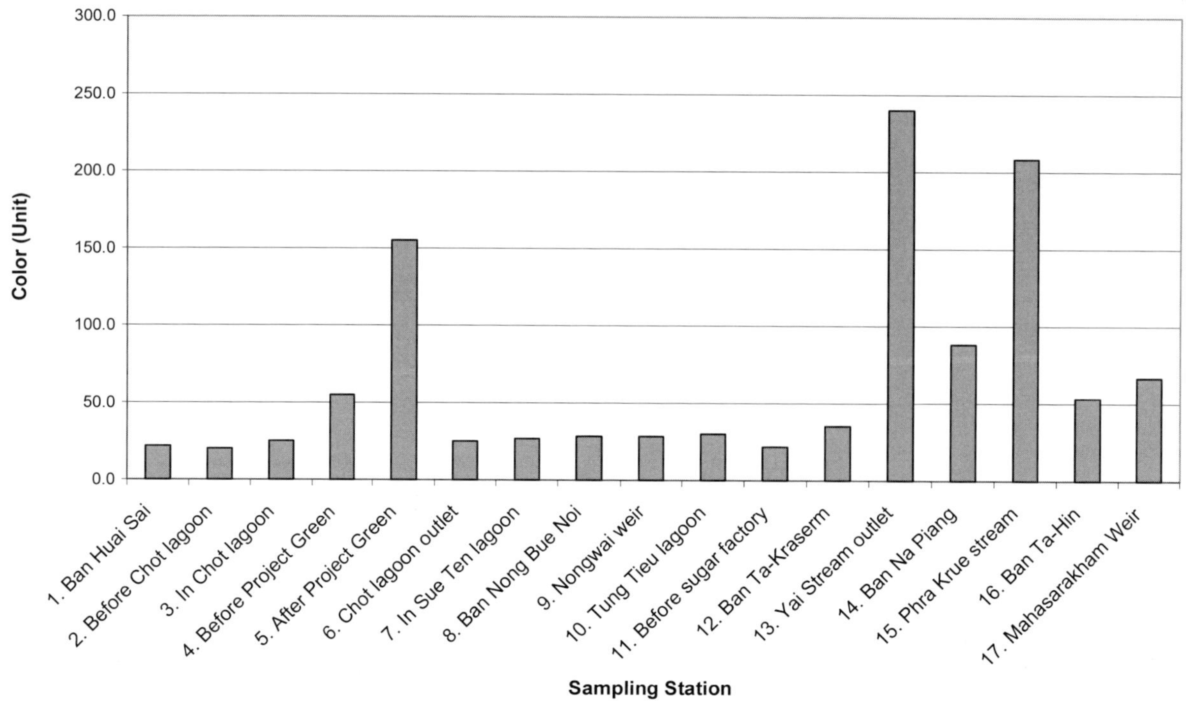

Figure 3. Results of color analyses at 17 sampling stations in the Pong River from Ubonrattana Dam to Mahasarakham Weir

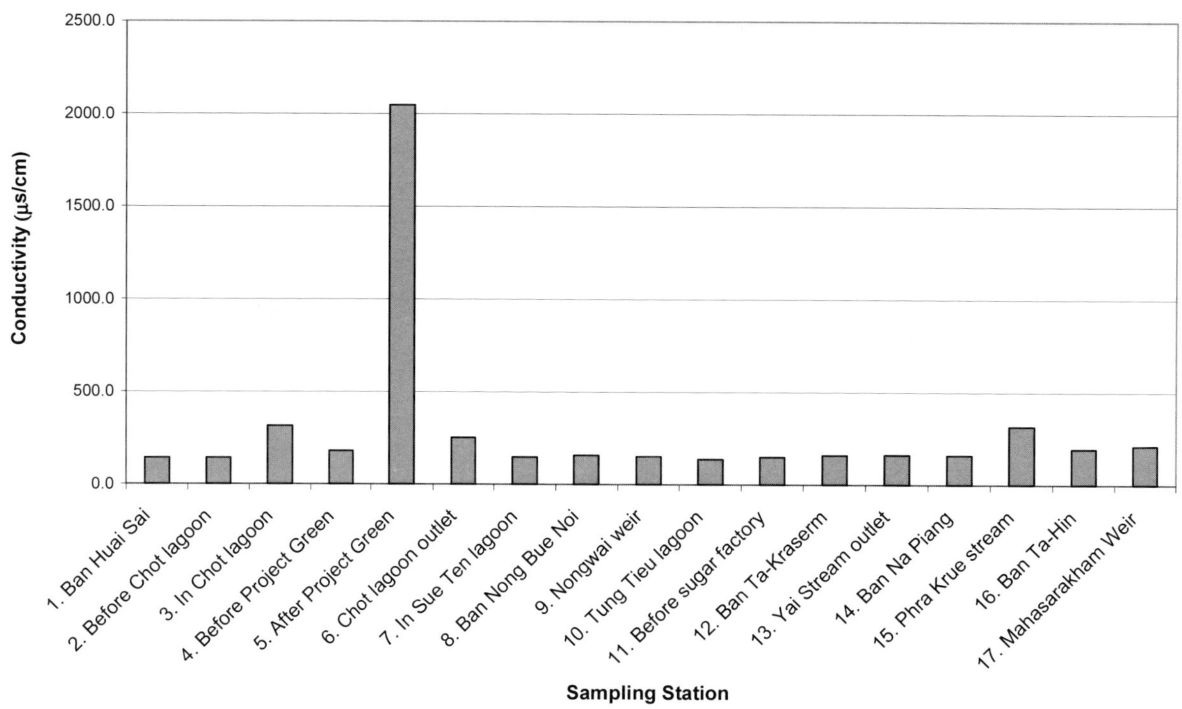

Figure 4. Results of conductivity analyses at 17 sampling stations in the Pong River from Ubonrattana Dam to Mahasarakham Weir

Chemical analysis: Generally the Pong River water quality, based on the analysis of DO and BOD content, could be classified as type 3 of the surface water classification in accordance with the Stand of Pollution Control Department, except for station 15 (Phra Krue stream) which presented as class 4. At station 15, indications of the impact of domestic wastewater TKN, NH_3-N and PO_4-P concentration would indicate contamination from agricultural and domestic waste. However, the analysis results of these parameters were still low and were not varied among the stations. The impact of agriculture on water quality was not yet obvious (Figures 5 and 6).

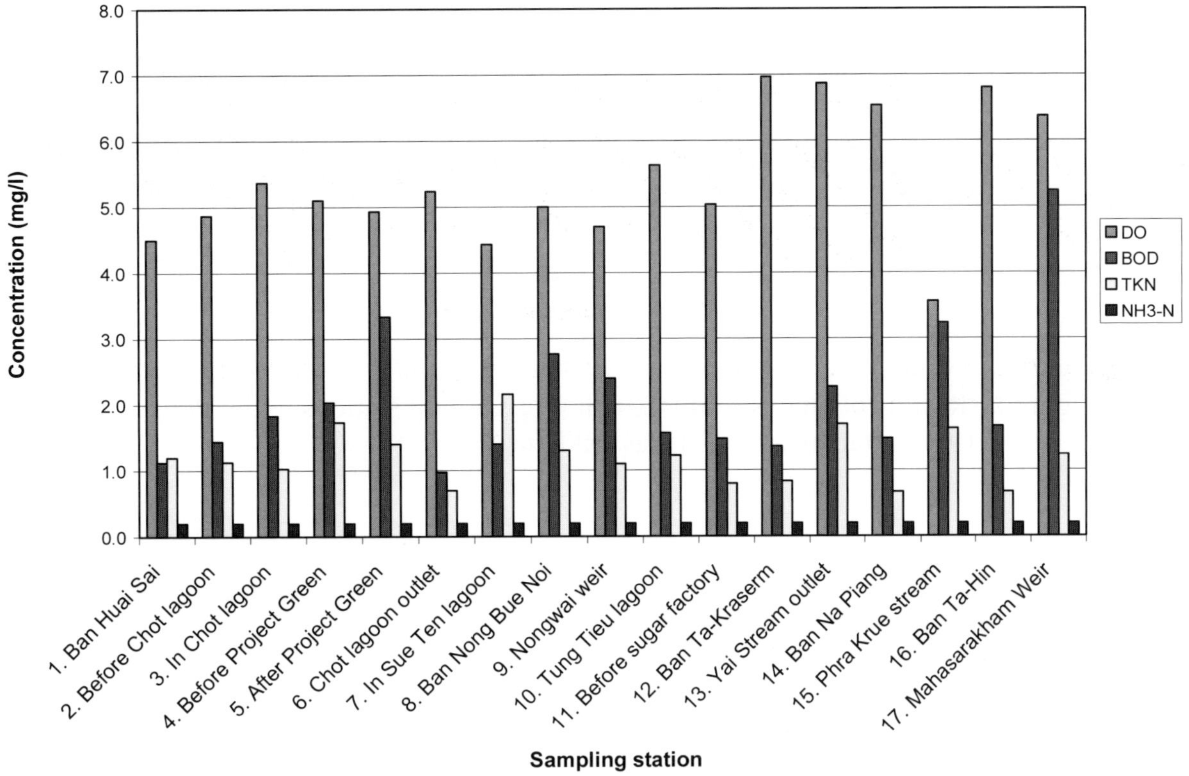

Figure 5. Results of analyses of DO, BOD, TKN, and NH_3-N at 17 sampling stations in the Pong River from Ubonrattana Dam to Mahasarakham Weir

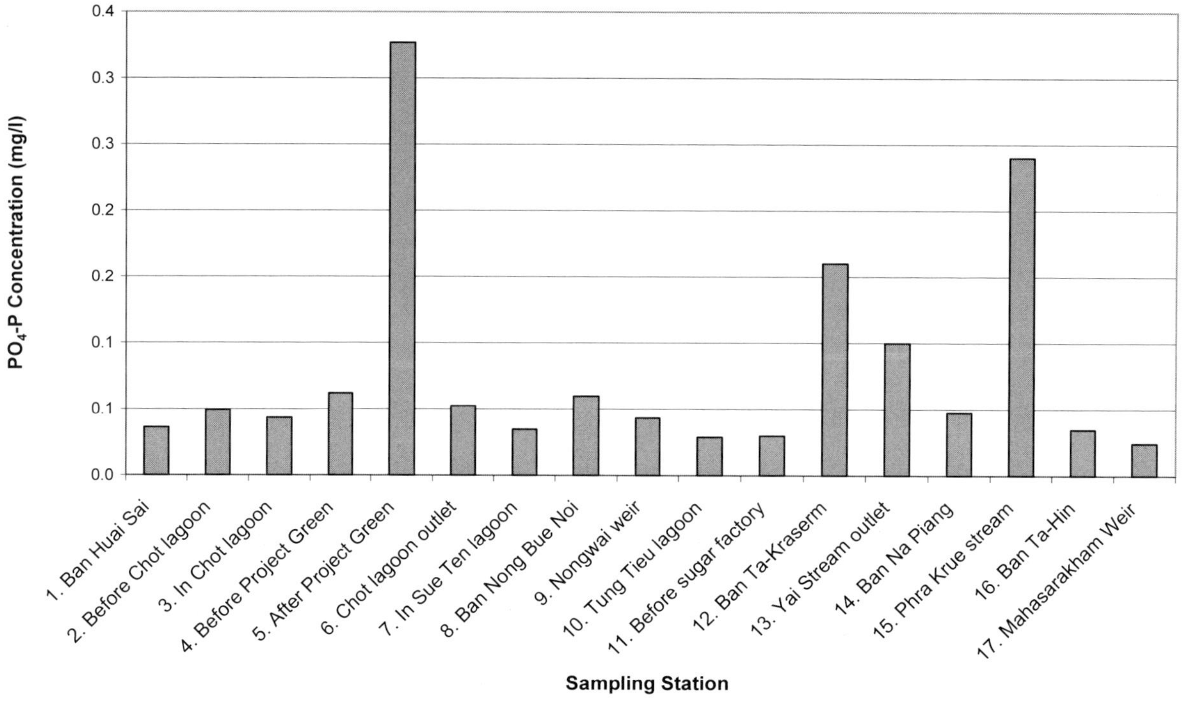

Figure 6. Results of analyses of PO_4-Pat 17 sampling stations in the Pong River from Ubonrattana Dam to Mahasarakham Weir

It was noted that station 5, where the river passes through the project green, showed higher COD (964 mg/l) and chloride (355 mg/l) concentrations than other stations where COD and chloride were less than 27 mg/l and 16 mg/l, respectively (Figure 7 and Figure 8). This indicated that there was some level of impact generated from the seepage from land treatment of the pulp and paper industry.

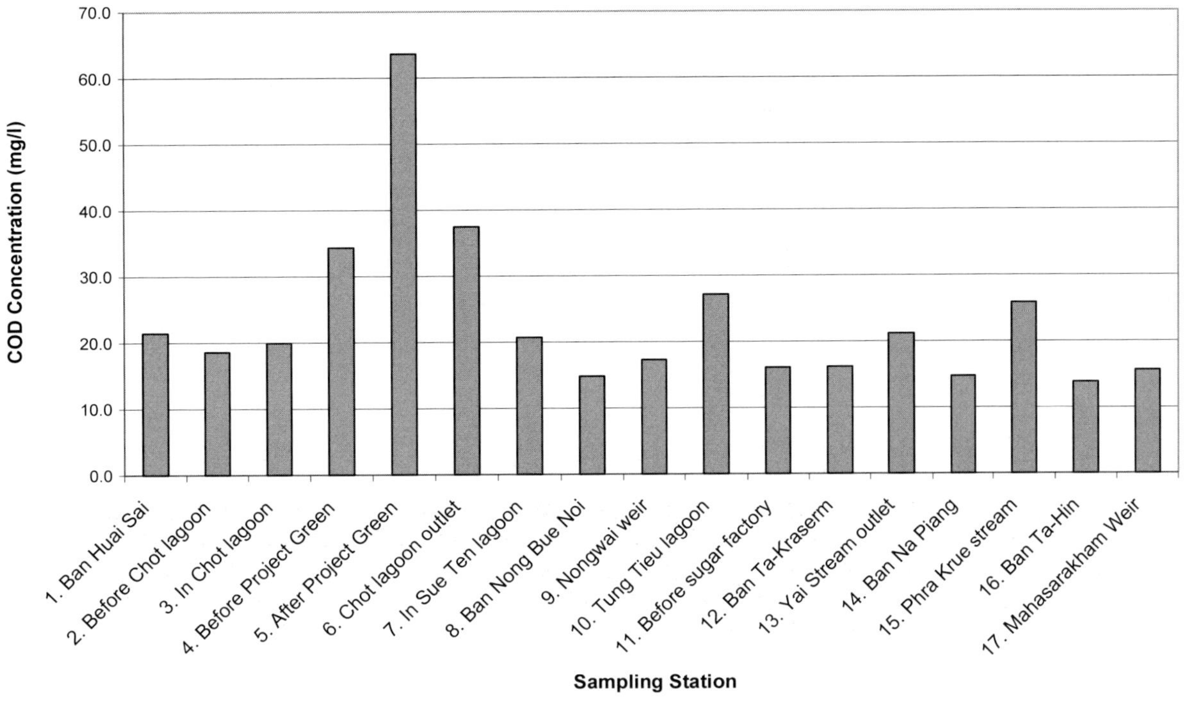

Figure 7. Results of COD analyses at 17 sampling stations in the Pong River from Ubonrattana Dam to Mahasarakham Weir

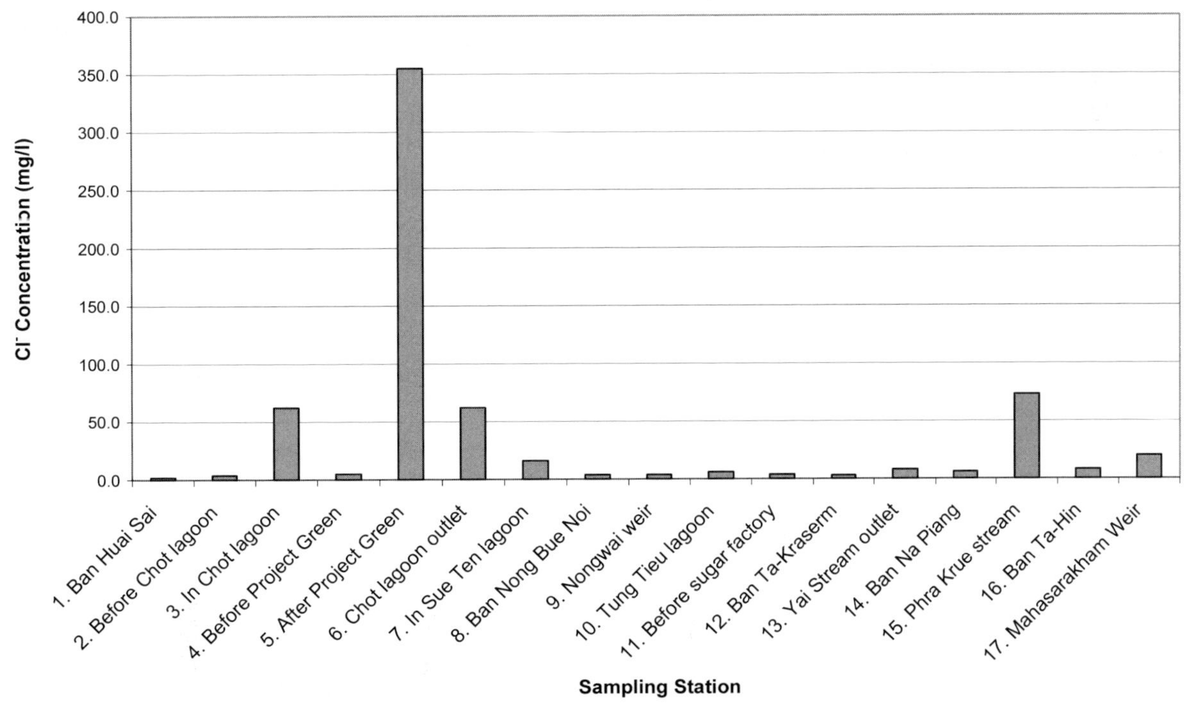

Figure 8. Results of Chloride (Cl^-) analyses at 17 sampling stations in the Pong River from Ubonrattana Dam to Mahasarakham Weir

Heavy metal analyses: Hg contamination in the river was found at station 3, 5, 6, 7, 8 and 15 with average concentrations of 0.49, 0.873, 0.60, 0.703, 0.767 and 2.8 μg/l, respectively (Figure 9). The Hg contamination at station 3, 5 and 6 could originate from the use of pesticides in the area. However, the source of the Hg contamination requires further study. At stations 7, 8 and 15 the contamination might be from domestic waste. Another possible source of Hg could be from the leachate of solid waste dumped onsite at the villages. Other heavy metals (Pb, Cd, Cr, Zn and Cu) were detected at very low concentrations which were accepted according to the surface water quality standards with the exception of station 16 which presented a high Zn concentration of 0.477 mg/l (Figure 10).

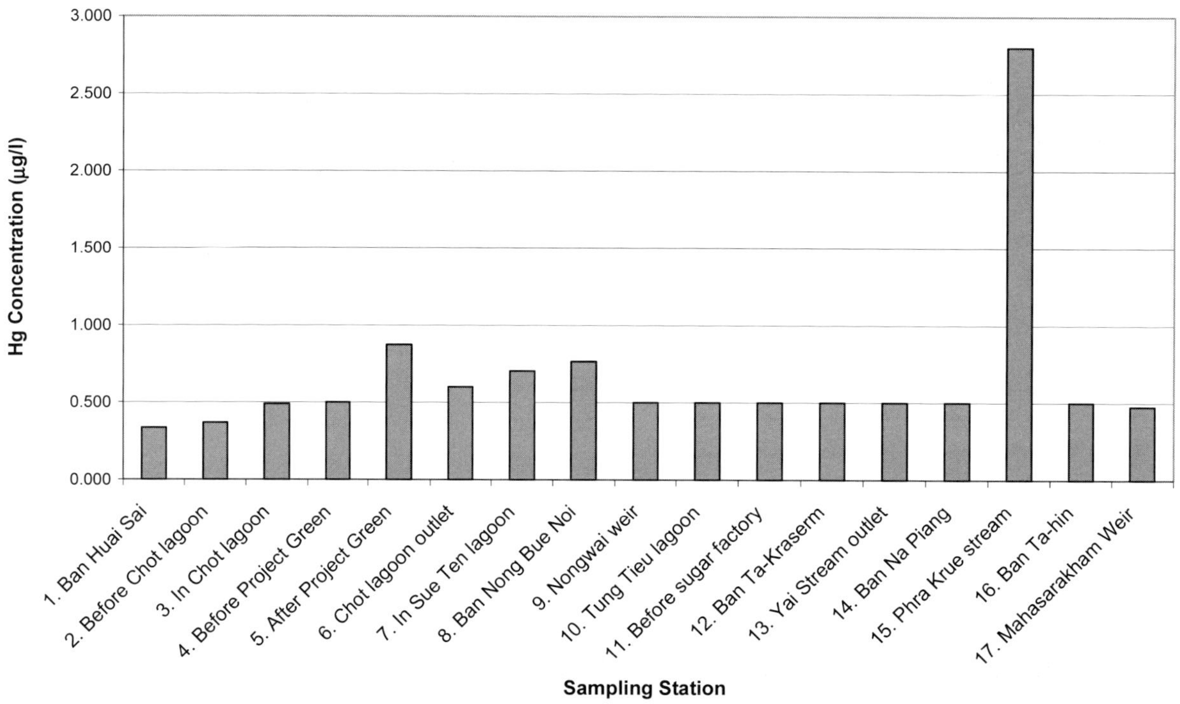

Figure 9. The results of Hg analyses at 17 sampling stations in the Pong River from Ubonrattana Dam to Mahasarakham Weir

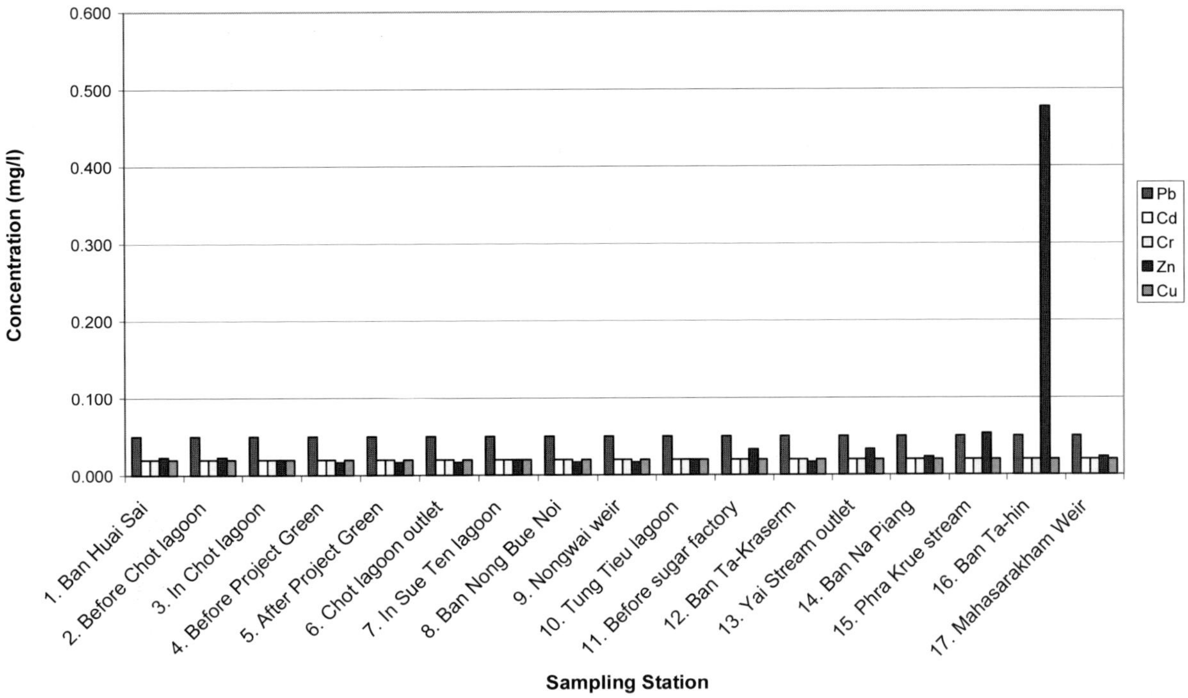

Figure 10. The results of Heavy Metal analyses at 17 sampling stations in the Pong River from Ubonrattana Dam to Mahasarakham Weir

Mutagens test: Since the cost of Mutagens analysis is quite expensive, the sampling took place only at stations 1, 3, 6, 7, 8, 10, 12 and 15 where likely potential pollution sources exist. At stations 7 and 8, where the floating basket aquaculture farm is located and at station 15, which is a community area, results were positive. Whilst at station 3, where the water body receives effluent from the pulp and paper industry, positive results showed when enzymes of the test organism were activated. It is likely that there are mutagen like substances at station 3 which could be changed to mutagens when activated by an enzyme.

Toxic chemical analysis: Organochlorine pesticide was not found in all sampling stations. This might indicate a low amount of pesticide use or that pesticides exist on land and not yet in the river. At stations 4, 5, 13 and 14, phenols were detected in the dry season, at concentrations of 0.0002, 0.042, 0.002 and 0.032 mg/l, respectively. The detection of phenols showed the possibility of chlorophenol contamination from the pulp and paper industry and also the tapioca industry. The contaminated compounds will be analyzed in a further study to find out how dangerous they are to the aquatic life and food chains in the area.

3. Conclusion

3.1 Both point (particularly the industrial source) and non-point sources showed some certain parameters that affect the Pong River water quality. This includes not only the change in physical and chemical characteristics of the river water, but also the possibility of the accumulation of heavy metals in the food chain and a potential risk from mutagens to the off-spring, as well as a cancer risk from phenols which were detected in some of the sampling stations.

3.2 The monitoring of the Pong water quality is necessary since it is the main source of water supply of Khon Kaen province and it is also upstream of the Chi and Mun Rivers, which are the main source of water supply for the other upper Northeastern provinces.

It should be noted that this is the preliminary study investigation of water quality in terms of toxic contamination. The results show a potential impact of different pollution sources. Further intensive study is necessary to identify the actual impact. Once the pollution sources are known, mitigation measures can be comprehensively recommended.

4. Acknowledgement

The authors would like to thank Khon Kaen University for providing the financial support for this project and would also like to express their great appreciation to the Environmental Engineering Department, Khon Kaen University for the use of their laboratory equipment and facilities. As this research project is a multidisciplinary study and without the strong effort of the study team who are from different faculties including the Faculties of Pharmacy, Science and Agriculture, this project could not be possible. All their efforts are much appreciated.

5. References

Khon Kaen University (1995). Khon Kaen province, Study project on action plan for Pong river water quality Rehabilitation and treatment.

Khon Kaen University (2001). National Research Council, Research on " Environmental Management of the Pong River".

Section 2

Dialogue between environmental science and policy making

Estimation of the Amount of Nitrate Originated from Nitrogenous Fertilizer in a Global River Model

Y. Suga[1)], S. Kanae[2)] and T. Oki[3)]

1) Institute of Industrial Science, The University of Tokyo
4-6-1, Komaba, Meguro, Tokyo 153-8505, Japan
e-mail: yoshito@iis.u-tokyo.ac.jp
2) Institute of Industrial Science, The University of Tokyo
4-6-1, Komaba, Meguro, Tokyo 153-8505, Japan
e-mail: kanae@iis.u-tokyo.ac.jp
3) Research Institute for Humanity and Nature
335 Takashima-Cho, Kyoto 602-0878, Japan
e-mail: oki@chikyu.ac.jp

Abstract

Fertilizer is one of the major sources of nitrate pollution in water resources. Nitrogenous fertilizer consumption is increasing rapidly because of the need to increase food production so that adverse impact on the water quality is a concern. In this paper, we estimated the distribution of fertilizer consumption in the global scale using 0.5 degree grid and in monthly time scale. Using a nitrogen behavior model, we calculated nitrogen uptake by crops, ammonia volatilization, nitrification of ammonia to nitrate, denitrification of nitrate and nitrate leaching. This nitrate leaching is regarded as a pollutant in water resources. Then we calculated the nitrate discharge from world river channels using a global river network model. This calculation was done under 4 assumptions; nitrate inflows directly into the river channel; the flow velocity is constant at 0.5 m/s; there are no lakes or reservoirs on the route; and there is no self purification in the river channel. After that, we focused on the calculation results of the discharges from Mekong River and Chao Phraya River. Not all nitrate leaching from cropland inflows to river channels so that it may be an overestimate, but we found that the nitrate concentration can reach a certain level.

Keywords: fertilizer, leaching, nitrate, nitrogen, water quality

1. Introduction

A number of research studies on world water resources assessment have been done so far. This is because now a water shortage crisis is occuring in many parts of the world and more water dificit in the future is possible for the reasons of population growth, lifestyle change, climate change and the like. But most of those studies focus just on the water quantity, and water quality is unconsidered although we would have to assess that an area is water-resource-poor if water is too polluted or the water purification cost is too high. Thus, we believe that global water resources assessment should include consideration of water quality.

When we think about water pollution, one of the key pollutional materials is nitrate. It is a globally widespread substance and its amount is enormous. But most nitrate is released from non-point sources, for instance, fertilizer sprinkle, livestock waste and domestic waste water. To begin with, we treated nitrate originated from nitrogenous fertilizer in this paper.

2. Distribution of nitrogenous fertilizer consumption

There are three processes for estimating global distribution of fertilizer consumption on a monthly time scale. We calculated the distribution in a 0.5 degree grid in order to utilize the following calculations with the same spatial resolution. First, we obtained a cropland distribution using Global Land Cover Characteristics (GLCC) data which is provided by the US Geological Survey (USGS). Next, we calculated the distribution of annual fertilizer consumption using the cropland distribution and country-by-country annual nitrogenous fertilizer consumption data of the year 2000, provided by Food and Agriculture Organization (FAO) of the United Nations. Finally, we divided the annual amount of fertilizer consumption into 12 months using a crop calendar calculated by Hanasaki *et al.* (2003).

2.1 Cropland distribution data

To obtain the global distribution of cropland, we used Global Land Cover Characteristics (GLCC) data of US Geological Survey (USGS). From the 30-arc-second gridded GLCC data, we extracted only the grid cells of cropland and summed up the area of the grid cells within each 0.5 degree grid, and we obtained 0.5 degree gridded cropland distribution data expressed by area ratio. This global distribution of cropland is shown in Figure 1.

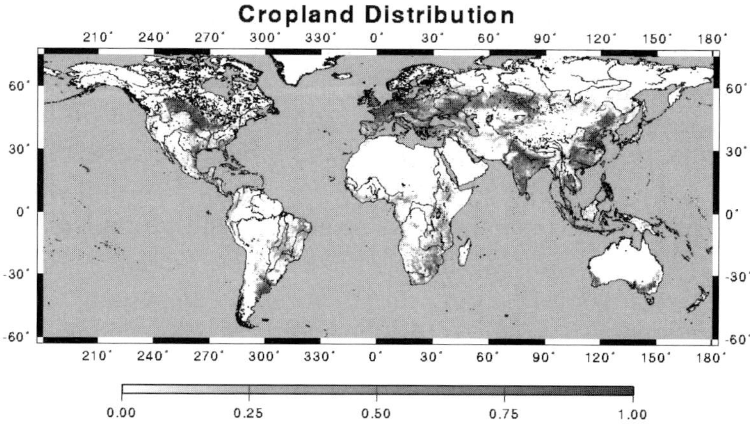

Figure 1. Global distribution of croplad

2.2 Annual fertilizer consumption

We used country-by-country annual fertilizer consumption data provided by FAO to estimate global distribution of annual nitrogenous fertilizer consumption. We calculated one country's cropland area at a time, ammonium nitrogen and nitrate nitrogen separately. Firstly, we summed up the total area of cropland of a country. Then, we divided the annual fertilizer consumption amount by total cropland area of the country and we obtained the fertilizer consumption ratio per unit

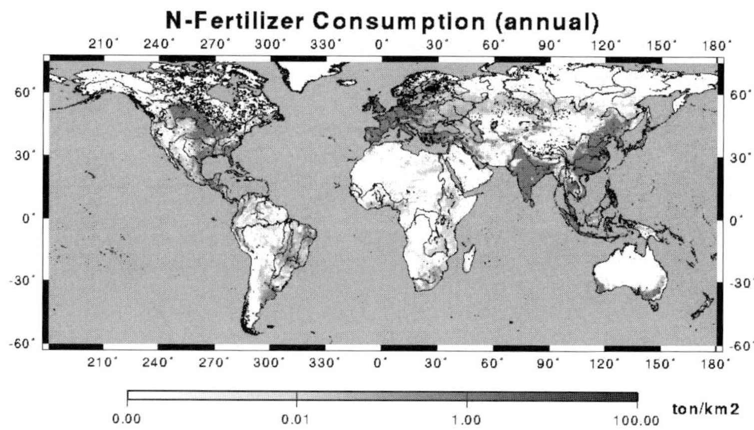

Figure 2. Global distribution of nitrogenous fertilizer consumption (annual).

cropland area under an assumption that the ratio is constant anywhere in the same country. Finally, we multiplied the fertilizer consumption ratio and the area of cropland of each grid cell and we obtained global distribution of nitrogenous fertilizer consumption which is shown in Figure 2.

2.3 Dividing into 12 months

We applied a crop calendar (Hanasaki *et al.*, 2003) in order to divide the annual amount of fertilizer consumption that we obtained in the previous section. The crop calendar was calculated by the method of point adding scoring according to the air temperature and river runoff data of each grid which are critical factors for cultivation. The duration of cultivation was fixed to 5 months.

In general, fertilization can be separated into basal fertilization and additional fertilization. We assumed that the basal fertilization is done in the first month of the 5 months and the additional fertilization in the third month. Then, we allocated 60 percent of annual nitrogenous fertilizer consumption to the first month of the crop calendar as the basal fertilization and 40 percent to the third month as the additional fertilization. Thus, we divided the fertilizer consumption into 12 months. Figure 3 shows the distribution of February, May, August and November of the year 2000.

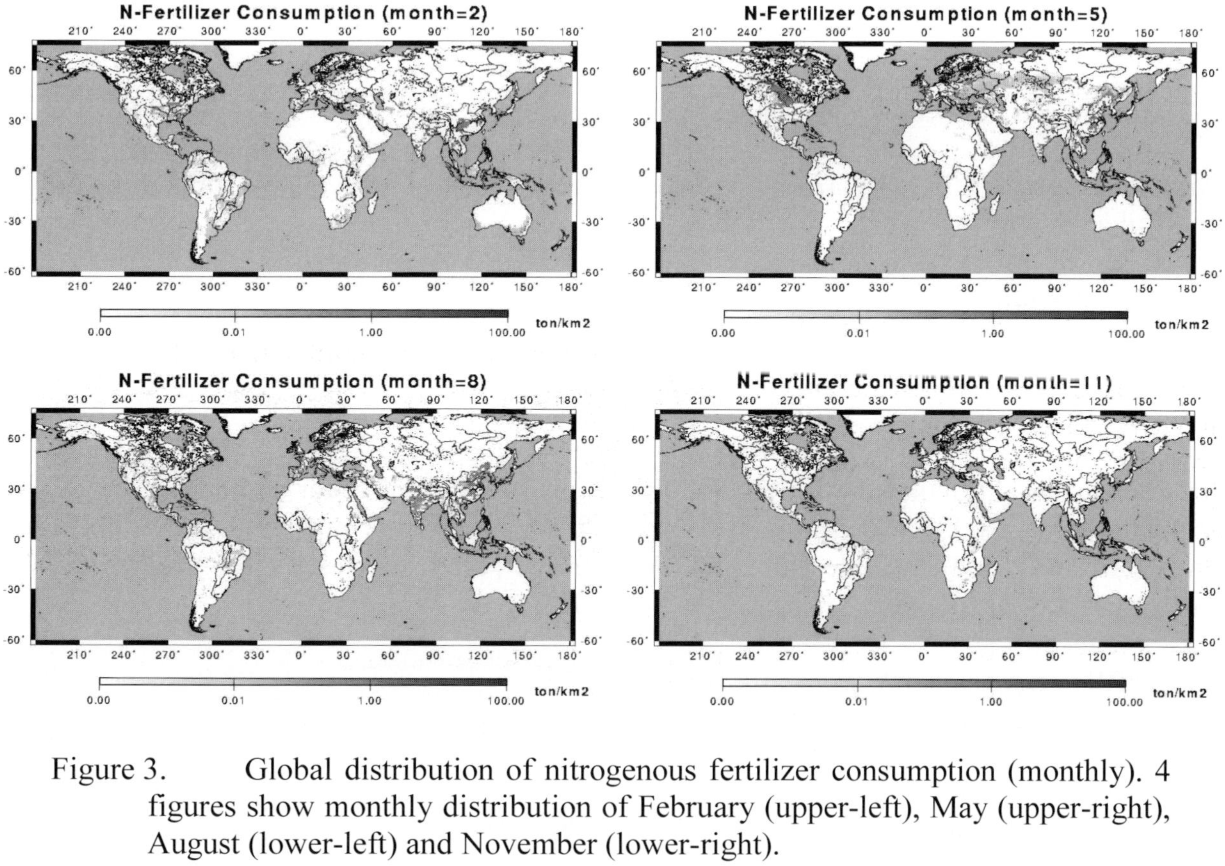

Figure 3. Global distribution of nitrogenous fertilizer consumption (monthly). 4 figures show monthly distribution of February (upper-left), May (upper-right), August (lower-left) and November (lower-right).

3. Estimation of nitrate leaching

We estimated nitrate leaching from cropland by referring to a global biogeochemical nitrogen-cycle model (Lin, 2000). We used the result of the calculation in the previous section as the input of monthly fertilizer consumption. The model consists of 5 components, uptake of ammonia and nitrate by crops, ammonia volatilization, nitrification of ammonia to nitrate, denitrification of nitrate and nitrate leaching as hereinafter detailed.

3.1 Nitrogen uptake (Raich *et al.*, 1991)

Nitrogen uptake (N_{UPTAKE}) by crops was modeled as a function of available nitrogen in soil (Nav), soil moisture (W_S), and temperature (T). N_{max} is the maximum rate of nitrogen uptake, $K_{s,uptake}$ is a parameter accounting for relative differences in the conductance of the soil to nitrogen diffusion. k_{UPTAKE} is the concentration of nitrogen at which uptake proceeds at one-half of its maximum rate.

$$N_{UPTAKE} = \frac{N_{max} \times K_{s,uptake} \times Nav}{k_{UPTAKE} + K_{s,uptake} \times Nav}$$

$$K_{s,uptake} = 0.90(W_S / W_{FC})3 + 0.10$$

$$k_{UPTAKE} = 90[gm^{-2}]$$

3.2 Ammonia volatilization (Martin *et al.*, 1984)

Ammonia volatilization (Vola) was assumed to be a function of available ammonia (Amm) and soil acidity as following.

$$Vola = 5.8 \times 10^{pH-10}(Amm)$$

3.3 Nitrification (Lin *et al.*, 2000)

Nitrification was calculated with the following function of soil-water content ($f_{NIT,SW}(W_S)$), temperature ($f_{NIT,T}(T)$) which was assumed with a $Q_{10,NIT}$ of 2 and optimum temperature $T_{opt,NIT}$ of 20℃, and available ammonia (Amm). The parameter μ_{NIT} (day^{-1}) represents the rate constant of nitrification determined to be 0.003.

$$NITRIF = \mu_{NIT} f_{NIT,SW}(W_S) f_{NIT,T}(T)(Amm)$$

$$f_{NIT,T}(T) = Q_{10,NIT}^{((T-T_{opt,NIT})/10)}$$

$$f_{NIT,SW}(W_S) = \begin{cases} 1.17 W_S / W_{FC} + 0.165 & (W_S < W_{FC}) \\ 1.0 - (0.1 / W_S / W_{FC}) & (W_S \geq W_{FC}) \end{cases}$$

$$\mu_{NIT} = 0.003\left[day^{-1}\right]$$
$$Q_{10,NIT} = 2$$
$$T_{opt,NIT} = 20[°C]$$

3.4 Denitrification (Lin *et al.*, 2000)

Denitrification was assumed to be a function of soil-water content ($f_{DENI,SW}(W_S)$), temperature ($f_{DENI,T}(T)$) which was calculated with a $Q_{10,DENI}$ of 3 and optimum temperature of 25℃, and available nitrate in the soil. μ_{DENI} (day^{-1}) is the rate constant of denitrification which was defined as 0.0015.

$$DENITR = \mu_{DENI} f_{DENI,T}(T) f_{DENI,SW}(W_S)(Nit)$$

$$f_{DENI,T}(T) = Q_{10,DENI}^{((T-T_{opt,DENI})/10)}$$
$$f_{DENI,SW}(W_S) = \begin{cases} 0 & (W_S \leq W_{FC}) \\ 1.0 W_S / W_{FC} & (W_S > W_{FC}) \end{cases}$$
$$\mu_{DENI} = 0.0015\left[day^{-1}\right]$$
$$Q_{10,NIT} = 3$$
$$T_{opt,NIT} = 25[°C]$$

3.5 Nitrate leaching (Lin *et al.*, 2000)

Nitrate leaching was calculated using following function. K_θ is hydraulic conductivity calculated using porosity (φ), water content(θ), residual water content (θ_r), pore-size distribution (λ) which were decided by soil type. Availabe nitrate in soil (Nit), water content (W_S), field capacity (W_{FC}) were also used.

$$LEACH = K_\theta \frac{(Nit)}{2\phi W_S / W_{FC}}$$
$$K_\theta = K_{s,leach}\left(\frac{\theta - \theta_r}{\phi - \theta_r}\right)^n$$
$$n = 3 + 2\lambda$$

3.6 Result

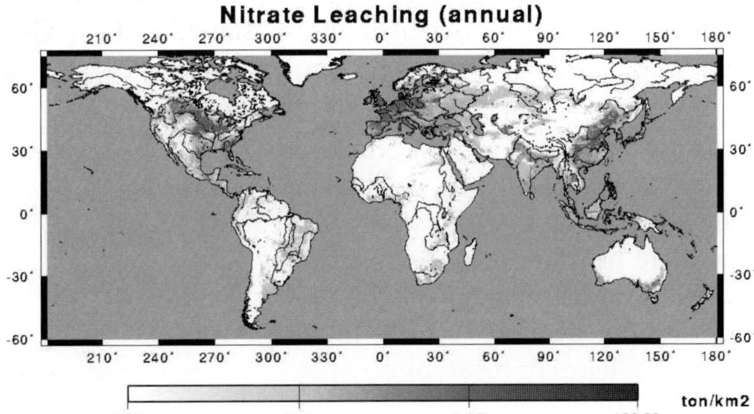

Figure 4. Simulation result of nitrate leaching

Figure 4 shows the calculation result, distribution of nitrate leaching. 80,900,000tons of nitrogenous fertilizer was used in the year 2000 and 14,500,000tons of nitrate leached according to the simulation. The proportion of leaching to fertilizer consumption is 18 percent.

4. Simulation of global nitrate discharge

We calculated the nitrate discharge by rivers around the world. We used the calculation result of nitrate leaching in the previous section and applied it to a global river network model under several assumptions. The target areas were the Mekong river basin and the Chao Phraya river basin. In Figure 5 the basins of the 2 rivers in the river network model are shown with a black border (the Mekong river basin is the right and the Chao Phraya river basin is the left). The annual total nitrate leaching amount in the Mekong river basin is 128000tons and in the Chao Phraya river basin, 39000tons, according to the calculation in the previous section.

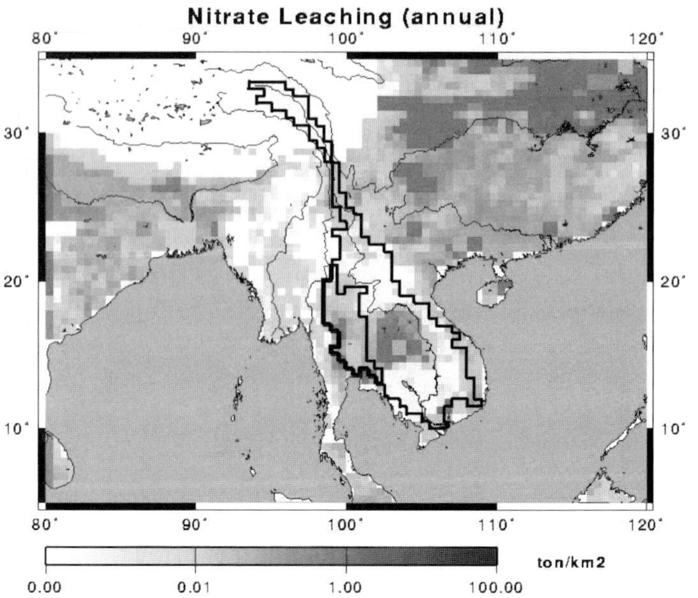

Figure 5. the Basins of Mekong river (right) and Chao Phraya river (left) in TRIP and estimated nitrate leaching

4.1 TRIP, Total Runoff Integrated Pathways

TRIP, Total Runoff Integrated Pathways has been developed by Oki and Sud (1998), Okada (2000) and Hanasaki *et al.* (2003). It is a routing model of global river network. Each grid cell has flow direction that is one of the proximate 8 grids, as shown in Figure 6. Figure and it makes network. 7 shows the river network of TRIP in Asian region.

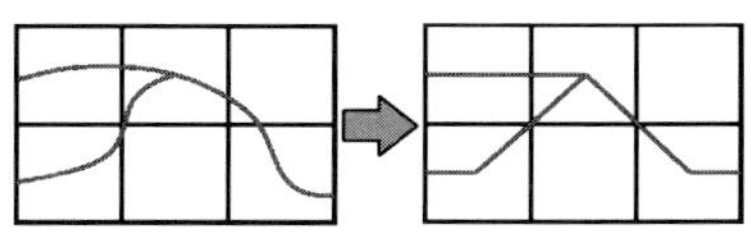

Figure 6.Desitalization of river routing

4.2 Assumptions

The following is what we assumed in order to make the calculation brief. We assumed that nitrate leaching from cropland inflows directly into the river channel; the flow velocity is constant anywhere at 0.5 m/s; there are no lakes or reservoirs on the route where nitrate remains for a long time; there is no purification effect in the river channels, lakes or reservoirs.

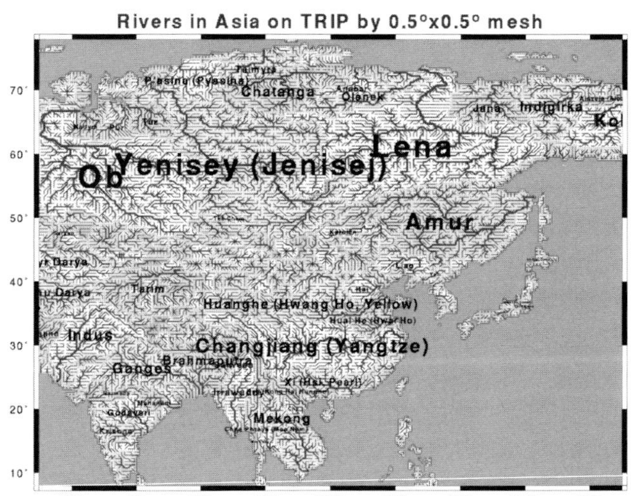

Figure 7. TRIP of Asian region

4.3 Result

Figure 8 shows the calculation result of Mekong river, and Figure 9. Chao Phraya river. The upper graphs are the monthly amount of nitrate discharge at the river mouths and its concentration. Lower graphs are river water discharge simulated using TRIP (Hanasaki, 2003). The seasonal variation of nitrate discharge can be recognized, however the concentration is highly dependent on the river runoff. The concentration reaches the 3.5mg/l at Mekong river mouth and 5.5mg/l at Chao Phraya river mouth.

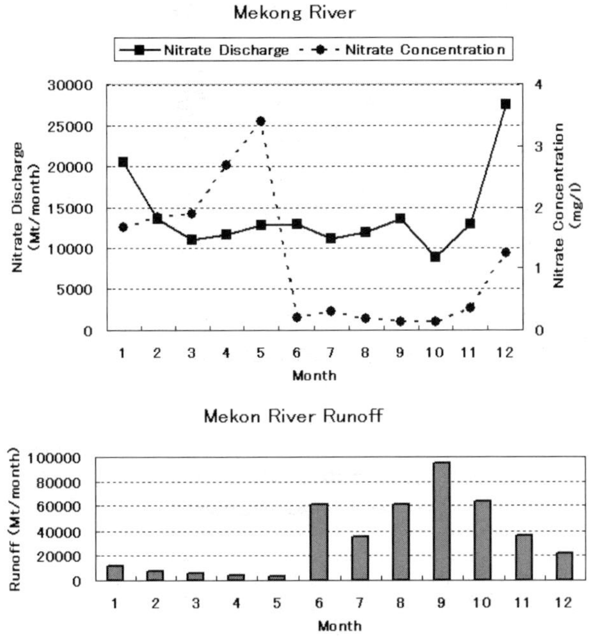

Figure 8. Simulation result of Mekong river. Upper graph shows nitrate discharge and its concentration. Lower graph shows the runoff (Hanasaki *et al.* 2003).

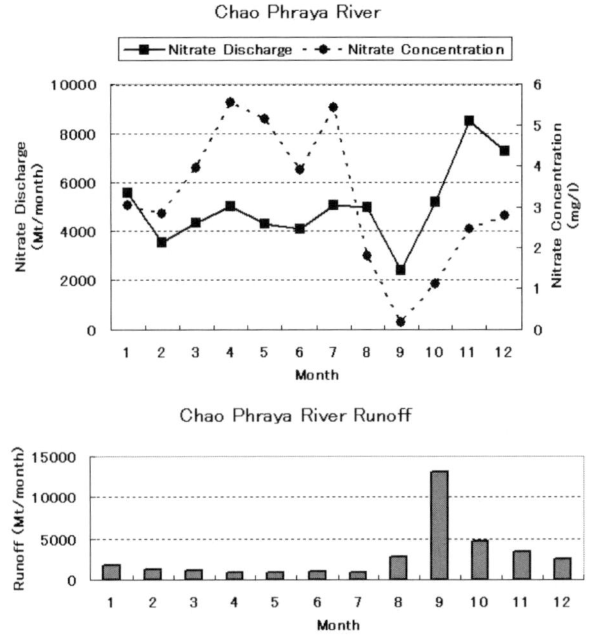

Figure 9. Simulation result of Chao Phraya river. Upper graph shows nitrate discharge and its concentration. Lower graph shows the runoff (Hanasaki *et al.* 2003).

5. Conclusions

In this research, we estimated global distribution of nitrogenous fertilizer consumption on a monthly time scale, calculated the nitrate leaching amount, and simulated the nitrate discharge from river channels around the world.

The total nitrate discharge in the Mekong river basin is 128000tons and in the Chao Phraya river basin, 39000tons. The annual total of the nitrogenous fertilizer consumption amount in the Mekong river basin is 707000 tons and in the Chao Phraya river basin, 324000tons. It means that in the basin of Mekong river, 18 percent of nitrogen leached in the form of nitrate and in the Chao Phraya river basin, 12 percent leached.

The simulated concentration of nitrate at the river mouth of the Mekong river reaches to 3.5g/l and at the Chao Phraya river, 5.5g/l. The standard value of WHO is 11.3mg/l so that the simulated value is not too high but cannot be ignored. We assumed that nitrate leaching from cropland inflows directly into the river channel and it may be an overestimate, however, nitrate pollutes either surface water or ground water so that this water pollution should be a concern. Taking into account how much nitrate inflows to the river channel, an approximation of the effect of other sources must be our future task.

6. References

FAO yearbook, Fertilizer (2000). Vol. 50, Bernan Assoc.

Hanasaki N., Kanae S., Oki T. and Musiake K. (2003). Simulating the discharge of the Chao Phraya River taking into account reservoir operation, *Water Resources Systems – Hydrological Risk, Management and Development*, 281, pp. 215-223.

Lin B.L., Sakoda A., Shibasaki R., Goto N. and Suzuki M. (2000). Modelling a global biogeochemical nitrogen cycle in terrestrial ecosystems, *Ecological Modelling*, 135(1), pp. 89-110.

Martin J.F. and Reddy K.R. (1997). Interaction and spatial distribution of wetland nitrogen processes, *Ecological Modelling*, 105, pp. 1-21.

Okada Y. (2000). Simulating discharge of major rivers in the world through development of a global river channel network at 0.5° resolution, Master thesis, Department of Civil Engineering, The University of Tokyo.

Oki T. and Sud Y.C. (1998). Design of total runoff integrating pathways(TRIP) - a global river channel network. *EarthInteractions*, 2.

Raich J.M., Rastetter E.B., Melillo J.M., Kicklighter D.W., Steudler P.A., Peterson B.J., Grace A.L., Moore B. and Vorosmarty C.J. (1991). Potential net primary productivity in South America : application of a global model, *Ecological Applications*, 1(4), pp. 399-429.

Riedo M., Grub A., Rosset M. and Fuhrer J. (1997). A pasture simulation model for dry matter production and fluxes of carbon, nitrogen, water and energy, *Ecological Modelling*, 105(1), pp. 141-183.

Measurement of N_2O Emission from Tropical Paddy Fields

T. Iida [1)], S. Kumar Deb [2)] and R. G. Kharbuja [3)]

1) Department of Bioproduction, Faculty of Agriculture, Yamagata University
1-23 Wakaba, Tsuruoka, Yamagata, 997-8555, Japan
e-mail: tiida@tds1.tr.yamagata-u.ac.jp
2) Soil Physics and Soil Hydrology Laboratory, Graduate School of Agricultural & Life Sciences, University of Tokyo
1-1-1 Yayoi, Bunkyo-ku, Tokyo 113-8657, Japan
e-mail: sanjit@soil.en.a.u-tokyo.ac.jp
3) Department of Hydrology and Meteorology, Hydrologist Engineer
Babarmahal Kathmandu. P.O. Box 406 Nepal
e-mail: rgkharbuja@yahoo.com

Abstract

N_2O emission from paddy soil to the atmosphere is attracting great interest from the point of the global environment. In order to characterize N_2O emission from paddy fields in tropical regions, in-situ measurement at an experimental paddy field was carried out for a whole cultivation period of rice, employing the C_2H_2 method. The measurement revealed that the inundation over the field was a major controlling factor for N_2O emission from the paddy soil, probably due to the drastic change in the soil conditions in respect of the availability of oxygen. High soil ORP values suggested that the expanded N_2O emission during absence of ponding water was caused by nitrification. The low emission around 10 μg-N m^{-2} h^{-1} was observed with ponding water of 2 cm in depth, indicating that N_2O emission can be mitigated by a shallow film of ponding water. A substantial fraction of the denitrification nitrogen loss was emitted as N_2O. The possible accounts for the high proportion of N_2O might be low soil temperature during the ponding period and low soil pH. N_2O emission was accurately correlated with soil temperature and had close relation with NO_3-N concentration in soil water in a plow layer.

Keywords: acetylene method, denitrification, nitrous oxide, paddy field, paddy soil

1. Introduction

The rice paddy field under wetland conditions is one of the predominant land uses in Asia. Ponding water on the soil surface obstructs diffusion of atmospheric oxygen into soil and brings about the existence of aerobic and anaerobic soil layers simaltaneously. Under this unique soil system, inorganic nitrogen in the soil is likely to be lost as nitrous oxide (N_2O) or dinitrogen (N_2) to the atmosphere through denitrification. Although the last step of denitrification reactions is reduction of N_2O to N_2, both N_2 and N_2O can be emitted from the soil.

N_2O is known as one of the greenhouse effect gases. Moreover, concern about depletion of stratospheric ozone by oxides of nitrogen has caused great interest in the sources of atmospheric N_2O. Paddy fields were recognized as one of the substantial sources of atmospheric N_2O. Thus, the rate and intensity of N_2O emission from paddy fields is an important issue from the point of the global environment.

Although a lot of studies regarding denitrification have been conducted so far on the various soil conditions, efforts to directly determine denitrification N losses from actual paddy fields have been hampered mainly by difficulties in distinguishing the traces of emitted N_2 from huge amount of atmospheric N_2. The observation that C_2H_2 in soil inhibits reduction of N_2O to N_2 (Balderston *et al.*, 1976; Ryden *et al.*, 1979) provided the theoretical basis for direct measurement of denitrification N losses by the C_2H_2 method. Since C_2H_2 inhibits N_2O reduction, the gas that would be produced by denitrification in C_2H_2 treated soil would all occur as N_2O. Therefore, total potential denitrification rates can be obtained by measuring N_2O emission from C_2H_2 treated soil.

The objective of this study is to characterize N_2O emission from paddy fields in tropical regions. The observed variation of N_2O emission from an experimental paddy field was analyzed. In order to obtain a clearer understanding of the characteristics of N_2O emission, the C_2H_2 method was applied and the ratio of N_2O emission to the total denitrification N loss was evaluated.

2. Materials and methods

2.1. The experimental field

The study was conducted at the research farm of the Asian Institute of Technology, which is located in the central plain of Thailand (14°04′N, 100°37′E) at an elevation of 2.3 m above the mean sea level. The mean annual rainfall in this region is about 1300–1400 mm and the average monthly temperature ranges about 19 - 36°C.

A rectangular experimental field with the size of 20 m × 40 m was prepared for this study. The soil in the field is Isohyperthermic Sulfic Tropaquept belonging to Rangsit Series and is classified as fine-textured clay. Soil physical and chemical properties at the beginning of the study are shown in Table 1. The groundwater level was observed to be at 0.3 to 1.47 m in depth from the soil surface during the study period.

Table 1. Soil phyisical and chemical properties.

Depth (cm)		0-40	40-60	60-90
Colour		Black	Reddish brown	Yellow
Bulk density (g cm^{-3})		1.42	1.46	1.38
Particle density (g cm^{-3})		2.72	2.81	2.85
Porosity		0.48	0.48	0.51
Particle size distribution	Sand (%)	9.5	2.0	7.0
	Silt (%)	13.8	23.8	18.8
	Clay (%)	76.8	74.3	74.3
Hydraulic conductivity (cm s^{-1}, vertical)		1.63×10^{-4}	5.56×10^{-5}	6.39×10^{-6}
NO_3-N content (μ g g^{-1})		0.38	0.27	0.93
NH_4-N content (μ g g^{-1})		1.94	1.46	3.91
Organic matter (%)		8.3	8.3	8.7
Organic carbon (%)		4.4	3.3	3.5
pH		5.45	4.50	4.25

Tillage operation by a power tiller up to 20 cm in depth and subsequent puddling were performed on October 5, 2001. Seeds of local hybrid Suphanburi-1 paddy rice (*Oryza sativa* L.) were broadcast onto puddled soil on October 9. 30 kg-N ha^{-1} ammonium metaphosphate, $(NH_4)_2HPO_4$, with 16% N was applied as basal fertilizer at the sowing. Experimental instruments were set up between October 15, 7 days after sowing (7 DAS), and October 19 (11 DAS). 57.5 kg-N ha^{-1} urea with 46% N was applied as top dressing on November 13 (36 DAS). The rice plants had followed the typical growth pattern in this region. The rice was harvested on January 25, 2002 (109 DAS), with average grain yield of 6,213 kg ha^{-1}. The field was intermittently irrigated to keep the ponding water depth between 0 and 10 cm from October 28 (20 DAS) to January 18 (102 DAS).

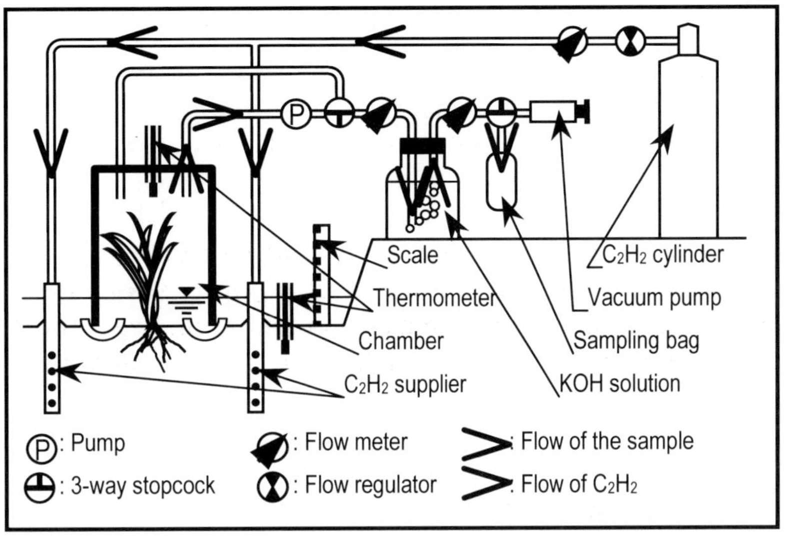

2.2. Experimental installation

The closed chamber method was applied to measure gas emission. Two chambers were set up at the C_2H_2 treated site, while a chamber was set up at the control site. The experimental set-up is schematically shown in Figure 1. The chamber boxes have the size of 0.3 m × 0.3 m in the base and 0.6 m in height and are made of 6 mm thick clear Plexiglas. Gas samples were led to the series of two CO_2 traps filled with 2N KOH solution before they were collected in an air-tight Cole-Parmer Teflon FEP bag. The gas sampling was carried out three times on each sampling day at 8:00, 12:00, and 16:00 at both sites.

C_2H_2 was supplied to the soil at the C_2H_2 treated site through six polyethylene tubes with 0.42 cm ID inserted up to 1 m in depth from the soil surface around each chamber. The tubes were perforated with 0.1 cm holes at every 5 cm over 90 cm of their length from the tips. The two sets of a chamber and C_2H_2 supply tubes were installed side by side at the C_2H_2 treated site and they were used in turn to avoid possible adaptation of microorganisms to C_2H_2. Soil water samplers made of a PVC pipe and a porous cup at the tip of the pipe were installed at 20, 40, and 60 cm in depth from the soil surface.

Soil temperatures were measured by soil thermometers installed at 15 cm in depth at both sites. The depth of ponding water in the field was continuously monitored by gauge scales. Oxidation-reduction potential (ORP) values were measured in-situ by oxidation-reduction potentiometers with platinum Eh electrodes at the soil surface and various depths up to 40 cm from the soil surface through holes with 30 mm in diameter and 60 cm in depth.

2.3. Analysis

Soil water pH was immediately measured at the site using a portable pH meter with a glass electrode. The gas samples for N_2O and C_2H_2 concentrations were analyzed by gas chromatograph (Shimadzu C-R4A). Soil samples were analyzed for NO_3-N, NH_4-N, pH, organic matter, and organic carbon content following the standard methods. Soil organic matters were determined by the standard dry ash method. The percentage of the decreased weight after the digestion at 350°C for 8 hours was taken as organic carbon content and it was converted into organic matter content. Soil water samples were analyzed within 24 hours from the sampling with Cadmium Reduction and Diazotization methods for NO_3-N and NO_2-N, respectively, using HACH DR/2010 spectrophotometer.

3. Results and discussions

3.1. The effect of inundation on N_2O emission

On 16, 104 and 107 DAS, when there was no ponding water on the soil surface, N_2O emissions were much higher than those during the ponding period. N_2O emissions at the C_2H_2 treated site and the control site on 16 DAS were as high as 1130.8 and 1002.8 μg-N m^{-2} h^{-1}, respectively. N_2O emission sharply decreased when ponding water inundated the field. Under ponding conditions, N_2O emission was kept at quite low level. The average daily N_2O emission fluctuated between 5.7 and 16.6 μg-N m^{-2} h^{-1} at the C_2H_2 treated site and between 5.0 and 14.7 μg-N m^{-2} h^{-1} at the control site. The average N_2O emissions under ponding conditions from 20 DAS to 102 DAS at the C_2H_2 treated site and the control site were 11.8 and 10.3 μg-N m^{-2} h^{-1}, respectively. N_2O emission abruptly increased when ponding water disappeared from the soil surface. On 104 DAS, N_2O emissions at the C_2H_2 treated site and the control site were 298.9 and 243.0 μg-N m^{-2} h^{-1}, respectively. N_2O emissions slightly decreased to 64.3 and 48.3 μg-N m^{-2} h^{-1}, respectively, on 107 DAS. The observation revealed that whether ponding water is present or absent on the soil surface was a major controlling factor for N_2O emission from the paddy soil to the atmosphere. This view is in essential agreement with Cai *et al.* (1999).

Figure 2 shows the variation of the observed ORP values of the soil. Before 22 DAS, the ORP values stayed in the positive range execpt at 10 cm from the soil surface. The ORP values at all the layers gradually decreased after the field was inundated. The ORP values at 10 cm depth reached as low as -329 mV on 90 DAS, indicating the strong anaerobic conditions in the plow layer. At 104 DAS, when the ponding water disappered, the ORP values at all the layers took a turn for marked increase.

High ORP values in the soil during absence of ponding water support that the expanded N_2O emission during absence of ponding water was caused by nitrification. As Bremner and Blackmer (1981) pointed out, nitrification in soil especially just after long submergence can cause high N_2O emission in a relatively short term with presence of enough amount of nitrogen. Under ponding conditions, on the other hand, the anaerobic state restrains nitrification and activates denitrification. As the observed ORP values in the plow layer under ponding conditions indicated strongly anaerobic conditions, most of the observed N_2O emission during the ponding period was considered to be produced by

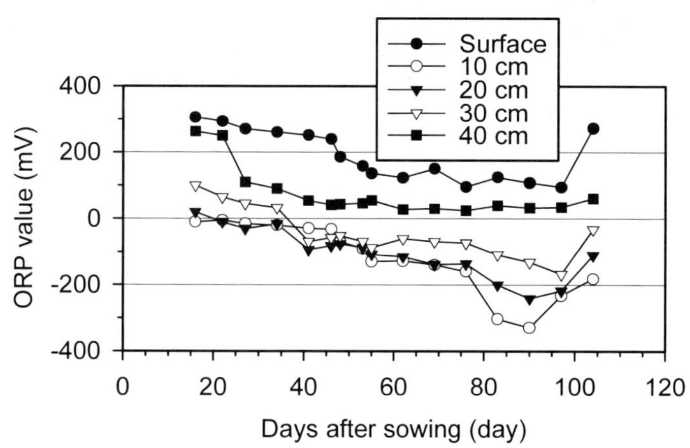

Figure 2. ORP values of the soil.

denitrification. Thus, the considerable difference of N_2O emission depending on whether ponding water is present or not could be attributed to the drastic change in the state of the soil in respect of the availability of oxygen.

One of other possible reasons for lower N_2O emission under ponding conditions may be absorption of N_2O by ponding water, because N_2O is quite soluble in water. Smith *et al.* (1978) gave the solubility of N_2O in water as 0.74 at 20°C. Minami and Fukushi (1984) reported that concentration of N_2O in ponding water was in the range between 0.65 and 10.4 μg L^{-1}. It is suggested that N_2O produced in the soil would immediately dissolve in the existing water and be further oxidized to NO_3^- by dissolved oxygen in the water. The NO_3^- in the water will become available again for leaching, plant uptakes, and other nitrogen transformation processes in the soil. Thus, ponding water and percolating water can act as a trap for N_2O emitting from the soil to the atmosphere. The low emission around 10 μg-N m^{-2} h^{-1} was observed with ponding water as shallow as 2 cm in depth, indicating that N_2O emission can be mitigated considerably by even shallow ponding water on paddy fields. Another possible explanation for lower N_2O emission under ponding conditions can be suppression of soil temperature by ponding water.

3.2. The ratio of N_2O emission to the total denitrification N loss

C_2H_2 concentrations in the chamber at the C_2H_2 treated site stayed in the range between 0.3 and 1.2 % (vol vol^{-1}), which was much higher than the minimum concentration to inhibit N_2O reduction to N_2 in soil suggested by the former studies such as Balderston *et al.* (1976) and Ryden *et al.* (1979). As C_2H_2 was supplied through six tubes inserted into the soil outside the chamber and the air in the chamber was completely isolated from the atmosphere during the C_2H_2 supply, C_2H_2 detected in the chamber must have travelled through the soil beneath the chamber. So, it is reasonable to consider that the C_2H_2 concentration in the soil was high enough to inhibit N_2O reduction. Based on the assumption that N_2O reduction to N_2 at the last stage of denitrification processes would be completely inhibited at the C_2H_2 treated site, it is considered that N_2O emission at the C_2H_2 treated site indicates the total denitrification N loss. Consequently, N_2 emission by denitrification can be calculated by subtracting the N_2O emission at the control site from the N_2O emission at the C_2H_2 treated site.

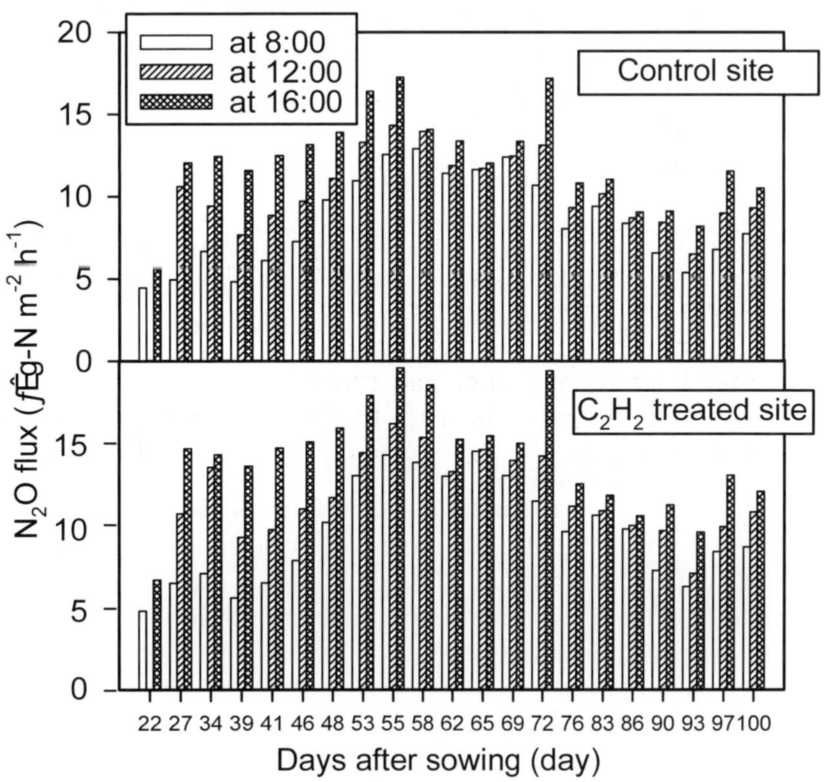

Figure 3. N_2O emission during ponding period.

Figure 3 presents the diurnal and seasonal variation of N_2O emissions during the ponding period at the C_2H_2 treated site and the control site. The minimum N_2O emission was observed at 8:00 on 22 DAS and the maximum was observed at 16:00 on 55 DAS at both sites. N_2O emissions in the morning were always lowest and those in the evening were always highest in a day, accurately corresponding to the diurnal variation of the soil temperature shown in Figure 4.

The total denitrification N loss in the form of N_2O+N_2 was calculated at 0.251 kg-N ha^{-1} for the ponding period, while the total denitrification N loss in the form of N_2O was calculated at 0.220 kg-N ha^{-1}. Consequently, the ratio of N_2O emission to N_2O+N_2 emission for the ponding period was obtained to be 87.6 %. This ratio indicated that a substantial fraction of the denitrification N loss was emitted in the form of N_2O gas from this experimental field.

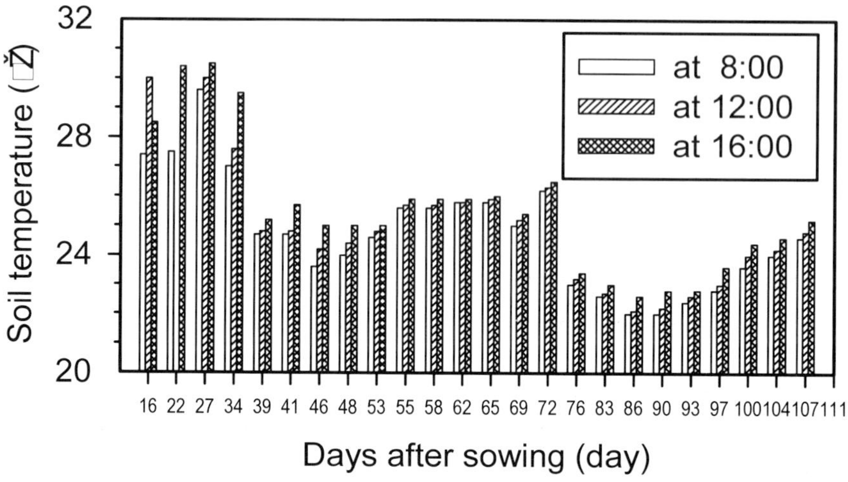

Figure 4. Soil temperature at 15 cm in depth.

One explanation for the high proportion of N_2O might be low soil temperature during ponding periods. Rolston (1981) presented that N_2O emission expanded when soil temperature was lower than around 30 °C while N_2 was predominant in higher temperature. The fact that most of the observed soil temperature was quite lower than 30 °C helps account for the high proportion of N_2O to the total denitrification N loss in this experimental field.

Another possible cause of the high proportion of N_2O might be low pH in the soil. It has commonly been observed that the proportion of N_2O occurring as the product of denitrification increases as pH decreases so that N_2O frequently appears as the dominant product in acidic soils (Fillery, 1982). The acidity of the soil in this experimental field could provide an account of the high proportion of N_2O observed in this study.

It is strongly suggested that in acidic soil at low soil temperature under ponding conditions the activities of microbes reducing N_2O to N_2 were lowered and consequently the proportion of N_2O to the total nitrogen emission from the field was elevated, which might contribute to the subsequent environmental degradation. It can be said that the control of soil acidity might be one of effective means to minimize emission of N_2O gas from the soil to the atmosphere also under ponding conditions.

3.3. Other factors affecting N_2O emission

Table 2 presents correlation coefficients between N_2O emission and NO_3-N contents in the soil water and soil mass at various depths. N_2O emission was highly correlated with NO_3-N concentration in soil water extracted at 20 cm from the soil surface and in soil mass at 0-20 cm in depth. It should be noted that this depth approximately corresponds to a plow layer of this experimental field. Close relation between N_2O emission and NO_3-N concentration in soil water in a plow layer was recognized. It is quite understandable that the amount of NO_3-N in a plow layer governs the N_2O emission from the soil because NO_3-N is used as a principal energy source by denitrifying bacteria.

Table 2. Correlation coefficients between N_2O emissions and nitrate contents during the ponding period.

		N_2O emission at the C_2H_2 treated site	N_2O emission at the Control site
NO_3-N in soil water	Ponding water	0.42	0.48
	At 20 cm in depth	0.86	0.86
	At 40 cm in depth	0.52	0.52
	At 60 cm in depth	0.50	0.48
NO_3-N in soil mass	0 – 20 cm in depth	0.69	0.73
	20 – 40 cm in depth	-0.13	-0.10
	40 – 60 cm in depth	0.30	0.30

Despite our expectation, no significant correlation was recognized between N_2O emission and other routinely observed parameters such as NH_4-N concentration in soil mass and NO_2-N concentration in soil water. The effects of these factors on N_2O emission might be relatively small or compensate each other under the condition of this experiment. Fillery (1982) suggested that the increase of soil organic matter and carbon content caused the decrease of the proportion of N_2O to total nitrogen emission. In our experiment, however, the relation between N_2O emission and soil organic matter was vague because the amount of soil organic matter in our experimental field was quite stable.

4. Conclusions

In-situ measurement of N_2O emission from an experimental paddy field by the chamber method revealed that whether ponding water is present or absent on the soil surface was a major controlling factor for N_2O emission from the paddy soil to the atmosphere. High ORP values in the soil during absence of ponding water support that the expanded N_2O emission during absence of ponding water was caused by nitrification. Thus, the considerable difference of N_2O emission depending on whether ponding water is present or not could be attributed to the drastic change in the state of the soil in respect of the availability of oxygen. The low emission around 10 μg-N m^{-2} h^{-1} was observed with ponding water as shallow as 2 cm in depth, indicating that N_2O emission can be mitigated considerably by even shallow ponding water on paddy fields.

It was reasonably considered that the required C_2H_2 concentration to inhibit reduction of N_2O to N_2 was achieved in the soil in this study. The ratio of N_2O emission to N_2O+N_2 emission for the ponding period was obtained to be 87.6 %, showing that a substantial fraction of the denitrification N loss was emitted in the form of N_2O gas from this experimental field. It is supposed that the high proportion of N_2O might be brought about by low soil temperature during the ponding period and low soil pH.

N_2O emissions in the morning were always lowest and those in the evening were always highest in a day, accurately corresponding to the diurnal variation of the soil temperature.

Close relation between N_2O emission and NO_3-N concentration in soil water in a plow layer was recognized. However, no significant correlation was recognized between N_2O emission and other routinely observed parameters such as NH_4-N concentration in soil mass and NO_2-N concentration in soil water. The relation between N_2O emission and soil organic matter was vague.

5. Acknowledgement

This study was partly supported by a research fund of GMS-JG project (Japanese Government) and a scholarship from DANIDA (Danish Government). Authors are grateful to the donors for their kind support to our research activities.

6. References

Balderston W. L., Sherr B. and Payne W. J. (1976). Blockage by acetylene of nitrous oxide reduction in pseudomonas perfectomarinus, *Appl. Environ. Microbiol.*, 31, pp. 504-508.

Bremner J. M. and Blackmer A. M. (1981). Terrestrial nitrification as a source of atmospheric nitrous oxide, In: *Denitrification, nitrification, and atmospheric nitrous oxide*, C. C. Delwiche (Ed.), John Wiley and Sons, Hoboken, NJ, pp. 151-170.

Cai Z. C., Xing G. X., Shen G. Y., Xu H., Yan X. Y., Tsuruta H., Yagi K. and Minami K. (1999). Measurements of CH_4 and N_2O emissions from rice paddies in Fengqiu, China. *Soil Sci. Plant Nutr.*, 45, pp. 1-13.

Fillery I. R. P. (1982). Biological denitrification, In: *Nitrogen in agricultural soils*, F. J. Stevenson (Ed.), Monograph No. 22, Am. Soc. Agron./Soil Sci. Soc. Am., Madison, WI, pp. 33-63.

Minami K. and Fukushi S. (1984). Methods for measuring N_2O flux from water surface and N_2O dissolved in water from agricultural land, *Soil Sci. Plant Nutr.*, 30, pp. 495-502.

Rolston D. E. (1981). Nitrous oxide and nitrogen gas production in fertilizer loss, In: *Denitrification, nitrification, and atmospheric nitrous oxide*, C. C. Delwiche (Ed.), John Wiley and Sons, Hoboken, NJ, pp. 127-150.

Ryden J. C., Lund J. and Focht D. D. (1979). Direct measurement of denitrification loss from soils: I. Laboratory evaluation of acetylene inhibition of nitrous oxide reduction, *Soil Sci. Soc. Am. J.*, 43, pp. 104-110.

Smith M. S., Firestone M. K. and Tiedje J. M. (1978). The acetylene inhibition method for short-term measurement of soil denitrification and its evaluation using nitrogen-13, *Soil Sci. Soc. Am. J.*, 44, pp. 611-615.

A Scientific Approach for Allocating Water to Competing Sectors

Mukand Singh Babel[1)], Ashim Das Gupta[2)] and Dhiren Kumar Nayak[3)]

1) Water Engineering and Management, Asian Institute of Technology
P.O. Box - 4, Klong Luang, Pathumthani-12120, Thailand
e-mail: msbabel@ait.ac.th
2) Water Engineering and Management, Asian Institute of Technology
P.O. Box - 4, Klong Luang, Pathumthani-12120, Thailand
e-mail: adg@ait.ac.th
3) Water Engineering and Management, Asian Institute of Technology
P.O. Box - 4, Klong Luang, Pathumthani-12120, Thailand
e-mail: dhiren_ait@hotmail.com

Abstract

The present study develops an Integrated Water Allocation Model (IWAM), which optimally allocates limited water supply from a source to different user sectors taking into consideration socio-economic, environmental and technical aspects. IWAM comprises three modules - a Reservoir Operation Module (ROM), an Economic Analysis Module (EAM) and a Water Allocation Module (WAM). The model can optimize the water allocation for two different objectives individually or together. The two objectives considered are the maximization of the satisfaction and the maximization of the net economic benefit by the demand sectors. Single objectives are optimized using the linear programming technique whereas the Simultaneous Compromise Constraint (SICCON) Technique is applied for solving the multi-objective function. The model applicability is demonstrated with an illustrative example.

Keywords: multi-objective decision-making, net economic return, optimization, reservoir operation, simultaneous compromise constraint technique, water allocation

1. Introduction

Rapid population rise along with improved standards of living, urbanization and industrial growth have led to increased demand, competition and conflicts among different water use sectors. While traditional water demands for agriculture, domestic and industries remain critical, the newly emerging demands for wetlands and environmental protection and conservation, fishing, navigation, tourism are being given increased importance. As long as water availability is enough, all demands can be met without conflicts. However, this is not true, with the increase in demand the conflicts between user sectors intensify and increase in frequency and the water allocation in such circumstances becomes an important issue.

Water allocation has received considerable attention in the recent past by the scientific community. Most of these studies are area-specific and based on the economic criteria for the allocation of water (Bielsa and Duarte, 2001; Quba′ *et al.*, 2002). Few modeling studies on water allocation in Aral and in Gediz basins have been reported (Voogt *et al.*, 2000; McKinney and Cai, 2002). However, not much work has been carried out in developing water allocation models considering socio-economic, environmental and technical aspects together.

Ideally, the allocation of water should be economically efficient and technically feasible as well as socially fair. However, these objectives are frequently in conflict with one another. While the economically efficient allocation tends to distribute water maximizing the economic returns, allocation with social equity, on the other hand, tends to distribute water in an attempt to protect the interests and fairness of allocation across economically desperate groups without caring for the economic efficiency. So, there is a need for a proper water allocation system that will consider water as an economic as well as a social good.

The present study aims to develop a simple and user-friendly water allocation model with Microsoft Excel in the Windows operating system using the Visual Basic language for allocating water to different user sectors under scarcity situation. The model optimizes either two single objectives independently using a linear programming technique or two objectives together applying a multi-objective compromise technique.

2. Model development

2.1 Conceptual framework

In the study the actual demand of a sector (i) is referred to as the normal demand (Dn_i). While allocating water the normal demand may or may not be satisfied. The bare minimum amount of water required by a sector that must be satisfied is defined as the minimum demand (Dm_i).

The developed model named as Integrated Water Allocation Model (IWAM) consists of three modules: Reservoir Operation Module (ROM), Economic Analysis Module (EAM) and Water Allocation Module (WAM) and runs on monthly time step. ROM is based on the standard reservoir operation algorithm and requires reservoir characteristics, inflows and operating rules as an input and the output of the module is the amount of Available Water (AW), depending on the normal demand, water availability and technical feasibility, for allocating to different demand sectors. EAM calculates the net economic return (NER) from one cubic meter of water used by sectors. The WAM allocates water among various competing sectors depending on the user-defined objective(s) among the two single objectives, maximization of satisfaction (OF_1) and maximization of NER (OF_2) and these two objectives together (OF_{12}). The model considers six water users sectors, namely agriculture, domestic, industry, hydropower, recreation and environment.

The conceptual framework of the IWRM is depicted in Figure 1. Depending on the AW and the sectoral demand, the model allocates water among various competing users. If the AW is more than the total normal demand, the water is supplied according to the Dn_i to all the sectors. When the AW is less than the total minimum demand of the sectors, the supply is equity-based, priority-based or under stressed conditions. Priority-based supply may be according to the choice given by the user or based on the doctrines in practice; namely riparian doctrine, prior-appropriation doctrine or the user-defined priority. The user-defined priority is suitable for the local conditions and can be either single priority or multiple priorities allocation. In case of multiple priorities the user has the privilege to assign the parts of Dm_i to be allocated with high priority and low priority. If the AW is in between the total Dm_i and the total Dn_i, the allocation is optimized using suitable optimization technique.

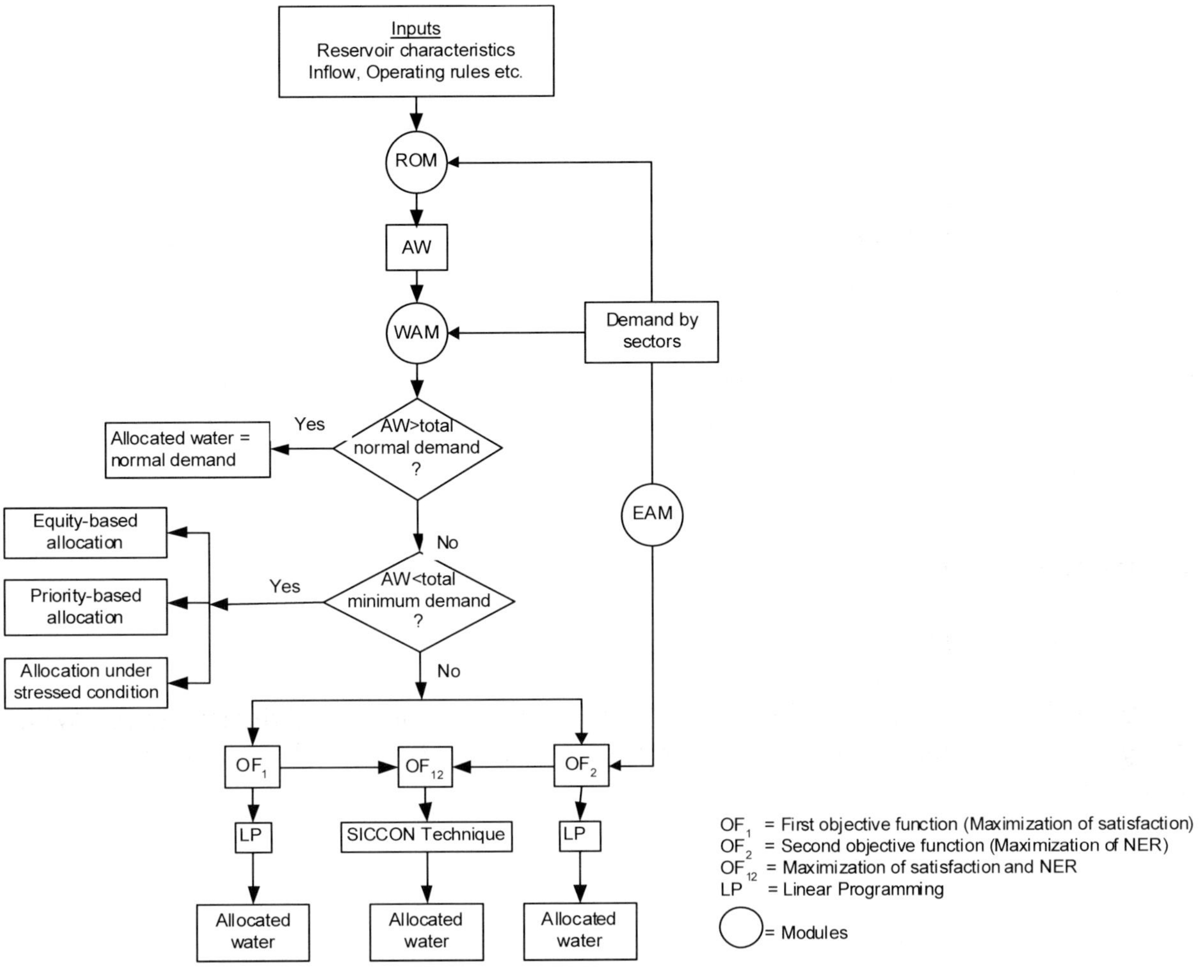

Figure 1: Framework of the integrated water allocation model

2.2 Optimization model

Objective functions

As stated earlier, the IWAM considers two single objective functions (OF_1 and OF_2) and a multi-objective function (OF_{12}), which are described as below:

(i) To maximize the satisfaction of various demand sectors: Satisfaction, in the present study, is defined as the ratio of the water allocated over the normal demand by a sector. Hence, OF_1 is the maximization of the total satisfaction and is given by the following equation:

$$OF_1 = \frac{1}{n}\sum_{i=1}^{n}\frac{S_{a_i}}{D_{n_i}} \tag{1}$$

Where, OF_1 = First objective function (maximization of satisfaction); Sa_i = Water allocated to sector i; Dn_i = Normal water demand of sector i; and n = Number of demand sectors. The value of OF_1 lies between 0 and 1.

(ii) To maximize the net economic return: This objective function is defined as the ratio of the total economic value (as represented by the summation of the water allocated multiplied by the *NER* of each of the sectors) to the maximum achievable total economic value. The objective function is represented as:

$$OF_2 = \left[\frac{\sum_{i=1}^{6} S_{a_i} * NER_i}{AW * NER_{max}} \right] \tag{2}$$

Where, OF_2 = Second objective function (maximization of NER); NER_i = Net economic return per unit volume of water from sector i (US$/m^3); AW = Available water (m^3); and NER_{max} = Maximum net economic return among the sectors considered (US$/m^3). Again, the value of OF_2 lies between 0 and 1.

(iii) To maximize the satisfaction as well as the NER of various demand sectors: This is defined as the sum of the two single objective functions multiplied by the respective weightage given to them less the sum of deviational variables and is given as:

$$OF_{12} = w_1 * OF_1 + w_2 * OF_2 - \lfloor \sigma_{12}^+ + \sigma_{12}^- \rfloor \tag{3}$$

Where, OF_{12} = Maximization of satisfaction and maximization of NER together; w_1 = Weight given to the first objective ($0<w_1<1$); w_2 = Weight given to the second objective ($0<w_1<1$); σ^+_{12} = Positive deviation from the supposed to be "zero value" of the compromise constraint developed between objectives OF_1 and OF_2; σ^-_{12}= Negative deviation from the supposed to be "zero value" of the compromise constraint developed between objectives OF_1 and OF_2.

Constraints

(i) Water availability constraint:

$$\sum_{i=1}^{n} S_{a_i} \leq AW \tag{4}$$

(ii) Water demand constraint:

$$D_{n_i} \geq S_{a_i} \geq D_{m_i} \tag{5}$$

(iii) Water supply constraint:

$$\sum_{i=1}^{n} S_{a_i} \leq \sum_{i=1}^{n} D_{n_i} \tag{6}$$

(iv) Non-negativity constraints: $\quad S_{a_i} \geq 0, D_{n_i} \geq 0, D_{m_i} \geq 0 \quad$ (7)

In addition to the above constraints, which are common to all the three objective functions, an additional compromise constraint developed between OF_1 and OF_2 is introduced for solving the multi-objective function OF_{12}. This is presented as below:

(v) Compromise Constraint:

$$w_1 * \left[\frac{1}{n} \sum_{i=1}^{6} \frac{S_{a_i}}{D_{n_i}} - OF_1 \right] - w_2 * \left[\frac{\sum_{i=1}^{6} S_{ai} * NER_i}{AW_{allocated} * NER_{max}} - OF_2 \right] + (\sigma_{12}^+ - \sigma_{12}^-) = 0 \tag{8}$$

Optimization techniques

The two single objective functions, OF_1 and OF_2, are optimized using built-in solver function in Microsoft excel and for OF_{12} a multi-objective optimization method called SICCON Technique is employed. Tabucanon (1976) developed the technique for decision-making in industrial systems with several conflicting objectives. Nayak (2003) recently applied the technique for modeling water allocation in competing demand.

2.3 Calculation of Net Economic Return

In the IWAM the *NER* from the six sectors is calculated by EAM. This is an input for OF_2 and OF_{12}. The *NER* calculation procedure is very briefly described here. Nayak (2003) presents details on the equations and procedure for estimating *NER*.

Agriculture sector: The *NER* per unit volume of water used by agriculture sector is calculated as the total crop benefit minus the total cost of production multiplied by the ratio of monthly water use to the total annual water use. It gives the *NER* on a monthly basis.

Domestic sector: The *NER* per unit volume of water from domestic sector is estimated using the inverse demand function for water (Ringler, 2001). This is calculated as water use benefit minus water supply cost on monthly basis.

Industry sector: The monthly *NER* from industry sector is also estimated using the inverse demand function for water. This is calculated as water use benefit minus water supply cost.

Power sector: The *NER* from power production is calculated as the ratio of power production multiplied by the difference between power selling price and power production cost to the total discharge from the plant.

Recreation sector: The model developed in this study may take into account various types of recreational activities, such as swimming pool, water sport, skiing, golfing, lake resort, park, boating, rafting, fishing, picnic areas and tourist places. The *NER* from recreation sector is calculated on a monthly basis.

Environment sector: The *NER* from the use of water for environmental preservation and protection is difficult to estimate. In the absence of well-established approach in the literature, the *NER* from environmental sector has been taken as average of all other sectors.

2.4 Input data requirement

Data required for ROM includes the monthly inflows to the reservoir, reservoir characteristics (elevation-area-volume relationships) and operating rules, rainfall and evaporation, percolation, installed capacity, tail water level, turbine efficiency and monthly water demand by the six sectors, etc. EAM requires data related to estimation of *NER* for the six sectors as described in Section 2.3. Normal and minimum water demands by the six sectors, AW, specified priority of allocation, *NER* (if OF_2 or OF_{12} is considered) and weights to be given to OF_1 and OF_2 (if OF_{12} is considered) are the input data for WAM.

3. Model application

Although the IWAM can be used for allocating water in complex allocation problems with conflicting social and economic objectives and where priorities and economic benefits may vary with time and space, due to space constraint, the applicability of the IWAM is demonstrated through a hypothetical case where the effect of the minimum water demand on water allocation is analyzed. The example considers six water user sectors: agriculture (A), domestic (D), industry (I), hydropower (H), recreation (R) and environment (E) competing for water from a storage reservoir.

The Dn_i by the six sectors and the monthly total normal demand are shown in Figure 2. A constant flow rate of 0.386 m^3/s is assumed as the environmental demand downstream of the reservoir for maintaining river system. The *NER* (US$/$m^3$) from agriculture, domestic, industry, hydropower and recreation sectors are calculated using the Dn_i by these sectors. *NER* from the environmental sector is assumed to be the average of *NER* of all other sectors.

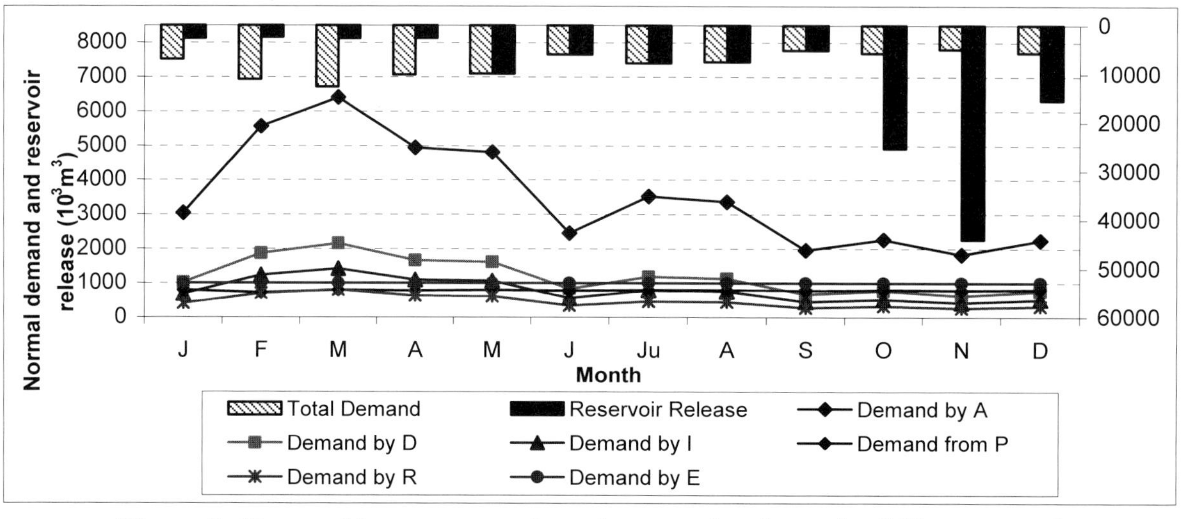

Figure 2: Demand by sectors and total reservoir release in different months

4. Results and discussion

The reservoir simulation is carried out with the initial normal water level of 41m MSL. The ROM results indicate that the release (AW + spillway overflow) is more than or equal to the normal water demand by the six sectors in the wet season from May to December. However, the release (AW) is less than the normal demand in the dry season from January to April (Figure 2). This necessitates that the limited water should be allocated to the six sectors during the dry period. The model is applied for OF_1, OF_2 and OF_{12} with the minimum demand varying from 0 to 20% of the normal water demand for the six sectors. For describing the model performance, the results of water allocation are presented and discussed only for the month February.

Table 1 presents the results of water allocation to different sectors for OF_1. The AW for February is 2419 (10^3 m^3). In this case, the model allocates water to a sector with the lowest normal demand first and then to the next higher normal demand sector. With no minimum demand defined for any sector, the water is allocated to the hydropower, recreation and environment sectors, as their requirements are lower compared to other three sectors. When the minimum demand is taken as 20% of the normal demand for each sector, the model satisfies the minimum demand and then the remaining water is supplied to the sector with the lowest normal demand in order to maximize the objective. It is noted that the value of OF_1

decreases with the increase in the minimum water demand. Numbers in bold face in Table 1 show that the water is supplied to these sectors with 100% satisfaction.

Table 1: Water allocation to different sectors in February with OF_1

Sector	Dn_i (1000 m^3)	AW (1000 m^3)	*NER* (US$/1000m^3)	Allocated water with minimum demand of 0%	5%	10%	15%	20%
A	5563	2419	12	0	278	556	834	1113
D	1866		2324	0	93	187	280	373
I	1233		4361	0	62	123	185	247
H	726		917	**726**	**726**	**726**	268	145
R	702		54	**702**	**702**	**702**	**702**	342
E	1000		1423	991	558	125	150	200
OF_1				0.499	0.451	0.404	0.328	0.248

The results of water allocation for OF_2 with varying minimum demands by the six sectors are presented in Table 2. It is interesting to note that when the objective is to maximize the net economic return, the water is supplied to the sectors with higher *NER*, however, after satisfying the assigned minimum demand for each of the sectors. The industry sector gets the highest priority followed by the domestic sector in fulfilling the normal demand. Again, the value of OF_2 decreases with the increase in the minimum water demand.

Table 2: Water allocation to different sectors in February with OF_2

Sector	Dn_i (1000 m^3)	AW (1000 m^3)	*NER* (US$/1000m^3)	Allocated water with minimum demand of 0%	5%	10%	15%	20%
A	5563	2419	12	0	278	556	834	1113
D	1866		2324	1186	787	387	280	373
I	1233		4361	**1233**	**1233**	**1233**	941	448
H	726		917	0	36	73	109	145
R	702		54	0	35	70	105	140
E	1000		1423	0	50	100	150	200
OF_2				0.770	0.693	0.615	0.481	0.308

Table 3: Water allocation to different sectors using SICCON technique with OF_{12}

Sector	Dn_i (1000 m^3)	AW (1000 m^3)	NER (US$/1000m^3)	Allocated water with minimum demand of 0%	5%	10%	15%	20%
A	5563	2419	12	0	278	556	834	1113
D	1866		2324	299	196	187	280	373
I	1233		4361	**1233**	**1233**	1040	760	399
H	726		917	**726**	627	466	289	194
R	702		54	0	35	70	105	140
E	1000		1423	161	50	100	150	200
OF_{12}				0.524	0.484	0.420	0.346	0.262

When OF_1 and OF_2 are taken together, the model analyzes the multi-objective function (OF_{12}) using SICCON technique. Table 3 presents the results obtained for various minimum demands with equal weights ($w_1 = w_2 = 0.5$) given to both the objectives. The SICCON technique maximizes the OF_{12} with a compromise between the two individual objectives and hence the value of OF_{12} is in between OF_1 and OF_2. When the minimum demand is zero, the water is supplied to domestic, industry, hydropower and environment sectors in such a way to maximize the overall objective.

Comparing the three objective functions for the zero minimum demand (Tables 1, 2 and 3), the number of sectors supplied with water is 3 for OF_1, 2 for OF_2 and 4 for OF_{12}. The increase in number of sectors is because of the compromising nature of the optimization technique. The economic benefit in monetary terms is US$ 2,113,843 for OF_1, US$ 8,133,377 for OF_2 and US$ 6,966,834 for OF_{12}. As expected, the total economic benefit with compromise solution lies in between the two single objectives.

5. Conclusions

The IWAM is a user-friendly tool and is capable of allocating water among six sectors with the objectives of maximizing satisfaction, maximizing *NER* or maximizing both satisfaction and *NER*. It can be used to analyze various water allocation problems in water scarce areas. The developed model is, thus, highly useful to water managers and professionals in making decisions for allocating water among user sectors. The model is further developed to incorporate the objective of sharing the benefits among the economic sectors.

6. Acknowledgement

Financial support in the form of a scholarship provided by the Royal Netherlands Government to the third author of the article for the Master of Engineering Program at the Asian Institute of Technology, Bangkok, Thailand is highly acknowledged.

7. References

Bielsa J. and Duarte R. (2001). An Economic Model for Water Allocation in North Eastern Spain. *Wat. Res. Dev.*, **17**(3), 397-410.

McKinney D. C. and Cai X. (2002). Multiobjective Optimization Model for Water Allocation in the Aral Sea Basin: Stochastic Optimization. Working Paper, Environment and Water Resources Engineering Program, University of Texas at Austin.

Nayak D. K. (2003). *Modelling Water Allocation in Competing Demand.* Master's thesis, Asian Institute of Technology, Bangkok, Thailand.

Quba′ A. R., El-Fadel M. and Darwish M. R. (2002). Water Pricing for Multisectoral Allocation: A Case Study. *Wat. Res. Dev.,* **18**(4), 523-544.

Ringler C. (2001). Methodology for Integrated Economic-Hydrologic River Basin Model. *ZEF Discussion Paper on Development Policy*, Bonn.

Tabucanon M. T. (1976). *Multiobjective Linear Programming for Decision Making in Industrial systems*. PhD dissertation, Asian Institute of Technology, Bangkok, Thailand.

Voogt K. D., Kite G., Draggers P. and Murray-Rust H. (2000). Modeling Water Allocation between a Wetland and Irrigated Agriculture in the Gediz Basin, Turkey. *Wat. Res. Dev.*, **16**(4), 639-650.

Pollution Inflows to Mahaweli River from Kandy City Suburbs: a Case Study from Sri Lanka

G. Herath

Department of Civil Engineering, University of Peradeniya,
Peradeniya 20400, Sri Lanka
e-mail: gemunuh@pdn.ac.lk

Abstract

During the recent past, Mahaweli river is under heavy and increasing pressure from human activities since it receives large amounts of contaminated water from domestic, commercial and industrial sources. In spite of this pollution, Mahaweli water is still utilized by many people for various needs such as drinking, recreational purposes and irrigation exposing them to many health hazards. Therefore to evaluate the present status, water quality in three main tributaries of Mahaweli around Kandy was monitored through September 2000 to July 2002. Variations of temperature, pH, conductivity, BOD_5, and chloride were within acceptable levels. In Pinga Oya, average DO decreased from 5 mg/l to 4 mg/l and NO_3-N increased from 0.5 mg/l to 1.75 mg/l during the study. Further, fecal coliform levels as high as 7500 pfu/100ml was measured. Obtained results show a strong relationship between fecal pollution and human settlement indicating frequent biological contamination from human excreta. Further, using the fecal count and nitrate content, a linear relationship between them was obtained for a branch stream running through a densely populated area during dry weather.

Keywords: biological pollution, dissolved oxygen, eutrophication, fecal coliform, Mahaweli River, nitrate nitrogen

1. Introduction

Developing countries particularly who are industrializing, have most of the contemporary water quality problems of developed countries; eutrophication, ecosystem damage, acidification, toxicity and microbial pollution. While the developed world faced these sequentially, the developing world faces them simultaneously. With very little technical, scientific and economic resources available, they are under tremendous pressure to cope with the situation. According to the United Nations (1997) Global Water Assessment, large numbers of developing countries, especially in Asia are now facing water shortages, a problem exacerbated by severely degraded water quality (Ongley, 1998). Also with UNIDO's projections for the water quality situation in the developing world over the next 25 years, public health problems and a possible complete ecological collapse is inevitable unless quick remedial actions are introduced. Therefore to sustain, most countries in Asia need new strategies to cleanup and develop river basins or sub-basins to increase the quantity of usable water as the available and new water sources are very limited.

The present situation in Sri Lanka is no exception from the above even although located in the tropics. In the tropics it is commonly expected that relatively constant environmental characteristics such as water quality (Payne, 1986) will prevail. In Sri Lanka, rivers are frequently polluted by untreated waste discharges from domestic, commercial and industrial sources. Additionally uncontrolled solid waste dumping and direct sewage disposal too is a common threat. As economic issues are given the priority, very little attention has been paid to the environmental impacts and has remained largely ignored and unmeasured in Sri Lanka. As the root for most of this pollution is from headwater contamination, this study focused on the level of pollution into the Mahaweli from the suburbs of Kandy.

Streams identified for this study are around Kandy city, in the central high lands of Sri Lanka. Kandy the second largest city in the country is at 7^0 20' N and 80^0 45' E. Suburbs of Kandy are moderately populated and are mostly utilized for paddy, tea and mixed cultivations since ancient times. All selected streams are upper reach tributaries of Mahaweli river. Among river basins in Sri Lanka, Mahaweli is the largest draining 16% of the island land surface that is more than 10,000km^2. The Mahaweli river originates from the central hills and flows 335 km towards the northeast sea, draining about one seventh of the total annual runoff to the island. Most of its water is utilized for irrigation and hydropower, damming at several locations downstream. Previous studies on these dammed reservoir waters indicate high accumulation of organic pollutants and toxins (Piyasiri, 1991). In spite of all this pollution people are still continuing to utilize this water for various recreational activities exposing them to greater risks. Geisler (1967) was the first to report on water chemistry in running waters of Sri Lanka (Silva, 1998). Subsequently many investigations on several rivers and their headwaters for a limited number of parameters had being conducted (Weninger, 1972, Silva, 1998). However, no extensive study, on pollution in the Mahaweli headwaters has ever been done before, and has become a timely requirement to built up records and understanding of the pollution processes and trends in order to keep these water bodies away from the pollution.

2. Materials and methods

In recent times with increased population and urbanization, settling around townships especially on the banks of Mahaweli and its tributaries have shown a marked increase. Among these Kandy city area is believed to contribute most of pollution into the headwaters through a number of its tributaries around the city (Dissanayake *et al.*,1982, Weerasooriya *et al.*,1982, Silva, 1998). Therefore, three tributaries around Kandy area, the Pinga Oya, (P), Maha Oya (M) and Nanu Oya (N) were monitored for water quality through September 2000 to July 2002 (Figure 1). Although flows in these tributaries are considerably small, their runoff is supposed to be highly polluted with contributions from a large number of point and non-point sources. All tributary upper catchments are loosely populated compared to lower reaches that are densely populated with concentrated dwelling and commercial establishments, especially by the side of the streams. Further, Pinga Oya downstream water is almost stagnated as the Mahaweli river is dammed a few kilometers downstream. The geology of these catchments consists of three main rock types quartzite, hornblende gneiss and garnet biotite gneiss. Soil cover in the area is of reddish yellow latosolic soils and immature brown loams with steeply dissected hilly and rolling terrains. The average elevation of the area is 500–1000m above mean sea level. Further these catchments lie in the wet zone of Sri Lanka with tropical weather that have distinct rainy seasons. When the annual precipitation occurs there tends to be a seasonal cycle of discharge rates on the rivers and streams.

There is a common understanding that water quality programs in the developing regions collect wrong parameters, from wrong places at inappropriate sampling frequencies (Ongley, 1998). In this study, 1-2 samplings per month at selected locations was done. This frequency is well above the required minimum for an effective study (Blind and Aalderink 1998). Stream P was sampled at eight cross sections selected according to the observed level of pollution. As a smaller branch of Pinga Oya is believed to contribute a substantial load of pollutants, separate sampling was done. Other two streams, which are much smaller were monitored at two sections each at upstream and close to the river confluence.

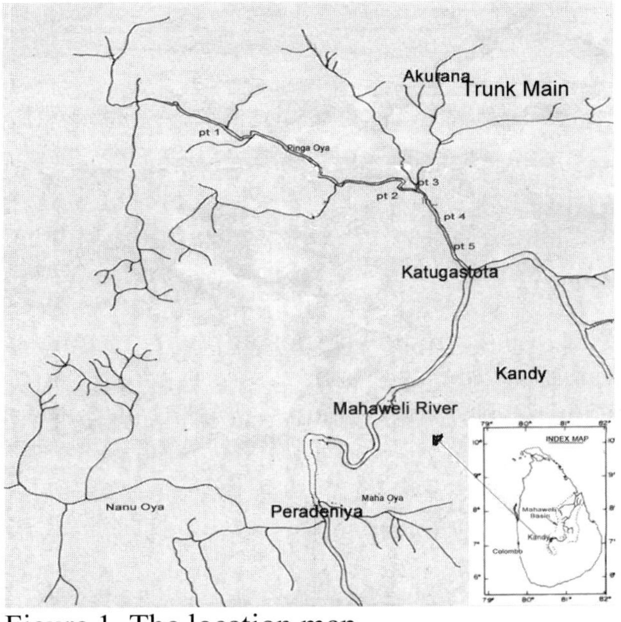

Figure 1. The location map

At all cross sections samples were taken from various depths across the river. Figure 1 illustrates the sampling locations along the main stream. Temperature, DO, pH, conductivity and turbidity was measured in-situ using the TOA WQC-22A Water quality checker. All samples were collected in pre-cleaned polythene containers following the Standard Methods (Eaton *et al.*, 1995). Suspended solids (SS), total solids (TS), BOD_5 (DO175 meter, HACH), COD (Chrome method), chloride (Argentometric titration with silver nitrate), nitrate nitrogen (Spectrophotometer HACH), Orthophosphate (Spectrophotometer) and fecal coliform (plate counting and membrane filtration with Difco m-FC broth) was tested in our laboratory.

3. Results and discussion

Although there was a significant variation in monthly temperatures, Pinga Oya had almost a constant temperature along the stream (Figure 2). Observed temperature variations in the stream were between 22-32^{0}C. This variation is mainly due to seasonal fluctuations in ambient temperatures. Table 1 shows extreme values recorded during the study for some selected parameters. The pH in the steam was almost constant. Although streams with similar watershed characteristics in the neighborhood had a lower pH, 4.4 to 5.5 (Ponnamperuma, 1987), obtained values in Pinga Oya were around 7.

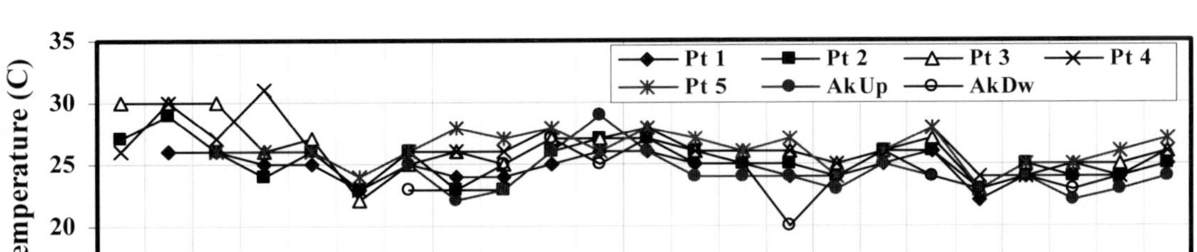

Figure 2. Temperature variation in Pinga Oya

Table 1. Minimum and maximum values recorded on Pinga Oya

Sampling point	T °C	pH	DO mg/l	Turbidity NTU	Nitrate mg/l	Coliform Pfu/100ml	Suspended Solid mg/l
Pinga Oya Pt. 1	22 - 26	6.98-7.66	3.78-6.2	1-185	<0.1-4.4	9-1,800	16-204
Pt. 2	23-29	6.88-7.40	3.05-5.60	1-257	<0.1-1.9	11-3,000	21-140
Pt. 3	22-30	6.90-7.63	1.7-4.9	1-342	<0.1-2.4	21-4,200	21-224
Pt. 4	23-32	6.70-7.67	1.92-5.29	1-390	<0.1-1.9	100-5,000	12-263
Pt. 5	22-29	6.85-7.64	1.48-5.11	1-112	<0.1-1.5	310-5,100	28-193
Trunk main AkUp	22-30	7.01-7.51	2.56-5.31	2-267	<0.1-3.3	98-7,000	47-190
AkDw	20-27	7.11-7.9	2.08-5.13	1-374	0.2-2.9	1000-7,500	62-214

The lowest Dissolved Oxygen (DO) concentration 1.48 mg/l was obtained at station 5 and highest 6.2 mg/l at station1. DO variation in the survey exhibits a reduction along the stream and with time (Figure 3). Average DO reduced from around 5mg/l in September 2000 to 4mg/l by February 2002. This indicates a continuous increase in pollution into the stream. Further the low DO levels observed in the Akurana trunk main (Figure 1), is a good example of heavy waste discharge. Frequent liquid and solid waste dumping including slaughterhouse waste from the dense households on riverbanks is responsible for this condition. As these wastes are accumulated down stream where the water is stagnant, it is approaching a possible failure on its self-purification capability leading to septic conditions. This is evident from the observed critical DO concentrations at station 5. However as there was a fair amount of algae and weeds visible in the down stream stagnant water especially during dry periods, their presence could also influence the DO levels at sampling stations 4 and 5.

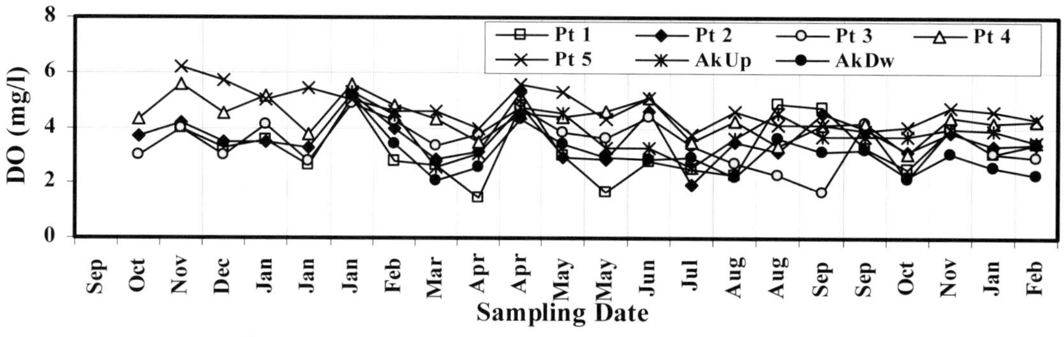

Figure 3. Dissolved Oxygen variations in Pinga Oya

Figures 4 and 5 shows the temporal variations in suspended solids and turbidity respectively on few selected days. These days were selected to make a representation of both rainy (day 7, day 19) and dry weather days. Very high turbidity and SS values are recorded on rainy days indicating high amounts of soil erosion in the sloping catchment. However even at low turbidities, comparatively high SS values were obtained. This is possibly due to the high bio solids in the stream. Pinga Oya conductivity variation on selected days is presented in Figure 6 below. Low conductivity (less than 30 mS/m) and chloride (7 to 23 mg/l) values were recorded in the main stream throughout the study. This implies that natural sources such as weathering of rocks contribute insignificant amount of free ions. Higher conductivity values reported in the trunk main is well understood as a large number of inlets bring in an array of effluents mainly from small industries, vehicle service stations and domestic. Silva (1998), observed similar results in the Kandy Lake.

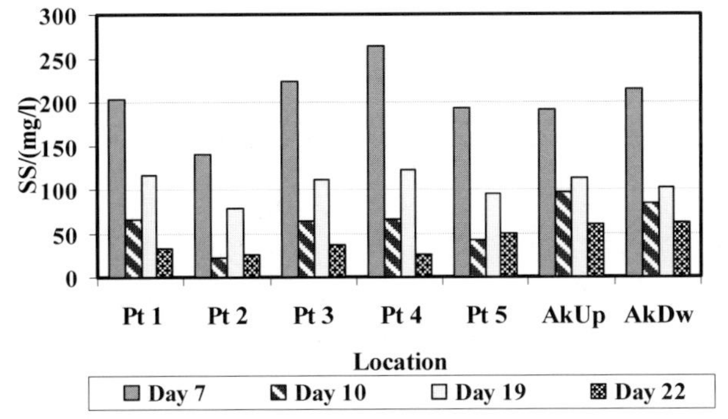

Figure 4. SS Variation

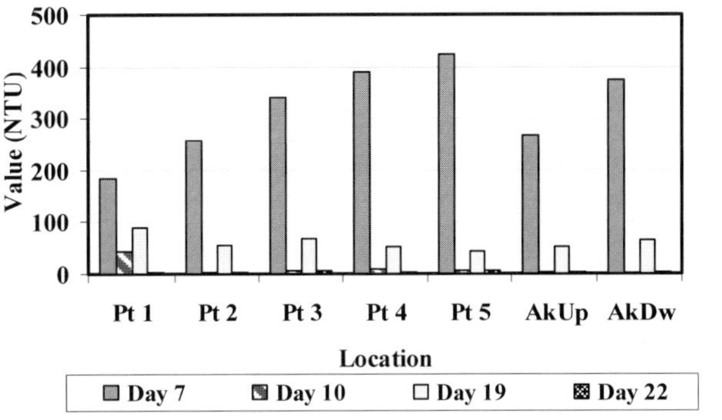

Figure 5. Turbidity Variation

Insignificant BOD_5 values indicate signs of good dilution and effective degradation (Table 1). However COD was relatively high during rainy days (Figure 7). This indicates that a fair amount of COD is washed into the stream with the runoff. The maximum COD, 132mg/l recorded is higher than the values reported in other water bodies around Kandy (Illaperuma 1998). It was noted that most of these highs are recorded near the vehicle service stations.

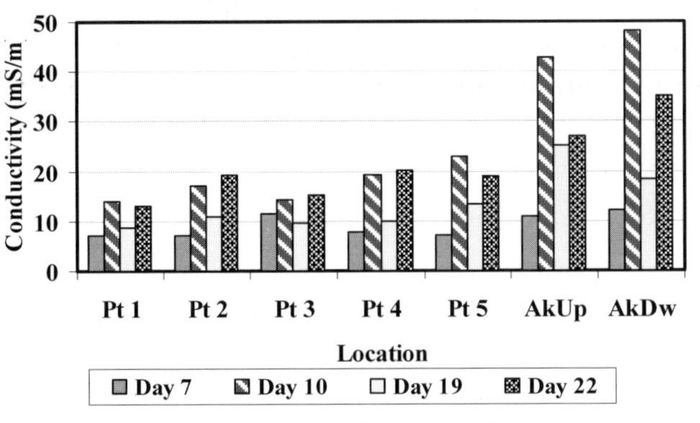

Figure 6. Conductivity Variation

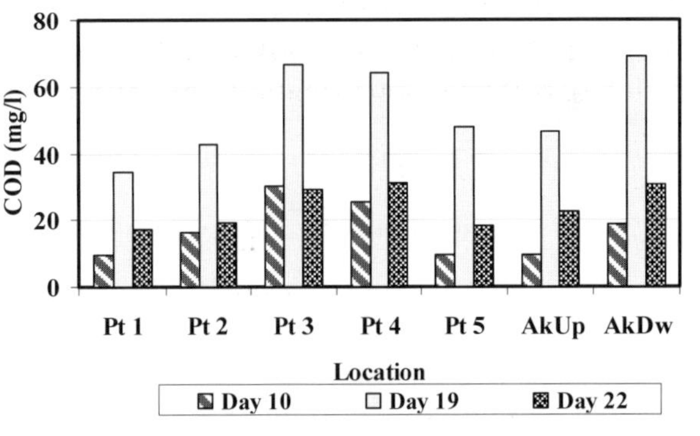

Figure 7. COD Variation

Nitrate nitrogen (NO^-_3 –N) is the most highly oxidized form of nitrogen usually abundant in streams. Geological, plant and animal breakdown products constitute the main natural sources of NO^-_3 –N in stream water. As nitrate is very mobile in soils, use of inorganic fertilizer, increased anthropogenic pollution can have a significant impact on surface water and ground water NO^-_3 –N. Throughout the study, fairly low NO^-_3 –N levels were observed (Figure 8). However, comparatively higher NO^-_3 –N concentration was obtained at upstream stations (Figure 9). Increased use of fertilizer washed away from agricultural land can be the main cause for this increase. Average nitrate nitrogen concentration increased from 0.5mg/l to 1.75mg/l during the 16 months study period. This increase was observed even during the base flow conditions, thus signifies the effect of land use pattern on the nitrate concentrations in surface water within the area. Correl, (1999) made similar observations during studies on Rhode river watershed. Organic nitrogen associated in surface runoff generated by biota is oxidized into nitrate in quick time and is also believed to be contributing to this increase too.

Phosphorus concentrations are generally low in streams as they readily bind with suspended particles and settle to the bottom. Even surface runoffs carry less phosphorous concentrations as a major portion is retained by soil. Phosphate concentrations obtained during the entire sampling period was below 2.58 mg/l. Majority of these were below 0.5 mg/l. Nevertheless these low concentrations are far adequate for eutrophication to occur (Muller, 1999).

Relatively higher phosphate concentrations are recorded on rainy days. This increase is expected as most phosphates associated in surface runoff easily settle to stream bottom with the sediments and are released to the water at a much slower rate due to anoxic conditions.

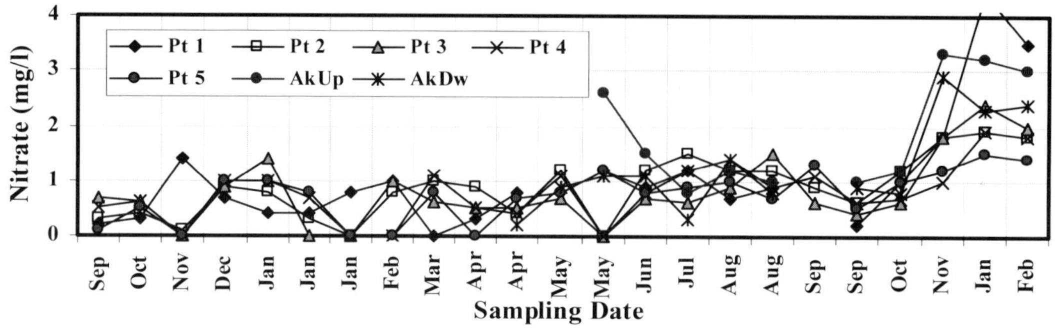

Figure 8. Nitrate Nitrogen variations in Pinga Oya

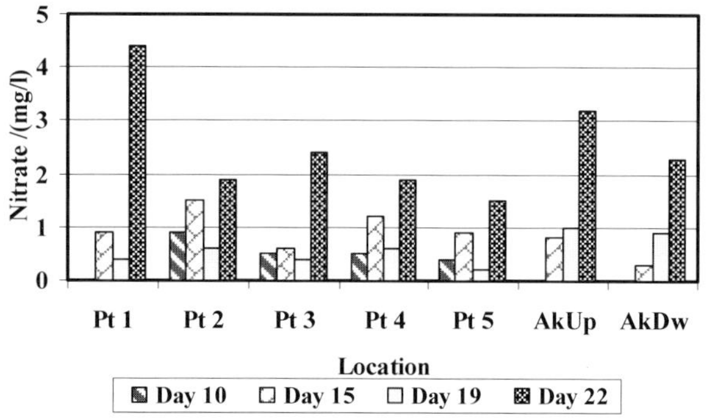

Figure 9. Nitrate Variation

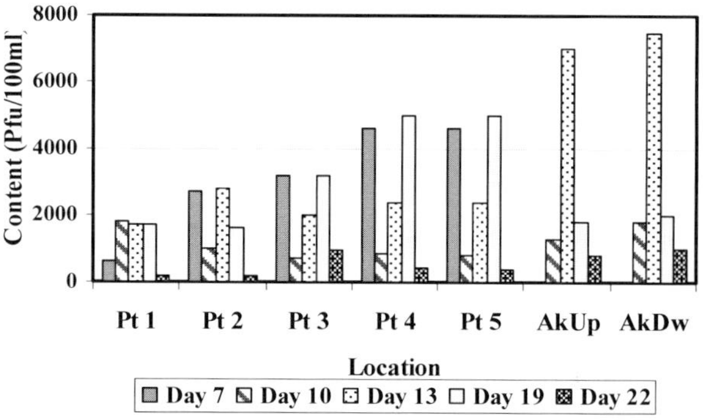

Figure 10. Fecal Coliform Concentrations

It is a common understanding that water bodies especially streams, rivers and lakes are at greater risk in developing countries as they are commonly exposed to sewage disposals (Dissanayake *et al.* 1999, Pearce *et al.* 1982). In this study too high levels of coliform were detected in all sampling points. Obtained fecal coliform values on selected days are shown in Figure 10. Values as high as 7500 pfu/100 ml was recorded during the survey (Table 1). Almost all recorded coliform levels are well in excess to the USEPA review total coliform criterion of 400pfu/100ml for drinking and recreational waters. Figure 10 show that the fecal concentrations during rainy days are much higher than that of a normal day. On rainy days as much as 35 times of normal day coliform levels were observed at upstream points. This increased levels after rain is probably caused by bacteria washing with runoff and on airborne particulates. Similar results were reported by many researches, Silverman and Erman (1979) who sampled Alpine lakes in Kings Canyon recorded 51 colonies per 100 ml after rain. As higher fecal pollution often occur in association with human waste, it can be assumed that high fecal matter on trunk main water, even during dry conditions is an indication of heavy human settlement in close proximity. Unlike in the main stream, coliform level during rainy weather did not show significant increase in the trunk main as the basin lag for this catchment is short.

It is of interest to note that Hutton and Lewes (1980) in a study on nitrate pollution in ground water, Botswana, obtain a strong association between water pollution and population density. However according to them, larger towns served mainly by surface water were unaffected. Though this study yields a strong connection between human settlement and coliform concentration, no distinct simple statistically significant relationship between nitrate content and coliform count was observed in the main stream (Figure 11). Brooks and Cech (1979) in their work on rural water supplies have made a similar comment but inferred that there could be a complex non-linear relationship, probably involving factors not well understood at present. A close look at the results obtained for trunk main shows a possible linear relationship between the fecal variation and the nitrate content during dry weather base flow conditions (Figure 12). According to the plot, with increasing nitrate, a decrease in coliform level is indicated. The reason for this may be the normal organic pollution that contains ammonia nitrogen takes few days to oxidize into nitrate and the deviation during rainy days may be due to the effect of dilution.

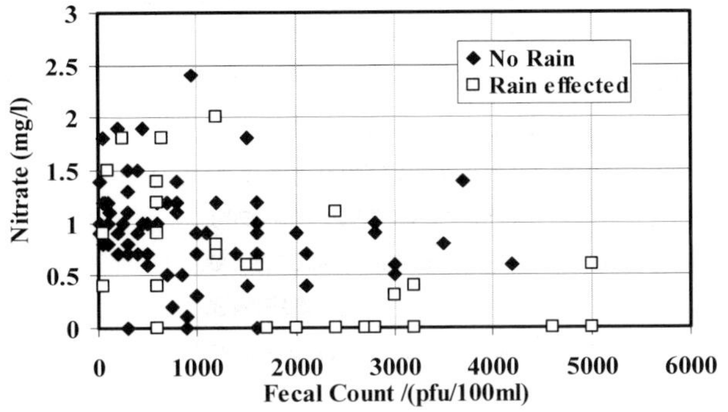

Figure 11. Coliform variation with Nitrate

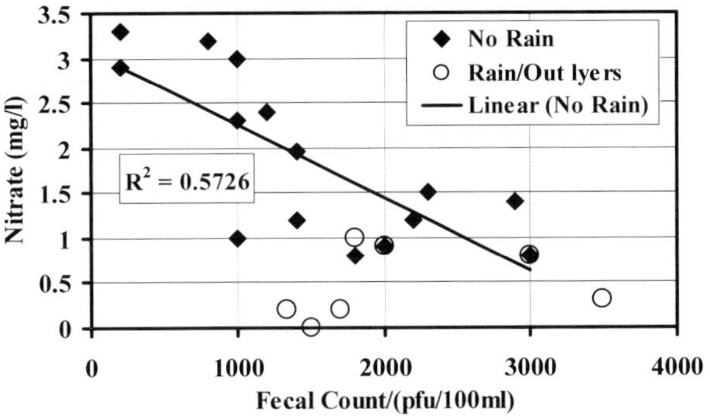

Figure 12. Coliform variation with Nitrate

Table 2. Minimum and maximum values recorded in Maha Oya and Nanu Oya

Sampling point	T °C	pH	DO mg/l	Turbidity NTU	Nitrate mg/l	Coliform pfu/100ml	Suspended Solid mg/l
Maha Oya Up.	23-29	6.32-7.70	4.4-6.21	3-66	0.45-1.3	50-850	15-78
Dw.	23-29	6.19-7.60	4.30-5.98	1-62	0.8-1.2	140-800	11-67
Nanu Oya Up.	25-31	4.89-7.45	3.83-5.76	1-19	1.3-1.8	50-1950	11-66
Dw.	25-31	5.01-7.69	4.12-5.78	1-11	0.9-1.5	10-1050	10-33

The recorded maximum and minimum water quality parameters in Maha oya and Nanu Oya are presented in Table 2. In total 10 sampling events each through February 2001 to July 2002 were done. This investigation was purely done for comparison purposes. The topography, geological formation, land usage and soil types are similar to that of Pinga Oya catchment. Compared to the Pinga Oya, lower temperature and pH value are observed in Nanu Oya. Occasional low pH value in the stream is understood as a large number of small scale metal finishing and foundry workshops with no treatment facility are located in the upper watershed of this stream. The observed turbidity and SS values indicate lower soil erosion in these catchments, as human disturbance is less compared to Pinga Oya. Further, high DO and the low coliform level too approves the above fact.

4. Conclusions

Pinga Oya, a stream in Sri Lanka city suburbs, exemplifies the problems encountered in water quality management, particularly in a suburban situation of a developing country. This study has clearly established that land use, particularly improper human settlements strongly influence the pollution outweighing the factors such as intensive agriculture and geological formation. Further, as in the case of most developing countries with very little large-scale industries, biological pollution due to human pollution is the predominant factor. Low DO values and high coliform counts indicate excessive biological contamination from human excreta. With the present rate of population increase, unless strict pollution control measures are implemented, eutrophic and septic conditions in certain stream stretches would be inevitable. Further, as most of these waters are diverted to the dry northern part of Sri Lanka to be used for drinking, recreational and irrigation purposes, unless hard control measures are implemented soon, this could turn out to be a source of serious health hazards, in the future.

5. References

Blind M. and Aaiderink H. (1998). Designing a monitoring network for long term trend detection, *Wat. quality international*, September/October, pp. 43-46.

Brooks D. and Cech I.(1979). Nitrate & bacteria distribution in rural domestic water supplies, *Wat Res.,* 13, pp. 33-41

Correll D.L., Jordan T. and Weller D.E. (1999). Transport of nitrogen and phosphorus from Rhode River watershed during storm events, *Wat. Re. Res*., 35, pp. 2513-2521.

Dissanayake C.B., Senaratne A., Weerasooriya S.V.R. and De Silva S.H.G. (1982). The environmental pollution of Kandy lake: a case study from Sri Lanka, *Env. Int*., 7, pp. 343-351.

Eaton A.D., Clesceri L.S. and Greenberg A.E., (1995). Standard methods for examination of water and wastewater: American Public Health Association (Eds), edition 19, United book press, Maryland, US.

Hutton L. G. and Lewis W. J. (1980). Nitrate pollution of ground water in Botswana, Sixth WEDC conference on water and wastewater engineering in Africa.

Illeperuma O.A. (1998). Report on water quality examination in the Kandy and Nuwara-Eliya districts, Postgraduate Institute of Science, University of Peradeniya, Sri Lanka.

Muller D. and Helsel D. (1999). Nutrients in the Nation's Water-Too Much of a Good Thing, US Geological Survey circular 1136.

Ongley E D (1998). Modernisation of water quality programmes in developing countries: issues of relevancy and cost efficiency, Wat. quality int., September/October 1998, pp 37-42.

Payne, A. I. (1986). The ecology of tropical lakes and rivers, John Wiley & Sons, UK, pp 301.

Pearce G.R., Chaudhry M.R and Ghulam S. (1998). A simple methodology for water quality monitoring: HR Wallingford in collaboration with International Waterlogging and Salinity Research Institute,Lahore.

Piyasiri S.,(1991). Limnology project at Mahaweli reservoirs:1. Some physical properties of Kotmale, Victoria and Randenigala reservoirs: *Vidyodaya journal of science*, 3, pp. 45-61.

Ponnamperuma, F. N. (1987). Toxic soils. Some aspects of the chemistry of the environment of Sri Lanka, Sri Lanka Association for Advancement of Science, pp. 165-168.

Silva E.I.L., (1998). Water chemistry of the Nilambe Oya, a tributary in the central Mahaweli valley of Sri Lanka 1, Some physicochemical characteristics: *Sri Lanka Journal of Aquatic Science,* 3, pp. 119-133.

Silverman, G. and Barica, J. (1979). Alpine lakes in Kings Canyon national park, California: baseline conditions and possible effects of visitor use, *J. Environ. Manage.*, 8, pp. 73-87.

Weerasooriya S.V.R., Senaratne A. and Dissanayake C.B., (1983). Environmental impact on polluted city canals-a case study from Colombo, Sri Lanka: *Env. Int.,* 9, pp. 401-407.

Weninger, G. (1972). Hydrochemical studies on mountain rivers in Ceylon, Bulletin of the fisheries research station, Sri Lanka (Ceylon) 23, pp. 77-100.

Technical Inefficiency in Intensive Fish Cage Production: the Importance of Understanding the Sources of Interdependencies in Abating the Nutrient Pollution in Taal Lake, Philippines

A. B. Vista

Department of Agricultural Economics, Michigan State University
200 Cook Hall, East Lansing, MI, 48824 USA
e-mail: vistaarv@msu.edu

Abstract

Production inefficiencies are inherent in the current fish cage culture technology but are further aggravated by the differences in policy implementation of each municipality surrounding Taal Lake and the ownership arrangements. Externalities abound in the current open-access policy structure and the inherent characteristics of the lake's resources. Forty-nine fish cage operators were surveyed and the production data elicited were used in the estimation of production coefficients employing the stochastic frontier analysis taking the Cobb-Douglas log linear form. Net revenue and a technical inefficiency index were measured for each operator and aggregated based on municipal jurisdictions and type of ownerships. Net revenue and technical efficiency were highest for sampled operators from Agoncillo and Talisay. Income and technical efficiency are higher for owner hiring caretakers. Low income and high index of technical inefficiency of sampled operators from Laurel and San Nicholas suggests future exit in the industry or the probability of transferring production into a lesser-concentrated and better water quality area. Nutrient pollution and decreasing biodiversity are inevitable in the current technology and institutional set-up.

Keywords: externalities, fish cage, interdependence, nutrient pollution, technical efficiency

1. Introduction

Philippine lakes are threatened by various stressors, such as watershed deterioration, household and agricultural pollution, unregulated open water fishing, and lately, intensive fish cage sprawl (Zafaralla, 1999). Recent fish kill occurrences in Taal Lake in the province of Batangas suggest a deteriorating water quality in cage areas. An excessive amount of nutrients accumulating from the cages led to nutrient enrichment, endangering the industry itself and the general lake ecosystem. Understanding the inherent nature of the lake and its interdependencies with the current institutional structures are imperative for abating the nutrient pollution.

The cage and its surrounding environment cater to two incompatible uses. Fish cage operators rely on water both as an input into the production and as a place to put waste. Using the water excessively for the latter decreases the productivity of the former. This situation arises for two reasons. First, the inherent characteristics of the lake make it difficult to assign efficient property rights, to implement laws, and to monitor compliance with regulations. Research results show that local government implementation of the open-access policy in Taal Lake tends to have negative effects on the lake's fisheries (Mutia and Magistrado, 1999). In theory, open-access policy results in ubiquitous externalities and has this problem of "free-ridership," a situation where others pay the cost of goods that will benefit everyone. Externality is defined as "...a by-product of some production process..." (e.g. air pollution) or "...the lost opportunity occasioned by the incompatible use of a good...." (Schmid, 2002). The lake's resource uses are high exclusion cost goods; hence, excluding other users is difficult and very costly. In the same way, the lake as a non-rival good implies the undesirability of exclusion wherein the marginal cost of an additional user is zero; so more persons can obtain benefits from using that good. Thus, a market situation will fail to produce an efficient quantity of the good. Second, the input-specificity associated with an intensive production system is in itself a polluting technology. The intensive production results in larger volumes of wastes generated (Handy and Poxton, 1993) and greater possibilities for the spread of disease (Naylor, 2003). It can disrupt the balance of plant and animal life (Howarth, 2000) resulting in a reduction in the species diversity (e.g. *Harengula tawilis*) and a dominance of the surviving organisms (e.g. *Oreochromis niloticus*) (Magistrado and Mercene, 2000). Other studies conducted indicate that the practice of aquaculture technology in many areas is on a destructive path that poses a threat not only to endemic fish populations but also to the industry's own long-term potential (Naylor, 2003). Many other areas in the Philippines are facing the same problem due to nutrient pollution (Dela Rosa, 1992; Sumalde *et al.*, 2002).

According to Lapido (1973), the main problem in the fishery sector is the allocation of scarce resources in an environment of complex interactions, among physical, social, economic, and political components, involving multiple and often conflicting values. By adopting the economic efficiency criterion, inefficient allocation of inputs can be eliminated. In the efficiency analysis and estimation of production coefficients, a stochastic frontier production model was used, which takes the log-linear Cobb-Douglas form. Stochastic frontier analysis was used because it can decompose the deviation from the frontier into its stochastic noise and technical inefficiency components. This paper also explores the differences in technical inefficiencies among the four municipalities. Finally, the paper looks into the ownership arrangement, i.e. owner as operator vs. owner hiring caretakers, in minimizing inefficiencies and other wastage, which is suspected to affect the net revenue and nutrient enrichment in Taal Lake.

2. The Taal lake environment and the fish cage industry

Taal Lake is located in the province of Batangas and lies approximately 60 km south of Manila, Philippines (Figure 1). It is a major tourist attraction and is one of the 10 priority freshwater wetlands in the Philippines. Its watershed has 38 tributary rivers draining into the lake. The only outlet, to Balayan Bay, is the Pansipit River. The lake covers an aggregate area of 24,236 ha, excluding the islands, and has an average depth of 60.1 m. It has a circumference of 120 km and is bounded by nine municipalities and two cities. The fish cages are limited to the barangays of four municipalities: Laurel, Agoncillo, Talisay, and San Nicholas. The municipal regulations for cage culture and lake utilization differ among the four municipalities (e.g. different rate of fees when securing an operator's permit).

Cage operators (classified either as owner/financiers or caretakers), open water fishermen, middlemen, and the different feed producers/ suppliers/ retailers are the key actors in the industry. Most of the financiers are non-residents, who hire labor to work as caretakers of their cages and significantly influence decision-making. However, the legal right to own at most five cages is vested in the local residents only.

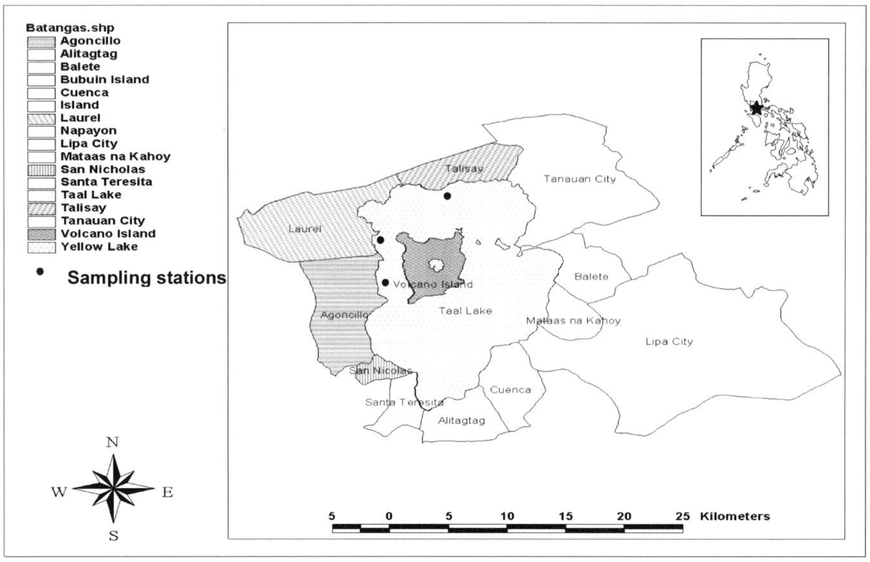

Figure 1. Map of Taal Lake, located in the province of Batangas, Philippines.

Fish cage culture has been practiced for the past two decades, and the present trend is towards increased stocking density and intensive feeding. Each fish cage measures an average of 10 x 10 x 6 m with a stocking of 65 -100 tilapia fingerlings per m^3. In 2001, Agoncillo had a total number of 2,362 registered cages, Laurel had 2,729, San Nicholas had 481s, and Talisay had 1,861. The fish are heavily fed with different types of commercial feeds, which comprise the highest proportion of production expenses (76 %). The culture period spans from 4-7 months, depending on the intensity of feeding and the quality of fingerlings. At present, the species cultured in cages are tilapia (*O. niloticus*), bangus (*Chanos chanos*), and maliputo (*Caranx ignobilis*). Among the three species, tilapia is the most commonly cultured. The estimated total fish production in Taal Lake between 1993 and 2002 is shown in Figure 2. Production from the open water fisheries dwindled after fish cage culture flourished in the lake. The lower productivity of endemic fish species may be attributed to the disturbance and displacement of their spawning ground brought about by tilapia dominance and fish cage wastes. Tilapia is an exotic and prolific breeder, and hence, a competitor of native fish species. Fish cage production plunged to a lower level starting 1998 due to massive fish kill occurrences, which were attributed to oxygen depletion during overturn and toxic poisoning from the suspected pollutants (NH_3, NO_2 and H_2S) (Rosana and Salisi, 2001).

In the Taal Lake watershed, excess nutrients are supplied to the system through point and non-point sources. Table 1 presents the estimated nutrient pollution loading from various sources in 2001. The fish cages mainly contribute to the point source pollution, which are mainly supplied by the discharges of nitrogen (N) and phosphorus (P) derived from uneaten feed, feces, and excretion via the gills and urine (Beveridge, 1993; Kibria, 1996; 1998). Beveridge (1993) estimated 76.56 % of N put into use is being lost into the cage and its surrounding waters, while Phillips *et al.* (1994) estimated 85 % for P. Consequently, for every 1000 kg tilapia harvested in 2001, an estimated 80.42 kg N and 17.46 kg P are being lost. This situation bears witness to high nutrient loading in the lake. In Taal Lake, *N* more than *P* seems to limit primary production (Zafaralla, 1999). This observation supports the findings in other freshwater lakes, where *P* has been demonstrated to be more critical in regulating water quality (Howarth, 2000; Zafaralla, 1999).

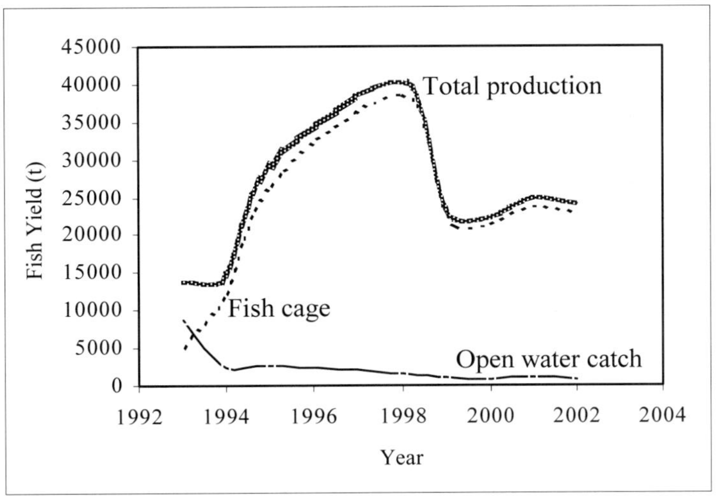

Figure 2. Estimated fish production in Taal Lake, 1993-2002 (after Mutia and Magistrado, 1999; BAS, 2003)

Table 1. Estimated nutrient input loading rates from various sources, 2001.

Source of Nutrients	Total N		Total P	
	(t)	%	(t)	%
Fish cages	3,757.73	75.68	815.88	81.85
Household sewage [a]	968.67	19.51	145.55	14.60
Agriculture/watershed [b]	239.15	4.81	35.43	3.55
Total	4,965.55		996.86	

[a] Adopted from Jacinto *et al.* (1998), assuming a 1.96 kg N per person per yr and 0.29 kg P per person per yr loss; [b] Adopted from Clemente and Wilson (2000), assuming a 6.48 kg N per ha per yr and 0.96 kg P per ha per yr loss and a 10% filtration of nutrients by the watershed vegetations.

3. Research methods

A fish cage operator survey was conducted in the four municipalities. A total of 50 respondents comprised the survey, with a varying number of fish cages and types of ownership. Aside from the data gathered through the survey questionnaire, auxiliary information was obtained from various sources. The Statistical Package for the Social Sciences (SPSS) software was used in the analysis of the data gathered. Financial analysis was done at Michigan State University. The physical relationship between inputs and output in the fish cage production was estimated following the Cobb-Douglas (C-D) production function, defined as:

$$Y = f(X_{SD}, X_F, X_D),$$

where Y = yield, kg per cage per cropping cycle;
X_{SD} = stocking density, number of fingerlings per cage per cropping cycle;
X_F = total feeding ration, kg per cage per cropping cycle; and
X_D = dummy variable for ownership arrangement,
= 1 if owner is the operator or = 0 if owner hire caretaker (s).

Data were transformed into their logarithmic values, which reduces the standard error of the estimated function. The initial ordinary least square (OLS) regression shows no evidence of multicollinearity based on the coefficient of determination (R^2) and t-statistics ratios. In the determination of the technical inefficiency index and the maximum productive capacity of fish production per cage per cropping cycle, a stochastic frontier production function [$f(X_i;\beta)$] was employed, as defined by Kumbhakar and Lovell (2000), which takes the log-linear C-D form: $\ln Y_i = B_o + \sum_n \ln X_{ni} + v_i - u_i$, where v_i is the two-sided "noise" component and u_i is the non-negative technical inefficiency component of the error term ($\varepsilon_i = v_i - u_i$). The idiosyncratic component v_i, which corresponds to things outside the operator's control, was assumed to be independently and normally distributed [$N(0,\sigma^2)$] over the observations. The u_i were assumed to be independently half-normally [$N^+(0, \sigma^2)$] distributed. The non-negative component u_i captures the technical inefficiency, which reflects the fact that each operator must lie on or below its frontier (Battese, 1992). A stochastic frontier production function was employed because it is assumed that not all operators succeed in utilizing the minimum inputs required to produce the output given the current technology (Kumbhakar and Lovell, 2000). It was also assumed that operators face a strictly positive vector of input prices and that they attempt to minimize the costs of producing the output Y. The parameters of the stochastic frontier production function and the technical inefficiency models were estimated using Stata version 8.0. Asymptotic t-ratio and generalized likelihood-ratio test were used to examine the significance of parameter coefficients of the model. Taking the partial derivative of Y with respect to X_i results in the marginal physical product of X_i (MPP_{Xi}) $= \frac{\partial Y}{\partial X_i} = \frac{\beta_i}{X_i}(Y)$. The optimum value of X_i was derived by equating MPP_{Xi} to the X_i input and output price ratio. The optimum yield can be derived from the optimum input levels obtained. Numerical solutions could be accomplished through optimization of the variables X_F and X_{SD} using Microsoft Excel Solver.

4. Results and discussion

Table 2 presents the comparative average income estimates of a grow-out cage per cropping cycle in the four municipalities in 2002. Talisay operators gained the highest net revenue compared with the operators from the other municipalities. The high cost of feed decreases the profit; that's why, while Agoncillo operators produced the highest yield per cage per cropping cycle, it ranked second to Talisay. A negative profit was obtained for sampled operators from Laurel because of low fish productivity and those of San Nicholas operators due to high feed cost. Notable to Laurel operators were the high stocking density, shortest rearing period, and the lowest yield. Some respondents believed that for 2002, many operators were still recovering from financial losses from the effects of massive fish kill occurrences and lower fish productivity that resulted in negative net return. It may be possible that some respondents understated the yield and overstated the costs.

Table 2. Estimated average incomes for a grow-out cage per cropping cycle in the four municipalities, 2002.

Item	Agoncillo	Laurel	San Nicholas	Talisay
Revenue	199,694	114,041	184,670	175,133
Operating costs [a]	166,651	130,336	189,755	122,130
Net Revenue (PhP)	33,043	-16,295	-5,085	53,002

[a] Include fees, fingerlings, feed, labor, transportation, depreciation, and miscellaneous expenses.

While some local residents have negative thoughts on the arrangement of 'owner hiring a caretaker', it has a higher net revenue compared with the 'owner as operator' (Table 3). The additional labor hours provided by the caretakers reflect an increased production. Feed wastage is minimized when a longer time is devoted to feeding, which is done manually. Contingent to these results are the proper cultural management practices, good interpersonal skills, and the paternalistic management style of owners and caretakers.

Table 3. Comparative estimated incomes of Talisay operators of a grow-out cage per culture cycle under two ownership arrangements, 2002.

Item	Owner as operator [a]	Owner hire a caretaker [b]
Revenue	141,205	204,384
Operating costs	126,101	138,172
Net Revenue (PhP)	15,104	66,212 [c]

[a] Six respondents; [b] Seven respondents; [c] Net revenue sharing scheme is 50 % to owner, 50 % to caretaker (= PhP 33,106).

The estimated frontier production coefficients and related statistics are shown in Table 4. The stocking density and ownership arrangement turned out to be not statistically significant; however, the two variables were included in the model based on *a priori* knowledge. The production coefficients for stocking density and total feeding ration have the expected positive sign while the ownership arrangement has a negative one. Testing the null hypothesis ($Ho: \sigma_u^2 = 0$) revealed that there is a technical inefficiency and that the frontier production function model employed was statistically and economically significant. A 10 % increase in stocking density will produce a 0.08 % increase in yield while a 10 % increase in the total feeding ration will produce a 0.36 % increase in yield. Since the sum of the production coefficients equal 0.35, a decreasing return to scale exists; a doubling of stocking density and feed inputs will decrease yield. One of the concerns in fish cage production is reducing the index of technical inefficiencies. The estimated frontier technical inefficiency shows that the sampled operators from Agoncillo and Talisay are more efficient in using the inputs (X_{SD} and X_F). An owner hiring a caretaker decreases the technical inefficiency by as much as 35 %. The data in Table 5 support the findings presented in Tables 2 and 3.

Table 4. Estimated frontier production coefficients and related statistics for a sample of 49 respondents in Taal Lake, 2002. Dependent variable: Yield (kg) per cage per cropping cycle

Variables	Parameter	Coefficient	SE	P value
Intercept	β_0	4.34	0.94	0.00 [a]
Stocking density (X_{SD})	β_1	0.08	0.07	0.26
Total feeding ration (X_F)	β_2	0.36	0.09	0.00 [a]
Ownership arrangement (X_D)	β_3	-0.09	0.13	0.45 [b]
Total error variance	σ_s^2	0.31	0.08	
Lambda	$\lambda = \sigma_v/\sigma_u$	3.27	0.13	
Log-likelihood function		-17.41		

[a] Statistically significant at 1 % level. $H_o: B_i = 0$.
[b] Initial OLS regression was found to be statistically significant at 5 % level.

Table 5. Estimated geometric mean of technical inefficiencies.

Item	Inefficiency value[a]
Location	
Agoncillo	0.26
Laurel	0.53
San Nicholas	0.44
Talisay	0.37
Ownership arrangement	
Owner as operator	0.56
Owner hires a caretaker	0.36

[a] Index value between 0 and 1.

The *MPP* of stocking density and total feeding ration were less than the price ratios (0.01 < 0.04; 0.21 < 0.30); hence, increasing these inputs are discouraged. The estimated optimum-stocking density was 6,415 fingerlings per cage per cropping cycle, apparently only 1/5 of the 2002 average stocking density in the area (32,649 fingerlings). The estimated optimum-feeding ration was 4,325 kg per cage per cropping cycle, about 7/10 of the 2002 level (6,220 kg). A 30 % reduction in feed would definitely be enticing to most cage operators. The estimated optimal yield was 3,134 kg per cage per cropping cycle. Given the optimum values of stocking density, feeding ration, and yield, and using the Talisay cost data, the estimated net return increased by as much as 62 %.

5. Conclusions

With the high technical inefficiency index of San Nicholas and Laurel operators in using the cage technology, it may be possible that more owners will exit or transfer production to the Talisay area. Moving away from the highly concentrated cage areas is promising for expanding the cage production. The role of caretakers is essential in reducing the technical inefficiencies in the use of inputs, specifically feed. Indeed, the current structure of right ownership and the inherent nature of the fish cage production technology influences the water quality in Taal Lake.

Externalities, such as nutrient pollution and decreasing biodiversity, are inevitable with the current fish production technology. The choice left is to accept a certain level of pollution as a trade-off for the benefits gained from use of the lake's resources and services. The enforcement of property rights is not sufficient to attain cage production efficiencies. As a result, exclusion through prices and transaction costs may be necessary. Finding a better cage culture technology that minimizes the production and transaction costs is promising and practical. However, a cheaper production process may not be that appealing to the current dominant operators who will lose their incentives once the least-cost technology succeeds. Of great importance also is the need to interpret and modify existing laws and regulations, determining who pays for fixed costs and whose preferences count? Developments in the institutional structures warrant future technological changes in the fish cage production system.

6. References

BAS. (2003). Fisheries statistics of the Philippines, 1997-2002, Bureau of Agricultural Statistics, Department of Agriculture, Available: http://bas.gov.ph, Retrieved March 12, 2003.

Battese, G. E. and Coelli, T J. (1992). Frontier production functions, technical efficiency and panel data: with application to paddy farmers in India, *The Journal of Productivity Analysis, 3*, pp. 153-159.

Beveridge, M. C. M., and M.J. Phillips. (1993). Environmental impact of tropical inland aquaculture, Paper presented at the Environment and Aquaculture in Developing Countries: ICLARM Conference Proceedings 31.

Clemente, R. S., and E.N. Wilson. (2000). Hydrologic and physico-chemical modelling of the watersheds draining into Lingayen Gulf, Quezon City: Environmental Science Program, College of Science, University of the Philippines Diliman.

Dela Rosa, M. (1992). Fish kill hits San Pablo, *Malaya, Vol. II.*

Handy, R. D. and M.G. Poxton. (1993). Nitrogen pollution in mariculture: toxicity and excretion of nitrogenous compounds by marine fish, *Review of Fish Biology, 3*, pp. 205-241.

Howarth, R., *et al.* (2000). Nutrient pollution of coastal rivers, bays, and seas, *Issues in Ecology*.

Jacinto, G. S., *et al.* (1998). N and P budget of Manila bay, Philippines, Quezon City, Philippines: Marine Science Institute, University of the Philippines.

Kibria, G., *et al.* (1998). Can nitrogen pollution from aquaculture be reduced? *NAGA, The ICLARM Quarterly, 21,* pp. 17-25.

Kibria, G., *et al.* (1996). Aspects of Phosphorus Pollution from Aquaculture, *NAGA, The ICLARM Quarterly, 19*, pp. 20-24.

Lapido, O. O. (1973). General system analysis and simulation approach: a preliminary application to Nigerian fisherie, Ph.D. thesis, Department of Agricultural Economics, Michigan State University.

Magistrado, L. S. and M.T.C. Mercene. (2000). Survey of migratory fishes in Taal Lake, Pansipit River and Balayan Bay: National Fisheries Biological Center, Bureau of Fisheries and Aquatic Resources.

Mutia, M. T. M. and L. S. Magistrado. (1999). Status of open fisheries and aquaculture productivity in Taal Lake, Unpublished Report, Taal, Batangas: Bureau of Fisheries and Aquatic Resources.

Naylor, R. L., *et al.* (2003). Effects of aquaculture on world fish supplies, *Issues in Ecology*.

Phillips, G., R. *et al.* (1994). The importance of sediment phosphorus release in the restoration of very shallow lakes (The Norfolk Broads, England) and implications for biomanipulation, *Hydrobiologia, 275/276*, pp. 445-456.

Rosana, M. R., and N.C. Salisi. (2001). Fish kill investigation in Taal Lake, Tanuan, Batangas: BFAR Inland Fisheries and Aquatic Resources.

Schmid, A. A. (2002). Institutional and behavioral economics: draft manuscript, Unpublished manuscript, East Lansing.

Sumalde, Z. M. *et. al.* (2002). Pollution-induced fish kill in Bolinao: effects of excessive aquaculture structures and overstocking, College: Department of Economics, College of Economics and Management, University of the Philippines Los Banos.

Zafaralla, M. T., *et al.* (1999). Taal Lake: limnological characterization and sources of ecological perturbations, Unpublished Report: IBS, CAS, University of the Philippines Los Banos.

How Good Intentions Turned Sour?

S. Rajchagool[1)] and C. Rajchagool[2)].

1) Intercountry Center for Oral Health
248 Chiang Mai- Lumphun Rd, Nong Hoi Muang, Chiang Mai, Thailand 50000
e-mail: sunsanee@icoh.org
2) Department of History, Faculty of Humanities
Chiang Mai University, Chaing Mai, Thailand 50200
e-mail: crajchagool@hotmail.com

Abstract

A well-intentioned policy of water provision for the rural population in Thailand has resulted in bringing a new kind of illness to over 8,500 communities. This amounts to around fourteen percent of the total deep-wells dug to serve as new water sources in the last two decades. This illness is caused by high level consumption of fluoride-contaminated water over a long period of time. It is known as fluorosis. The visible sign of fluorosis is the deformity of permanent teeth. Chronic intake of excessive fluoride can lead to severe and permanent bone and joint deformations, known as skeletal fluorosis. This paper tries to give an account of a village case study in northern Thailand, which is apparently indicative of community fluorosis. It then discusses the problems of what could be regarded as the maximum safety limit of fluoride in water. It also touches upon a controversial debate as to whether there can be such thing as 'the safety limit' set by international and national agencies alike. Underlying the tone of the paper, the authors, from recent research and growing scientific evidence, challenge the traditional idea of a possible benefit of systemic fluoride intake. In conclusion, the paper proposes ten points for collective actions for further consideration.

Keywords: fluoride, fluorosis, safe drinking water, WHO guideline for safe drinking water

1. Introduction

Many societies in their own respective ways coin a common wisdom that pertains to the idea of possible non-correspondence between intention and outcome. Due to partial knowledge, good intentions can bring about undesirable, and sometimes disastrous, consequences. Water provision in Thailand, in certain areas, is a concrete example of what was once a noble enterprise that has become a menace. This paper presents a version of history that the state's attempts to find water for people in rural areas have also resulted in an occurrence of a new illness previously unknown to local people. Now the people do not only get water but also the toxic element that comes with it. The element in question is fluoride, and it causes fluorosis.

This paper is divided into seven parts. The first gives a brief picture of the state policy of water supply in rural areas. It then follows with an unforeseen situation, which is initially perplexing and later well - understood as a serious health problem. It then goes on to give an account of fluoride, its deposits on a world scale and to discuss the knowledge about it and its harm to people's health. Finally the paper proposes ten ideas of actions which can mitigate the harmful effects of fluoride. It is hoped that the awareness of its harm will be appreciated more widely among professionals and the public alike. Thus we will all be in a better position to fight this water-borne disease.

2. Good intentions of the water provision policy

With the launching of the UN international campaign to provide water availability for the poor in developing countries known as the International Drinking Water Supply and Sanitation Decade (1981-1990), Thailand made serious efforts to comply. Various organizations under different ministries incorporated the policy into their rural development programmes. At the forefront were the Department of Mineral Resources (then under the Ministry of Industry), the Department of Health (under the Ministry of Public Health), the Department of Public Works and the Office of Accelerated Rural Development (both under the Ministry of Interior) The fact that different organizations were working in the same field to implement the same policy was not regarded as a problem, but as a sign of a unified plan at the top and of cooperation among implementing agencies. Those agencies, though working in different geographical areas, applied a common method, namely, the use of water pumps from deep wells to the pipeline systems. They dug deep wells, laid out the pipelines and acquired electric or manual water pumps for the villagers. As a result, a large percentage of the rural population in Thailand could enjoy a water supply similar to the system available to those in urban areas. With little money (less than 100 Baht, i.e., approx. US$ 2.5) per month, the villagers could have supposedly clean water from taps for domestic consumption. This novelty, hitherto unheard of, has been very popular. The Government allocated a large budget (many thousand million baht until recently) to this Safe Water Provision Programme. The water was assumed to be sanitary, and hopefully water-borne diseases of the population could be reduced. But as an adage goes, things are not what they seem.

3. The initial detection of a hitherto unknown symptom

Compared to the formerly “hard-to-obtain” situation, the piped water has been a real treat for the rural poor. Water ready at the turn of a fingertip became a new fact of life for drinking, cooking and washing. It did not occur to anybody that it might induce a problem of any kind. Six years after the use of water from the new source began, however, a primary school teacher in a village in northern Thailand noticed a hitherto unknown characteristic in newly erupted permanent teeth in every student from the village. It was opaque white patches on the dental enamel, which became stained yellow or a darker colour in some and the developing normal tooth structure was damaged, in more severe cases. Due to the as-yet-absence of other symptoms, these dental "physical" deformities were not perceived as a problem, or at least not a significant one. The perception was changed at a later age. When they, as grown-up kids and teenagers came into contact with other kids in other areas or in town, it was still regarded as an aesthetic problem, not a heath problem. They became increasingly reluctant to have their teeth seen. Lack of self-confidence in speaking and smiling gradually crept into their minds. They may have had different names for the problem, but the medical terminology is ‘dental fluorosis’. Its cause is from the excessive intake of fluoride into the body, or technically speaking, systemic fluoride consumption.

On the positive side, however, once they began to regard the symptom as a problem, they tried to make efforts to overcome it. In the beginning they tried to help themselves in several ways : by the application of charcoal, the use of sand paper and the employment of quack doctors. Sandpaper was used to erase the brown stain from tooth surfaces. A mobile "doctor" with certain elementary dental equipment went from one village to another to polish teeth for the villagers.(Lekswat, 1986) These methods undoubtedly could be very damaging, for they could cause sensitive teeth and lead to infection. A minority with sufficient economic means sought professional dental treatment, e.g. tooth extraction, which is medically both unnecessary and unethical, or crown restoration which is extremely costly.

In addition to the dental aspects, the doctors of the local hospital received a lot of complaints related to joint pains from young adults. The pains varied from vague pains most frequently in the small joints of the hands and feet to severe pains in the knees and spine. Some doctors may have mistakenly diagnosed them as rheumatoid arthritis or ankylosing spondylitis. And many prescribed analgesic drugs to relieve pain, which could be best reduced simply by fluoride-free water consumption. The statistics from this hospital showed that the usage of analgesic drugs had been unusually high. (Jumpolkul, 2002) The side effects of the overuse of analgesic drugs have not been assessed systematically.

This micro picture from a locality in northern Thailand, however, is merely a case in point. Speaking of the macro picture all over the country, a situation of a similar nature has been found in more than 8,500 spots scattered from north to south, (Athikomrangsun, 2002) but mainly in the northern part. The bad news is that it is worse than it seems at first. Worse still is that the problem is not confined to Thailand, but people in a large number of countries are being exposed to the menace of fluoride.

4. The global distribution of fluoride deposits

Fluoride exists fairly abundantly in the earth's crust and can enter groundwater by natural processes. (UNICEF, 2001) Fluoride in most ground waters occurs as the anion F^- Waters with high fluoride content are found mostly in calcium-deficient ground waters in many basement aquifers, such as granite and gneiss, in geothermal waters and in some sedimentary basins. (World Water Day 2001) Waters with high levels of fluoride content are mostly found at the foot of high mountains and in areas where the sea has made geological deposits. (World Water Day fact sheets 2001)

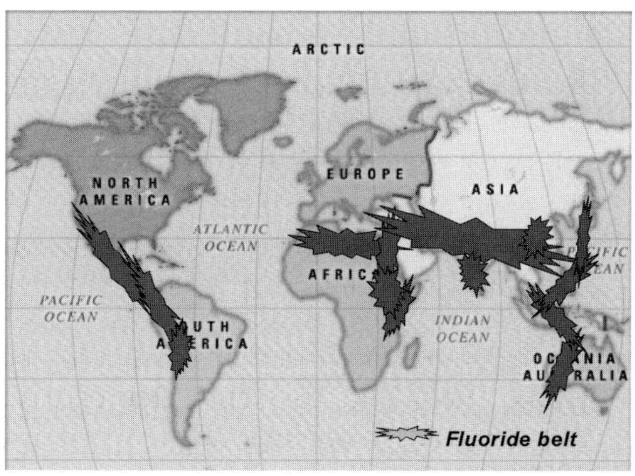

Geographically speaking, on a world map with an illustration of fluoride deposits, there are five strips. Known as global fluoride belts on land, these include: Belt 1: Turkey, Syria, Jordan, Egypt, Sudan, Somali, Ethiopia Kenya, Tanzania and South Africa. Belt 2: Egypt, Libya, Algeria, Morocco, Western Sahara and Mauritania. Belt 3: Turkey, Iraq, Iran, Afghanistan, Pakistan, India, North Thailand and (parts of) China. Belt 4: Sierra Nevada, USA Rocky Mountain, Central America, Colombia, Peru, Bolivia, Andel Mountains. Belt 5: Japan, Philippines and Volcanic Indonesia. (UNICEF, 2001, World Water Day 2001, World Water Day fact sheet 2001)

As a result of the abundance of fluoride all over the globe, the magnitude of its harm is considerably high in both actual and potential terms.

5. Fluorosis: medically speaking

There are many kinds of fluororsis depending on where and what kinds of cells are affected by fluoride. They can be dental, bone, tissue, plasma, etc. But some are known better than others. Tissue fluorosis, for example, is only in the initial state of research. Scientists with an interest in it are few and far between. Dental and bone fluorosis are more well known, and thus can be fairly delineated.

A judicious "cut and paste" below from WHO provides a detailed medical explanation and an appreciation of the harms of fluoride. "Dental fluorosis is (brought about by) a specific disturbance of tooth formation caused by excessive fluoride intake during the development of teeth. It is characterized by opaque white patches in the dental enamel. These patches may become stained yellow to a darker colour and in more severe cases normal tooth structure may be destroyed. (World Water Day: Oral Health 2001) When fluoride reaches the cells which make enamel, ameloblasts, it causes the ameloblasts to lay down irregular enamel. Instead of the regular hydroxyapatite, they will produce mottled, porous and thin enamel. At the same time the enamel is being mottled other hard and ligament tissues are being affected as well ... (Albright, 1978) Dental fluorosis is a clear indication of overexposure to fluoride during childhood when the teeth were developing. These effects are not apparent if the teeth were already fully grown prior to the fluoride overexposure; therefore, the fact that an adult may show no signs of dental fluorosis does not necessarily mean that his or her fluoride intake is within the safety limit." (UNICEF 2001)

"Fluoride can have a deleterious effect on bone growth and cause premature joint and ligament aging, as well as calcification of the inter osseous membranes and irregular bone formation. (Robins and J. L. Ambrus, 1982) ...the bone structure may change and ligaments may calcify, with resulting impairment of muscles and pain ...(World Water Day fact sheet 2001) Chronic intake of excessive fluoride can lead to severe and permanent bone and joint deformity known as skeletal fluorosis. In the earliest stages of skeletal fluorosis, the only complaints are vague pains... (Singh, et al., 1963, Smith, 1985). In the more advanced stages, there is obvious stiffness of the spine with limitation of its movements followed by kyphosis. Patients experience difficulty in walking, partly because of stiffness and limitation of movements of various joints and partly because of neurological defects in the advanced cases. (Jolly, 1968) Whereas dental fluorosis is easily recognized, the incipient skeletal involvement is not clinically obvious until the disease has advanced to the state of crippling. ... (WHO 1970) Acute high-level exposure to fluoride causes immediate abdominal pain, excessive saliva, nausea and vomiting. Seizures and muscle spasms may also occur ." (World Water Day fact sheet 2001)

The fact that fluoride is tasteless, odorless, colorless and totally soluble makes it undetectable for people who are exposed to it. It is, by analogy, an unknown and hidden "ingredient" imperceptibly damaging body organs. Its detection requires special laboratory equipment and specially trained personnel. (UNICEF 2001) Their limited availability, in effect, increases the danger to people. Fluorosis is irreversible and curative treatment of any kind does not yet exist. And to the best knowledge of the authors, no serious research in the field has been carried out anywhere. The preventive measure is pure and simple, i,e, not to let fluoride into one's body. Although this is the only possible and practical way to escape fluorsis, it is easier said than done. Since fluoride is ubiquitous in people's diets through water consumption, how one can avoid it is a big question. Admittedly it is unlikely that the people in fluoride-infested areas can obtain fluoride-free water. One important question then is what a safe level of fluoride is. This question has become a hot debate among the pro and anti-fluoride camps alike.

6. The problem of the safety limit of fluoride in water

The cynics from the anti-fluoride camp would argue that there is no such thing as a safety limit. There can only be more harm or less harm from fluoride intake. Not all from the camp share this view, though many are sympathetic to it. A common view among them and also of those who see some benefits of fluoride is that fluoride can be systemically taken up to a certain point beyond which fluorosis will occur. They differ on the question of its quantity: what amount of fluoride can be regarded as safe. This question can be phrased in some other ways depending on one's position in the polarized camps. The anti-fluoride camp would frame the question as 'what is the maximum permissible limit of fluoride?' Whereas the pro-fluoride camp would ask 'what is fluoride's optimum level to bring benefits to dental health?' The differences arise more often in the case of water fluoridation than that of defluoridation. It is stated here in passing for the sake of research methodology. Even though the differences are as wide as ever, a common ground can be reached, i.e., what is the upper limit at which fluoride can be taken into the body and still not cause fluorosis.

It is customary in most countries (particularly in developing ones) that the value 1.5 mg/l is set as the maximum level of fluoride for drinking water. Those countries have adopted the figure not because of their research achievements, but because of the lack of it. The figure is just accepted (unfortunately uncritically) from *WHO guidelines for drinking-water quality*, (1996). While the figure is in fact a guideline, it is mistaken both as a recommendation and as a standard. Since most of those developing countries have neither research facilities nor expertise of their own, they, by force of necessity and circumstance, simply adopt what they believe to be scientifically authoritative.

However, when the figure was reexamined in the light of research in a northern province, it goes far beyond the critical level at which fluoride could cause fluorosis to the entire community. The research results from ICOH carried out during 1981 - 1983 (Phantumvanit et al, 1984) clearly indicate that the narrow range between 0.4-0.65 mg/l is the borderline above which community dental fluorosis occurs. Yet it does not mean that dental fluorosis is absent under the borderline. It still does occur, but in a limited number of individuals. Hence, as far as the community fluorosis is concerned, the research team is of the opinion that 0.5 mg/l (which is only one third of the WHO guideline figure) should be set as the maximum tolerable level of fluoride in drinking water for Thailand.

The diagram shows the relationship between the fluoride level in water and the occurrence of community fluorosis in a comparative perspective. The number in front of each country is the temperature which is believed to be a factor, among many others, of the body absorption of fluoride. The higher it is, the more likely fluoride, through water consumption, is absorbed into body organs. Additionally the diagram shows the maximum tolerable level of fluoride in water above which fluorosis increases.

Besides the temperature factor, personal genes, age, physiological status, gender and nutro-pharmacological agents, influence fluoride biosusceptibility of individuals. There could well be many other factors, but at present our knowledge about fluoride and its interaction with body cells is limited. Much more research is required to supplement our rudimentary knowledge. (Clarkson *et al*, 2000)

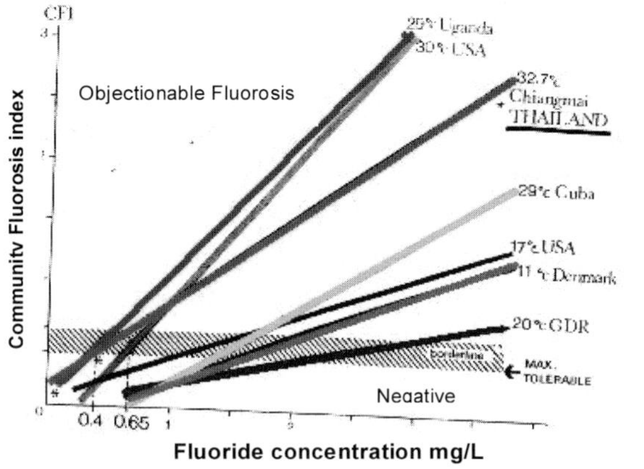

From the experience in northern Thailand, the figure from WHO is not of much value. WHO, however, cannot be directly held responsible for the (misleading) figure (1.5 mg/l). They are probably well aware that the figure is far from gospel. They explicitly state that the figure is in essence a guideline, and not a recommendation, let alone a standard. The publication also gives an additional explanation of the level 1.5 mg/l.

> "In setting national standards for fluoride, it is particularly important to consider climatic conditions, water intake, and intake of fluoride from other sources (e.g. from food and air)."(p. 236)

A precaution has been suggested, and it is true that it is up to each country to modify the figure. Ideally it is best and safest for each country to carry out its own research for its own safety limit of fluoride in water. But in view of inadequate research prospects in general and in poor countries in particular, with or without the precaution the end result cannot be much different. Most developing countries are not in a position to modify the figure through their own scientific experiments. WHO's intended objective is certainly beyond any doubt, it is simply another example of unfortunate consequences from initially good intentions.

7. What is to be done?

In view of the problems which are not peculiar to Northern Thailand alone, a set of considerations is here proposed for collective actions:

- The WHO guideline figure should be abandoned. There are two reasons for the abandonment. First it is too high, and it hence cannot serve a positive role in fluoride determination. Secondly since it is to be "modified", if not radically adjusted, the figure cannot be regarded as a point of reference. To continue to adhere to it is to mislead a large number of not-fully- informed people, both professionals and the public alike. As it has been, it is still taken by some health organizations (for example, Thai FDA) as the standard to go by. If the "standard" level, which is unsafe, remains, it naturally preempts all other needful preventive measures.

- Those countries capable of their own research need to have their own experiments. Since it is not possible to have one straightjacket for all people, each country must find its own maximum tolerable fluoride level in water. Moreover in a large country, which has cultural diversities, a figure for each locality is required. This is a challenging role for all concerned professionals, and also government officials, or even politicians.

- Those countries that are not yet well equipped, either in terms of personnel or of materials, could critically follow other countries with a similar condition. Mutual-help forums and/or an appropriate international assistance are to be encouraged.

- The determination of fluoride in water supply, particularly from wells, should be on the agenda. If an unsafe source is found, it should be discontinued for consumption in due time. Water well mapping is urgently needed, especially in high-risk areas. This could help avoid fluoride-infested locations for future well digging, and simultaneously find remedial measures for the actual as well as potential victims of fluorosis.

- Knowledge about the dangers from fluoride needs to be disseminated widely through mass media. The public's right to be informed is of top priority. Since the belief in fluoride as a beneficial element, despite ample contrary evidence, is still strong and persistent, an open channel for debates and the right to know more are necessary conditions for a higher public awareness of how harmful fluoride-contaminated water can be. Considering the fact that the view of "fluoride is good for you" is still dominant, the task for the awareness of adverse effects of fluoride is all the more pressing.

- For communities suffering from fluorosis, a "self-reliance" approach to defluoridation as a component of community health programmes is as important as a search for alternative sources of safe water. There have been experiments to find an appropriate technology for the approach, but a satisfactory degree of success is not within reach. More cooperation between professionals and the people is to be fostered and strengthened, if a people-centered development is to be the leading principle.

- A new definition of safe water is called for. The present understanding is too limited, as it does not take into account seemingly harmless elements (e.g. calcium, iron, magnesium, fluoride, etc.). An inadequate definition is not conducive to a higher goal attainment.

- A comprehensive review of underground water provision is essential not only for health reasons, but also for more effective water supply management. There has not been a policy of division of labour among water providing agencies. A number of organizations carry out the same functions, e.g., well-digging, water pump supply. Each agency has its own budget and its own area of coverage. The Department of Health, for example, would be better assigned to the job of water quality, and leave the job of drilling and digging to the Organization for Accelerated Rural Development. A similar point can also be said of the Department of Mineral Resources whose tasks should focus more on the concern for increasing use of underground water.

- More economical habits for more effective utilization of water are to be cultivated. This can be done at different levels, from the individual, household to community, and from the enterprise to town and city, so that a fairer distribution of safe water could be aimed at, and the long-term provision of water could be better managed.

- Last but not least, concerted efforts to put an end to the idea of fluoride as a supplement are called for. At the international front the tide has turned, from popularizing the use of fluoride supplements to outlawing the practice. Belgium is a case in point, and it is trying to have it implemented in EU countries (Vermeire and Cremers, 2002). France and certain states in the US have long realized the detrimental effects of fluoride. Scientists' admission of mistakes in the past has also become increasingly more prevalent. Despite the still strong support for fluoride supplement (be it water or milk fluoridation) from some quarters - many dentists are in the forefront – the message could not be clearer: systemic fluoride intake causes more harm than the benefits it is supposed to bring. A major task for all of us working in the areas of water provision and consumption can perhaps be best realized by continuous and concerted attempts among like-minded people. But the resistance to be expected from certain professional groups is strong and hard to overcome. One of humanity's predicaments since time immemorial is that a process of change requires time. It is more so with one's habits of mind, but there are no other ways than taking the first steps on a long journey.

8. References

Albright J. A.(1978). The effect of fluoride on the mechanical properties of bone, *Transactions of the Annual Meeting of the Orthopedics Research Society*, 3, pp. 98.

Athikomrangsun S. (2002). Fluoride in deep-well water, *Appropriate use of fluoride in Thailand,* Dental Health Division, Department of Health, pp. 139-154 (In Thai).

Clarkson et al. (2000). International collaborative research on fluoride, *J Dent Res* 79(4), pp. 893-904.

Jolly SS. (1968). An epidemiological, clinical and biochemical study of endemic, dental and skeletal fluorosis in Punjab, *Fluoride*, 1(2): pp. 65-75.

Jumpolkul W. (2002). VDO presentation, Skeletal Fluorosis in Ban Thi, (in Thai).

Lekswat P. (1986). Perception and health behavior regarding fluorosis of rural population in Chiangmai (dissertation), Bangkok: Chulalongkorn Univ. (in Thai).

Overview of disease fact sheets: World Water Day 2001: Fluorosis, August 3, 2002, http://www.worldwaterday.org/2001/disease/fluorosis.html.

Overview of thematic articles: World Water Day 2001: Oral Health, August 8, 2002, http://www.worldwaterday.org/2001/thematic/oral.html.

Phantumvanit P, Songpaisan Y, Hansawat S., Lekswat P. and Shamshula R.G. (1990). The relationship between fluoride in drinking water and community fluorosis index in Chiangmai, Thailand (1984), *The program and abstracts of the 10th Meeting on Science and Technology of Thailand,* Chiangmai University, Chiang Mai, pp. 456, (in Thai).

Robins J.C. and Ambrus J. L (1982). Studies on osteoporosis IX. effect of fluoride on steroid induced osteoporosis, *Research Communications inb Chemical Pathology and Pharmacology*, Volume 37, No. 3, pp. 453-461.

Singh A, et al. (1963). Endemic fluorosis, epidemiological, clinical and biochemical study of chronic fluoride intoxication in Punjab, *Medicine*. 42: pp. 229-246.

Smith GE. (1985). Repetitive strain injury, or incipient skeletal fluorosis? (Letter.) *New Zealand Medical Journal* 98: pp. 328.

UNICEF's Position on Water Fluoridation, August2, 2002, http://www.unicef.org/programme/wes/info/fluor.htm.

Vermeire C. and Cremers P., Away with the trade: Belgium prohibits - as the first country in the world - fluoride supplements, November.11, 2002, http://64.177.90.157/pfpc/html/hume.html.

World Health Organization. (1970). Fluorides and human health, pp 32, 239-240.

World Health Organization (1996). *Guidelines for drinking-water quality (2nd edition)* vol. 2: Health criteria and other supporting information, WHO, Geneva.

World Water Day (2001). Overview of thematic articles: naturally occurring hazards, August 3, 2002 http://www.worldwaterday.org/2001/thematic/natural.html.

Section 3

Emerging issues and environmental technologies

Treatment of Textile Wastewater by a Combination of Anaerobic and Aerobic Processes: A Denim Processing Plant Case

[1] T. Setiadi, Y. Andriani and M. Erlania

1) Department of Chemical Engineering, Institut Teknologi Bandung
Jl. Ganesa 10, Bandung, 40132, Indonesia
e-mail : tjandra@che.itb.ac.id

Abstract

In this study, the ability of a combination of anaerobic and aerobic processes to treat denim process wastewater and its application on an industrial scale was reported. Laboratory experiments were conducted with the varied hydraulic retention time (HRT) in anaerobic process, i.e. 12, 18 and 24 hours. In the aerobic process, the HRT were 12 and 24 hours, and the SRT (Solid Retention Time) was maintained at 20 days. All experiments were performed at room temperature. Each experiment was run for about 90 days. From the laboratory study, it was found that increasing HRT would increase color removal on the anaerobic processes, however no significance different for COD removal. However, the majority of color was removed in the anaerobic process, on the other hand the majority of COD was removed in the aerobic process. The combination of anaerobic and aerobic processes was able to reduce COD, BOD and color up to 91, 94 and 96 %, respectively. The color removal was consistently high and visually, no color has been observed in the final effluent.
Based on the laboratory results, the industrial scale required to treat the wastewater was established. The design flowrate was 20 m^3/hour. The biofilter was selected as an anaerobic process with the HRT of 24 hours. The aerobic process was determined extended aeration with the HRT and was also for 24 hours. The system has been completed in March 2002 and steady state condition was achieved in June 2002 with a performance not much different to the laboratory scale results. Although, slightly less COD and color removal was observed. From 4 months of operation, the COD, BOD and color removal was 87, 93 and 90 %, respectively. At the moment, the treated effluent was recycled back to the denim processing plant and used in washing machines.

Keywords: aerobic, anaerobic, decolorization, industrial scale, textile wastewater

1. Introduction

Textile industries have developed rapidly in Indonesia in the last ten years. The development of textile industries can be seen from the increasing value of textile (include textile products) exports. It was recorded at as much as US$ 2.02 billion in 1989 and has increased up to around US$ 7 billion in 2002. Thus, the textile industry becomes one of the most important sources of foreign exchange for Indonesia. However, textile industries have caused serious environmental problems because of their wastewater. Most textile industries produce wastewater which is relatively high in BOD, COD, suspended sludge and color. In addition, the wastewater may also contain heavy metal depending on the type of coloring substances used. In general, the objectives of the textile industry wastewater treatment are to reduce the levels of organic pollutants, heavy metal, suspended solids and color before releasing to the environment. At present however, the Indonesian government has not included color as one of the parameters in the regulation for textile industry effluents. In reality, wastewater containing color may cause environmental and aesthetic problems.

Color substances are used during dyeing and printing process. The wastewater from these two processes is the most polluted wastewater in the textile industry. Wastewater treatment technology either biological, chemical, physical or the combination of the three can be used to treat textile wastewater. Chemical coagulation is frequently practiced in Indonesian textile industries and sometime it is the only means to treat the wastewater. The main disadvantage of this method is that it produces a lot of sludge which needs to be disposed of. Moreover, according to the current Indonesian Government Regulation (Peraturan Pemerintah No. 18, 1999), textile sludge may be classified as a hazardous waste, therefore the sludge should be treated in a proper way. This means that the sludge disposal causes a substantial increase in the cost of wastewater treatment.

Furthermore, in the past decades, the use of synthetic compounds in the textile processes has increased. Thus, the amount of more complex compounds, which have a characteristic of being slowly biodegradable by aerobic processes, in the textile effluent has increased. On the other hand, aerobic processes have been widely used for treating textile wastewater. Therefore, to meet the stricter regulations, a more effective and efficient system to remove organic pollutants and color, and to reduce sludge production has to be found.

Several scientific research studies have reported that a biological process consisting of an anaerobic and aerobic process in series is a promising method for treating textile wastewater [Loyd *et al.*, 1991, Seshadri *et al.*, 1994 and Zaoyan *et al.*, 1994]. The anaerobic microorganisms reduced the color and increased the biodegradability of the waste [Setiadi *et al.*, 1997 and Minke and Rott, 1999]. This will render the effluent more amenable to subsequent aerobic treatment. Moreover, the sludge production will be reduced due to no chemicals being added to remove color. However, the report on the industrial application of these processes is very limited. Therefore, in this study, the ability of a combination of aerobic and anaerobic processes to treat denim process wastewater was investigated and the application on an industrial scale was reported.

2. Materials and Methods

Wastewater preparation. The wastewater was from a textile plant performing denim processing operations where only indigo dyes were used. The COD, BOD, color and pH range of wastewater was 400 – 978 (689) mg/l, 100 – 700 (504) mg/l, 12 – 300 (90) mg/l and 8 – 14 (10), respectively. Together with the above data, the average value was given in parentheses. A necessary amount of nitrogen and phosphorous was added to the wastewater to make the ratio of COD : N : P of 100 : 5 : 1. And also the pH of wastewater was adjusted to be neutral.

Reactors. The anaerobic and aerobic reactors were made of Plexiglas having working volumes of 6 and 7 litres, respectively. The anaerobic reactors were baffled reactors with the basic design of a rectangular box with internal vertical baffles alternatively hanging and standing. These baffles divided the reactor into 2 compartments. The aerobic reactors were a simple completely mixed activated sludge reactor where the box was divided into two chambers, i.e. aeration and settling chamber [Reynolds, 1982]. The experimental set-up is shown in Figure 1.

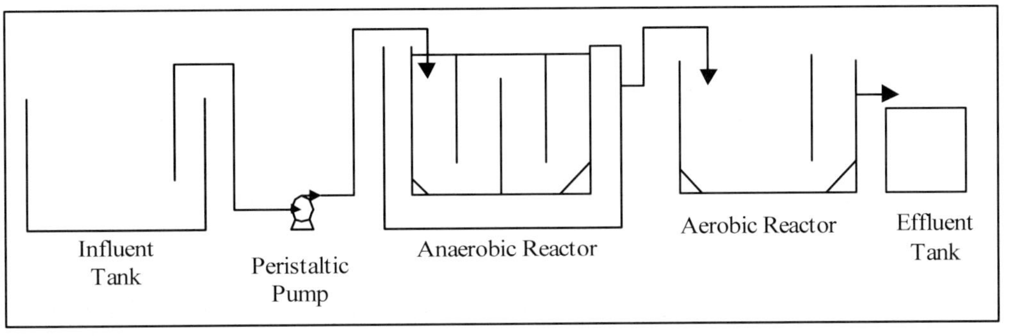

Figure 1. Experimental Set-up for Treating Textile Wastewater by a Combination of Anaerobic and Aerobic Process

Acclimatization and Reactor Operation. Sludge from an activated sludge process treating textile mill effluents was used as an initial inoculum for anaerobic reactors. Then, the cultures were developed anaerobically by feeding the textile wastewater with the necessary amount of nitrogen and phosphorous. Anaerobic conditions were achieved by overgassing the reactors with oxygen-free nitrogen for 30 minutes. This acclimatization period was carried out for about 3 months. The detailed procedure of acclimatization was presented elsewhere [Setiadi *et al.*, 1997].

For the aerobic reactors, the sludge from an activated sludge process was acclimatized by the textile wastewater in aerobic conditions. The necessary amount of nitrogen and phosphorous were also added. After the acclimatization period has been completed, the steady state performance was conducted with the different hydraulic retention time (HRT) at each reactor. The HRT variation of this study is given in Table 1 with sludge retention time (SRT) of the aerobic process maintained at 20 days for all experiments.

Table 1. HRT variation in the laboratory scale study

Denim process wastewater		
Run	Anaerobic	Aerobic
I	12	12
II	12	24
III	24	12
IV	24	24
V	24	48

Note: The HRT in each reactor has the unit of hours

Analytical procedure. Influents and effluents of each reactor were collected and analyzed regularly for the following parameters: MLSS (mixed liquor suspended solids), pH values, COD, and BOD. These parameters were measured in accordance with the Standard Methods [Standard Methods, 1992]. The supernatant was also analyzed colorimetrically at a wavelength of 486 nm using Spectronic 20. The color concentration was quantified by converting the absorbance figures to the equivalent dye concentrations using a calibration curve. Some of the samples were analyzed by a scanning spectrophotometer to identify the dye degradation products.

3. Results and Discussion

This section will be divided into two parts: a). results and discussion of laboratory studies, and b). results and discussion of industrial scale operations.

Laboratory scale studies

As has been mentioned above, the textile plant produced denim fabric with only indigo dyes used in the process, specifically Indigo C.I. Vat Blue 1 having a double bond of C=C. The COD, BOD and color removal in each process is presented in Tables 2 and 3 with the overall reduction shown in Table 4. The reduction of COD, BOD and color shown in Table 2 and 3 was calculated based on the influent and effluent of each step. Table 5 shows the effluent quality of the wastewater after treating through anaerobic and aerobic processes. A range and average values of final effluent are presented in the table.

Table 2. Reduction of COD, BOD and Color in the Anaerobic Process

Run	HRT (h)	COD reduction (%)	BOD reduction (%)	Color reduction (%)
I	12	44.4	31.2	57.0
II	12	45.5	32.3	53.0
III	24	45.0	29.4	84.5
IV	24	44.0	30.5	80.1
IV	24	40.1	33.9	78.2

Note: all the above value is an average value of steady state operation

Table 3. Reduction of COD, BOD and Color in the Aerobic Process

Run	HRT (h)	COD reduction (%)	BOD reduction (%)	Color reduction (%)
I	12	55.4	90.0	67.0
II	24	58.0	88.0	68.3
III	12	77.4	88.5	73.2
IV	24	83,0	91,5	78,7
V	48	78.5	90.5	76.4

Note: all the above value is an average value of steady state operation

Table 4. The Range and Average value of COD, BOD and Color Reduction in the Overall Treatment

	COD reduction (%)		BOD reduction (%)		Color reduction (%)	
Run	range	average	Range	average	Range	average
I	65 – 90	75.2	80 – 95	93.1	70 – 90	85.8
II	75 – 93	87.2	74 – 95	91.9	73 – 91	85.1
III	63 – 94	87.5	75 – 95	92.0	73 – 98	96.1
IV	74 – 98	90.5	83 – 99	94.1	79 – 99	95.8
V	74 – 98	87.3	76 – 99	93.7	80 – 88	95.0

Table 5. Final Effluent Quality

	COD (mg/l)		BOD (mg/l)	
Run	range	Average	range	average
I	83 – 210	170.8	20 – 75	34.7
II	90 - 187	157.7	25 – 75	40.9
III	33 – 176	85.9	20 – 60	40.9
IV	12 – 144	65.6	10 – 45	29.8
V	30 – 134	87.5	10 – 51	31.6

A typical absorption spectra of the sample is presented in Figure 2. In the influent sample (Figure 2.a), there were 10 peaks appearing with wavelengths of 220, 226, 244, 250, 486, 496, 510, 518, 552 and 560 nm. Then, only 2 peaks appeared at wavelengths of 202 and 486 nm after treating in the anaerobic reactor (Figure 2.b). Finally, there were 2 peaks left at wavelengths of 202 and 486 nm in the final effluent (Figure 2.c). Although, a peak of 486 nm was still recorded in the absorption spectra of anaerobic effluent, visually the color was completely decolorized. This finding confirmed that the anaerobic process has played a greater role in the decolorization and degradation of dyes than that of the aerobic one as has been pointed out by others [Carliell *et al.*, 1995 and Setiadi and Hasibuan, 1999].

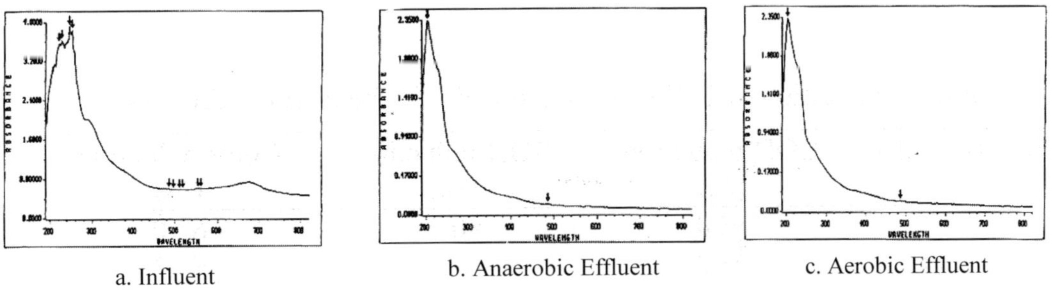

a. Influent　　b. Anaerobic Effluent　　c. Aerobic Effluent

Figure 2. Typical Absorption Spectra of Influent, Anaerobic Effluent, and Aerobic Effluent (the y-ordinate is absorbance value and the x-ordinate is wavelength)

The final effluent quality as shown in Table 5 revealed that the COD and BOD values for anaerobic HRT of 24 hours were consistently low. This value will meet the new Indonesian effluent standard for textile wastewater being 150 and 60 mg/l for COD and BOD, respectively. Therefore, the HRT of 24 hours in anaerobic and 24 hours in aerobic process was recommended for treating denim plant wastewater to meet the current Indonesian effluent standard.

Industrial scale operations

Based on the laboratory results, the industrial scale to treat the wastewater was established. The design flowrate was 20 m^3/hour. After discussion with the respected engineers in the factory and consultants, an upflow biofilter was selected as the anaerobic process. The biofilter was a fixed bed type of reactor and packed with plastic media, however the detail of the reactor is not able to be disclosed here. Basically, the HRT of 24 hours was determined in the anaerobic reactor having a diameter of 9.5 m and height of 9 m.

The aerobic process was determined extended aeration with the HRT and was also 24 hours with the length, width and depth of 12 m, 12 m and 3.5 m, respectively. The system was completed in March 2002 and steady state condition was achieved in June 2002 with a performance not much different to the laboratory scale results. Although, slightly less COD and color removal was observed. From 4 months operation, the average COD, BOD and color removal was 87, 93 and 90 %, respectively. At the moment, the treated effluent was recycled back to the processing plant and used in washing machines. Figure 3 gives an example of influent (A), effluent of anaerobic reactor (B) and final effluent of the aerobic process (C). Visually, the color in the final effluent was removed almost completely with an average value of removal of about 90 % from the average original concentration of dyes of 90 mg/l.

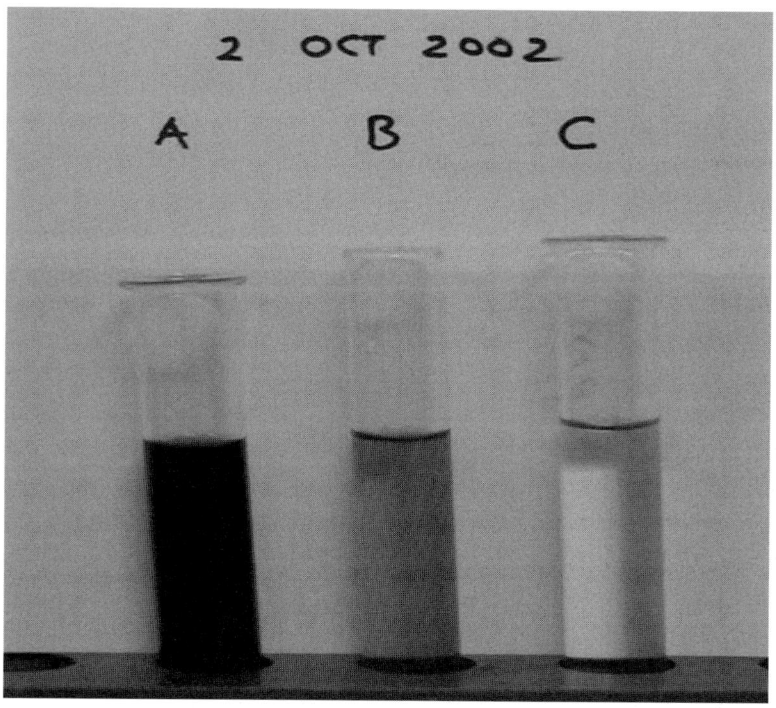

Figure 3. An example of influent (A), anaerobic process effluent (B) and final effluent (C) from the industrial scale

4. Conclusions

Results found in this study for treating denim plant wastewater with a combination of anaerobic and aerobic are as follows:

1. Increasing HRT would increase color removal on the anaerobic processes, however there was no significance different for COD removal.
2. The majority of color was removed in the anaerobic process, on the other hand the majority of COD was removed in the aerobic process.
3. The combination of anaerobic and aerobic processes was able to reduce COD, BOD and color up to 91, 94 and 96 %, respectively. The color removal was consistently high and visually, no color has been observed in the final effluent.
4. An industrial scale application has been established quite successfully with a performance not much different to the laboratory scale results

5. Acknowledgement

The first author would like to thank the students of the Department of Chemical Engineering ITB who have been involved in this research since 1994. Some of the names of the students are on the references. He would also like to express his gratitude to all institutions having provided financial support for this research, namely : OPF-ITB (1996/1997); the ESA-UNET Project, the European Union, Directorate General I, 1996-1997 : The Asahi Glass Foundation, Japan, 1998 – 1999, and also to PT Garuda Mas and PT Aquakultura Lestari for providing the industrial scale data.

6. References

Bhattacharya, S.K., Wang, S., Anggara, R.V., Kawai, T. and Bishop, F.D. (1990). Fate and Effect of Azo Dye on Anaerobic-Aerobic System, *Proc. 44th Purdue Industrial Waste Conference*, Lewis Publishers Inc., Chelsea, Michigan, pp. 295-297.

Carliell, C.M., Barclay, S.J., Naidoo, N., Buckley, C.A., Mulholland, D.A. and Senior, E. (1995). Microbial Decolourisation of a Reactive Azo Dye under Anaerobic Conditions, *Water SA*, Vol. 21(1), pp. 61-69.

Haug, W., Schmidt, A., Nortemann, B., Hempel, D.C., Stolz, A. and Knackmuss, H.J. (1991). Mineralization of the Sulfonated Azo Dye Mordant Yellow 3 by a 6-Aminonaphthalene-2-Sulfonate-Degrading Bacterial Consortium, *Applied and Environmental Microbiology*, Vol. 57 (11), pp. 3144-3149.

Loyd, C.K., Bordman G.D. and Michelson, D.L. (1991). Anaerobic/Aerobic Treatment of a Textile Dye Wastewater, *Pre-printed for Mid-Atlantic Industrial Waste Conference*, July 15-17, Morgantown, WV.

Minke, R. and Rott, U. (1999). Anaerobic Treatment of Split Flow Wastewater and Concentrates from Textile Processing Industry, *Wat. Sci. Tech.*, Vol. 40 (1), pp. 169-176.

Nigam, P., Banat, M.I., Singh, D. and Marchant, R. (1996). Microbial Process for the Decolorization Effluent Containing Azo, Diazo and Reactive Dyes, *Process Biochemistry*, Vol. 31 (5), pp. 435-444.

Reynolds, T.D. (1982). *Unit Operations and Processes in Environmental Engineering*, Broke/Cole Engineering Division, Monterey, California, pp. 549-550.

Seshadri, S. and Bishop, P.L. (1994). Anaerobic/Aerobic Treatment of Selected Azo Dyes in Wastewater, *Waste Management*, Vol. 14 (2), pp. 127-1317.

Setiadi, T., Ma'rup, F. and Khaqim, A. (1997). Effect of Temperature and pH on the Biodegradability Enhancement of Textile Mill Effluent by Anaerobic Processes, *Proc. Regional Symp. on Chemical Engineering 1997*, UTM, Johor, Malaysia, Oct. 13-15.

Setiadi, T., Partiwi, F.I. and Widyarsa, I.I. (1999). Treatment of Textile Wastewater Containing Reactive Azo Dyes in Anaerobic-Aerobic Processes (Indonesian), *Kongres Pertama Himpunan Toksikologi Indonesia*, Jakarta, 22-23 February.

Setiadi, T. and Hasibuan, R. (1999). Anaerobic Decolorization on Selected Azo Dyes: Effect of Carbon Sources and Mechanisms of Dyes Decolorization, *Presented at the 3rd SEA Regional Conference on Higher Engineering Education Network*, Manila, 27-29 October.

Standard Methods (1992), *Standard Methods for the Examination of Water and Wastewater*, 18th Edition, APHA, Washington.

Zaoyan, Y., Ke, S., Guangliang, S., Fan, Y., Jinshan, D. and Haunian, M. (1992). Anaerobic-Aerobic Treatment of a Dye Wastewater by Combination of RBC with Activated Sludge, *Wat.Sci.Tech.*, Vol. 26 (9-11), pp. 2093-2096.

Impacts of Wastewater Reuse in Agriculture and Aquaculture in Hanoi, Vietnam

D. T. Trang[1)], K. Mølbak[2)], A. Dalsgaard[3)], P. D. Cam[1)] and N. V. Hoa[4)]

1) Department of Bacteriology, National Institute of Hygiene and Epidemiology
1 Yersin Street, Hanoi 10000, Vietnam
e-mail: trangdt@cardvn.net; cam@fpt.vn
2) Department of Epidemiology, Statens Serum Institut
Artillerivej 5, 2300 Copenhagen S, Denmark
e-mail: krm@ssi.dk
3) Department of Veterinary Microbiology, Royal Veterinary and Agricultural University
Bülowsvej 13, 1870 Frederksberg C, Denmark
e-mail: ad@kvl.dk
4) Department of Clinical Microbiology, Hanoi Friendship Hospital
1 Tran Khanh Du Street, Hanoi 10000, Vietnam

Abstract

This study aimed to assess the risks associated with wastewater application in agriculture and aquaculture in a rural population in Hanoi, Vietnam. A cross-sectional study was conducted in October, 2002. Data were collected from a survey of 400 agricultural households, including information on the demographic, socioeconomic and sanitary conditions. Parasitological examination was conducted on single stool samples obtained from 807 individuals of the surveyed households, including 620 adults and 187 children under 5 years old. About 69% of the adults were engaged in wastewater-fed agricultural and aquacultural activities, and 60% of the household incomes derived from such activities. The prevalence of helminth infections, with *Ascaris,* hookworm, and *Trichuris*, was 46.5% in adults and 14.4% in children. The overall risk prevalence of helminth infections was highest in adult farmers with everyday exposure to wastewater (52%) and lowest in farmers with no contact with wastewater (22%). Female farmers had a higher risk of getting infected with helminths (RR = 1.46, 95% CI = 1.19-1.80). The application of wastewater has generated more incomes for the households, improving the family welfare and living conditions. Nevertheless, it poses some potential health risks to the users of wastewater and their families, in particular helminth infections. Adequate personal protective measures are highly recommended.

Keywords: benefits, health impacts, helminth infections, wastewater reuse

1. Introduction

Wastewater reuse has been a traditional practice in many countries around the world, especially in arid or semi-arid regions with scarce fresh water. The reuse of wastewater means the application of municipal and industrial wastewater for beneficial purposes, such as for crop irrigation and aquaculture. The most compelling reasons for wastewater reuse include job opportunities, increased crop yields, cheaper investment on additional nutrient sources for crops, and overall, improving incomes and livelihoods of poor people. In many low-income countries, wastewater reuse has also been seen as a low-cost wastewater treatment method.

Besides the undeniable benefits and efficiency that the reuse of wastewater has brought to agricultural and aquacultural production, possible negative impacts on human health of wastewater irrigation activities are of great concern to the public health sector. The majority of wastewater used in developing countries does not receive any conventional treatment before being applied directly to agricultural lands. Reused wastewater, especially raw and preliminarily treated wastewater, harbors several pathogens and toxic chemicals, which can cause diseases in humans when it is not managed and applied properly. Nematode infections are among the most common diseases contracted by the people working or living in the areas where wastewater is used, having constituted a major concern of health care in the community (Feachem *et al.,* 1983).

In Vietnam, the application of raw wastewater, including domestic sewage, industrial effluents and human excreta in agricultural and aquacultural production is widespread, especially in the northern Vietnam. In Hanoi, more than 400,000 cubic meters of wastewater is discharged daily, and a large amount of this wastewater is pumped or diverted for irrigation and aquacultural production. It is evidenced that wastewater application in agriculture, particularly in fish farming, has brought benefits to the sector as well as surplus profits to farmers from better harvest yields (Thanh, 1994; Hoan, 1996; Thang, 1996). However, little is known in Vietnam about the impact of wastewater on the health of farmers, especially those who are directly exposed to the wastewater.

This study is part of an ongoing research project entitled "*Sanitary Aspects of Drinking Water and Wastewater Reuse in Vietnam*", which includes activities on the assessment of health risks associated with wastewater reuse. The study's objectives were to identify and establish a cohort of eligible people for a one-year morbidity monitoring program as part of an epidemiological research, and to collect information about a farming population practicing wastewater irrigation, in terms of social, economic and health status. The assessment of health conditions was based on analyzing fecal samples obtained from the study population for helminth parasites.

2. Materials and Methods

Study area

The study was conducted in Yen So commune, a rural area in the southeast outskirts of Hanoi. The area has been known for a long time for its intensive use and reuse of wastewater in fish farming and irrigation of rice and vegetable cultivation, which are the major activities of the commune's agricultural economy. The main source of irrigation water is Kim Nguu river, which crosses the commune and is one of the main wastewater recipients, receiving almost one third of the city wastewater. The water quality of the river monitored by the Center of Environmental Engineering and Technology in Urban and Industrial Areas (CEETIA) showed a heavy pollution load carried by the river, with the values of COD and BOD of twofold to fivefold of the national standard for surface water. Toan (in preparation) also found a high concentration of heavy metals (lead, zinc, cadmium, mercury and arsenic) in the river water. In addition to the wastewater from the river, domestic wastewater from local households, and fresh or composted human excreta are frequently discharged to the fishponds and fields.

Study methods

A cross-sectional survey was conducted during October 2002. A total of 400 randomly selected agricultural households were included in this study. Local field workers were trained to carry out the survey. Data were collected by household interviews and parasitological examinations of single stool samples. In the interviews, each household was censused, and data were collected on the household's socioeconomic status, water supply, sanitation, and animal husbandry. In addition, preliminary information on the status and frequency of exposure to wastewater was also sought.

For the parasitological examination, stool specimens were obtained from people who met the inclusion criteria set out for the epidemiological research (i.e. adults from 15 to 70 years who were involved in agricultural work, and children less than 5 years old) and who were also willing to participate in the follow-up program. The fecal specimens were examined by microscopy for helminths, namely *Ascaris,* hookworm, and *Trichuris*. Upon the completion of the study, people identified with parasitic infections were provided with free treatment. Data were processed and analyzed using Microsoft Access and EPI INFO 2000, statistical software used in public health.

3. Results

The total censused population from the 400 households was 1723 people. A total of 1182 who met the inclusion criteria of the follow-up program were invited to participate in the study and asked to provide stool samples. The response rate was 68.3% (807 persons, including 620 adults aged 15 - 70 doing agricultural work and 187 children under 5 years old). There were more females in the cohort (62.9%), and the females made up 67.3% of the adult group.

A large proportion of the people in the working age were engaged in agricultural and aquacultural work (69%). This involvement was also reflected through the sources of households incomes, where the majority of households said that their incomes mainly came from agricultural activities, particularly terrestrial and aquatic vegetable farming (76.3%) and rice cultivation (73%). Less than 5% of the households had incomes from fish farming. This low figure could be due to the random selection of households. Additionally, fishponds were often contracted for farming by a limited number of households. The keeping of animals was rather widespread in the commune, with more than 66% of the households raising animals. The most common types of domestic animals were ducks, chickens and pigs.

All the surveyed houses were built from bricks, and the majority of them (79%) had a flat roof made from iron and cement, otherwise they were tiled roof. Few houses had earthen or cement floors, while more than 90% of the house's floors were paved with bricks or enameled tiles. Almost every family had a television and a bicycle, and about 50% of them possessed a motorbike. All households had access to a latrine, and 396 households had a latrine built in the house vicinity. The most common types of latrines were sanitary septic tank latrines (48.5%) and semi-septic tank latrines (43%). In comparison with other rural areas of the northern Vietnam, the traditional single or double vault composting latrines, in which human excreta was composted with ash, were less common in Yen So (less than 10% of the households). About 35% of the households composted human wastes (usually for a period of less than 6 months) which were then used as fertilizers for plants, and more than 40%

collected urine to water gardens and fields. Few households applied fresh human excreta on lands but discharged it directly to the fishponds.

All of the 807 people included in the follow-up program provided stool specimens for the parasitological tests. Overall, the prevalence of helminth infections (positive with at least one of the three helminths examined) in the cohort was 39% (315 persons). About 23.6% of the 315 persons were positive with two kinds of helminths, and 6% of them were positive with all of the three helminths. As shown in Table 1, the rates of parasitic infections were much higher in the adult group than in the children. There was a significant difference in the infections between the adult and the children groups, and the adults were likely to have a higher risk of being infected with helminths, especially hookworm, than the small children.

Table 1. Prevalence of helminth infections in Yen So commune, 2002

Helminths	Adult (*n = 620*)	Children (*n = 187*)	Overall (*n = 807*)	Relative risk (95% CI)
Overall	288 (46.5%)	27 (14.4%)	315 (39%)	**3.22** (2.25 - 4.61)
Ascaris	149 (24%)	24 (12.8%)	173 (21.4%)	**1.87** (1.26 - 2.79)
Trichuris	76 (12.3%)	3 (1.6%)	79 (9.8%)	**7.64** (2.44 - 23.94)
Hookworm	174 (28.1%)	2 (1.1%)	176 (21.8%)	**26.24** (6.57 - 104.75)

* CI - confidence interval

Figure 1 illustrates the prevalence of helminth infections among adults with different frequencies of exposure to wastewater. In general, the prevalence of parasitic infections was high in the first three groups, indicating that farmers doing agricultural work with exposure to wastewater had a higher risk of helminth infections. The risk was 1.46 times higher (95% CI = 1.02 - 2.09) in the people with contact to wastewater than those who did not contact with wastewater. There was a slight association between daily contact to wastewater and parasitic infections (RR = 1.24, 95% CI = 1.05 - 1.47).

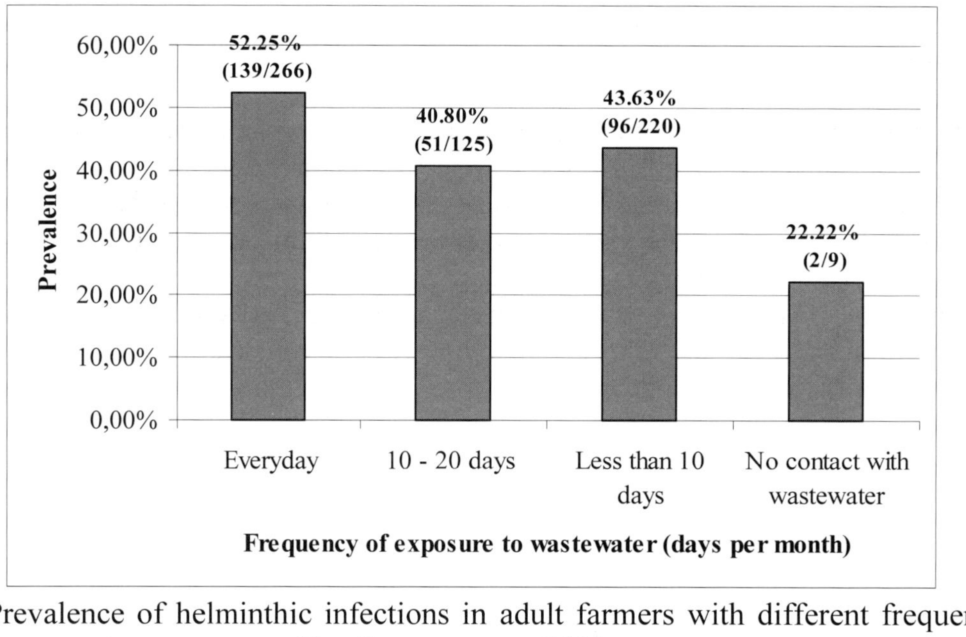

Figure 1. Prevalence of helminthic infections in adult farmers with different frequencies of exposure to wastewater, Yen So commune, 2002

No associations existed between parasitic infections and factors such as socioeconomic status, presence of animals in households or the existence of a sanitary latrine, but between the infections and gender. The prevalence of helminth infections was higher in female adults, and this group also had a higher risk of getting infected than the male group (RR = 1.46, 95% CI = 1.19 - 1.80). This was also consistent when the infection rates were compared between the female and male groups by the breakdown of single helminth (Figure 2).

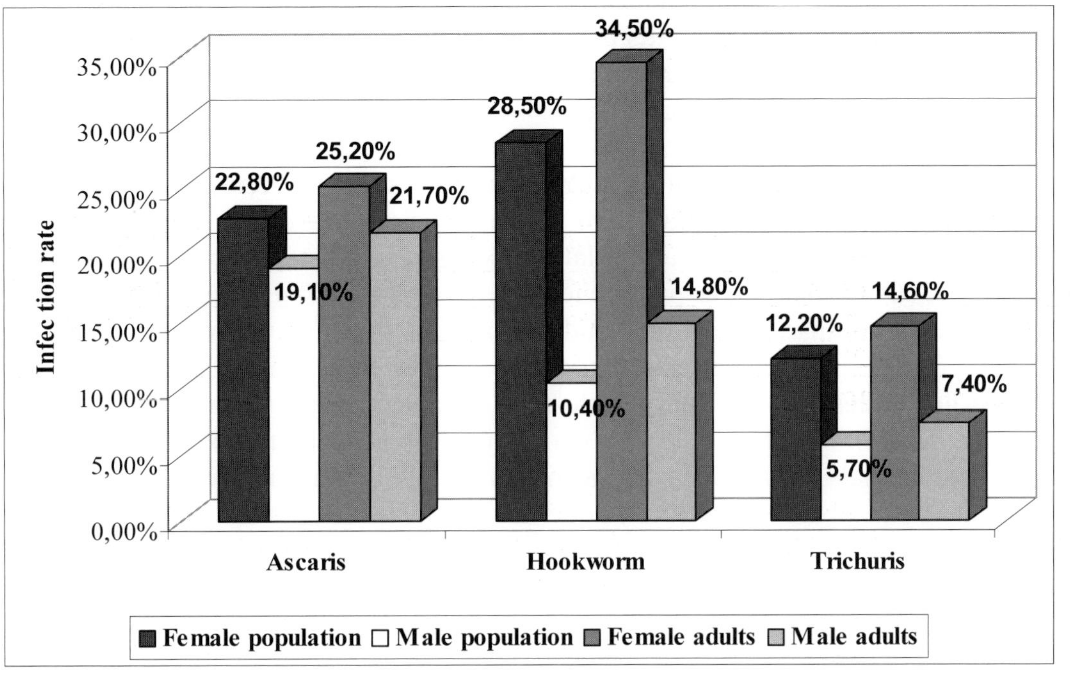

Figure 2. Prevalence of helminthic infections, stratified by sex and individual helminth, Yen So commune, 2002

4. Discussion

Results from the parasitological examination showed that irrigation with untreated wastewater posed a risk of gastrointestinal parasitic infections. The microbiological quality of the river supplying water for the communal irrigation and fishponds contained high concentrations of helminth eggs and fecal bacteria. The levels of fecal coliforms in the river ranged from 1.2×10^7 to 11.8×10^7 CFU/100ml, while the concentration of nematode eggs was of about 240-2840 eggs/1 liter (Toan, in preparation). These values clearly exceeded the WHO guidelines for wastewater reuse in agriculture and aquaculture, which propose that less than or equal to 1000 fecal coliform bacteria per 100ml and 1 human intestinal nematode egg per liter should present in the wastewater intended for use in agriculture (Mara and Cairncross, 1989).

The high risks for helminth infection observed in adult people, especially with hookworm, were obvious since parasitic infections were often age-dependent. It was also to some extent associated with the fact that most of the adults under examination (69%) were farmers and nearly everyone of them (611/620) had contact with wastewater. Direct contact with wastewater could introduce a higher risk of getting parasitic infections in adults than in children. In our ongoing case-control study, about 70% of the interviewed people claim that they often do the fieldwork manually and barefoot in wastewater irrigated fields, which is a risk factor for hookworm infection, since larvae may penetrate the skin of farmers. In addition, the application of fresh and partially composted human and animal excreta as well as the habit of eating raw vegetables irrigated with wastewater, such as morning glory, are more likely to make adults get helminth infections.

Gender is also a likely risk factor of parasitic infection. Female farmers have a higher risk of infection because they are more heavily involved in the fieldwork than the males. Most of the daily work in the fields (weeding, manuring, watering, etc.) are done by the females, while the males mainly provide significant inputs during the sowing or harvest of crops. In general, the men in the commune often have additional jobs, such as carpenters, builders, repairmen or salesmen, to increase their family incomes.

The prevalences of parasitic infections in Yen So are slightly different to what were found elsewhere where wastewater irrigation are practiced (Feenstra *et al.,* 2000; Habbari *et al.,* 2000; Blumenthal *et al.,* 2001). In a population exposed to wastewater irrigation in Pakistan, Feenstra *et al.* (2000) found lower prevalences of *Ascaris* and *Trichuris* infections (5.3% and 0.8% respectively) compared with our study. However the hookworm infection was rather high (38.6%). These differences could be attributed to the differences in study inclusion criteria, such as gender and age (only male farmers and children below the age of 12 were included for analysis). The rate of *Ascaris* infection found in adults beyond 15 years old in Mexico by Blumenthal *et al.* (2001) was also lower (4.5%) than in our study. In comparison with the results by Habbari *et al.* (2000) done in school children of 7-14 years old exposed to wastewater reuse, we reached a similar finding on the overall prevalence of *Ascaris* infection (21.4% compared with 20.5%). However when age difference was taken into account, our prevalence of trichuriasis was higher, while the rate of *Ascaris* carriers was lower.

Other investigations also discussed other potential health risks contracted by the farmers who irrigated their lands with wastewater, such as diarrheal diseases, nail problems, and trachoma (Thanh, 1994; Thang, 1996; and Feenstra *et al.,* 2000). In addition, gynecological diseases were of high prevalence in female farmers. In this cross-sectional study, we did not assess these diseases, however in our ongoing morbidity monitoring program, they are under surveillance. It is expected that after one year of follow-up, incidence density rates for these diseases would be obtained, and risk factors of the diseases could be identified for people who are exposed to the reuse of wastewater.

Besides the negative impacts of wastewater reuse, figures on the socioeconomic profiles of Yen So showed that the commune gained many benefits from the use of wastewater. All the major agricultural activities of the commune, such as rice, vegetable and fish farming, were largely dependent on the availability of wastewater from the Kim Nguu river. These activities constituted main incomes of the local farmers (more than 60% of local incomes), which also generated the major sources for the purchases of household expensive items such as motorbikes and televisions, as well as house building and renovation. Compared to other rural areas of Vietnam where thatched or mud houses are still present, the house conditions in Yen So have been much improved with brick buildings and paved floors. Access to latrines have been easier, and more families have upgraded their latrines to a sanitary latrine with septic tank. The Unicef wells also became more popular in the commune, replacing the traditional shallow dug wells.

The application of wastewater in agriculture also resulted in increased productivity. As suggested by Hoan (1996) and Thang (1996), the annual yield of fish raised in wastewater was more than 3 tons/ha, while the yield of fish without wastewater application was only 500-800 kg/ha. In Yen So, the yields are different between wastewater-fed fishponds located in the upstream and downstream of Kim Nguu river. Experience from local farmers shows that the upstream fish yield are twice higher than the downstream yield, where the wastewater is less strong and nutrient-rich. The yield of rice grown in wastewater almost doubles that grown in natural water. At the same time, farmers can save the cost of fertilizers and additional foods for plants and fish because they can rely on the nutrient supply from the wastewater.

The efficiency gained through the reuse of wastewater in Yen So can also be seen in the integration of fish farming and the rearing of domestic animals such as pigs, chickens and ducks. In the commune, the application of the garden-fish farming-livestock breeding model (the VAC model) is rather widespread, especially in those households with proximity to the fish ponds supplied with wastewater. Water from the ponds is often used for irrigation of the households' gardens as well as surrounding rice and vegetable fields. The rearing of pigs, chicken and ducks has contributed to the increase of fish productivity because of surplus feeds and manure used for aquaculture, at the same time, sub-standard fish could be supplemented as extra foods for the animals.

5. Conclusions and Recommendations

The application of wastewater in agriculture and aquaculture has created many economic benefits for farmers. However, the use of wastewater without any proper treatment poses potential health risks to the farmers and their children. The local farmers do not recognize such potential health risks and few if any protective measures have been introduced.

At present, little is done in Vietnam to enforce the existing laws and guidelines regarding the reuse of wastewater in agriculture and aquaculture. Especially there is a lack of regulations addressing the health impacts of such reuse to the users. Although Vietnamese environmental standards specify the quality requirements for discharged wastewater to be used for agricultural purposes, there are no standards on the microbiological quality, except for the numbers of total coliforms. So far, the WHO guidelines on the allowable limits of nematode eggs and fecal coliforms in wastewater for irrigation have not been incorporated into the national standards. These limits seem difficult to be obtained in Vietnam, where wastewater producers have encountered problems in treating their wastewater to comply with such regulations, and physiochemical parameters of the water often receive more attention than the microbiological ones. While appropriate treatment of the wastewater is not yet attained, local authorities of the areas where untreated wastewater has been applied should encourage farmers to wear protective measures such as shoes, boots, and gloves when working in the fields irrigated with wastewater. Awareness on the health impacts of wastewater and improvement of hygiene behaviors should be raised in the community so that farmers can protect themselves. Local governments can also partially support the farmers with periodical health check and treatment of the diseases possibly associated with the reuse of wastewater.

6. Acknowledgement

The Danish International Development Agency (DANIDA) is thanked for the financial support provided to the research capacity building project entitled "*Sanitary Aspects of Drinking Water and Wastewater Reuse in Vietnam*", in which this study was conducted. We wish to thank the staff at the Division of Enteric Infections (Department of Bacteriology, National Institute of Hygiene and Epidemiology), and the Hanoi Friendship Hospital for their hard and dedicated work. Special thanks are also given to the field workers and the People's Committee of Yen So commune for their cooperation and support.

7. References

Blumenthal, U.J., Cifuentes, E., Bennett, S., Quigley M. and Ruiz-Palacios, G. (2001). The risk of enteric infections associated with wastewater reuse: the effect of season and degree of storage of wastewater. *Transactions of the Royal Society of Tropical Medicine and Hygiene*, Vol. 995, pp. 131-137.

CEETIA (2002). *Monitoring data of Kim Nguu River - Hanoi* (unpublished).

Feacham, R.G., Bradley, D.J, Garelick, H. and Mara, D.D (1983). *Sanitation and Disease: Health Aspects of Excreta and Wastewater Management*, John Wiley & Sons Editions, New Yorks.

Feenstra, S., Hussain, R. and van der Hoek, W. (2000). *Health Risks of Irrigation with Untreated Urban Wastewater in the Southern Punjab, Pakistan* (Report No. 107), Institute of Public Health, Lahore and International Water Management Institute, Pakistan Program, Lahore.

Habbari, K., Tifnouti, A., Bitton, G. and Mandil, A. (2000). Geohelminthic infections associated with raw wastewater reuse for agricultural purposes in Beni-Mellal, Morocco. *Parasitology International.* **48** (3), 249-254.

Hoan, V.Q. (1996). *Wastewater Reuse through Aquaculture in Hanoi: Status and Prospects.* MSc. thesis. Asian Institute of Technology, Bangkok.

Mara, D. and Cairncross, S. (1989). *Guidelines for the Safe Use of Wastewater and Excreta in Agriculture and Aquaculture.* World Health Organization, Geneva.

Thang, V.Q. (1996). *Effects of Sewage Utilization on Fish Farming and Irrigation (Vietnam)*, Final Technical Report prepared for the International Development Research Center (IDRC), Center of Resources and Environmental Studies, Hanoi.

Thanh, T.T. (1994). Health impacts of sewage use in livestock breeding, *Proceedings of workshop on Gender, Environment and Development in Vietnam*, Hanoi, 9-11 November.

Toan, T.Q (in preparation). *Efficiency of Wastewater-fed Fishponds as Low-cost Wastewater Treatment Systems*, MSc. Thesis, "*Sanitary Aspects of Drinking Water and Wastewater Reuse in Vietnam*" project.

Water Reused in Dairy Milk Factories by Clean Technology Approaches

W. Wirojanagud [1)] and N. Tantemsapya[2)]

1) Environmental Engineering Department, Khon Kaen University
Khon Kaen, 40002, Thailand
e-mail: wanpen@kku.ac.th
2) Environmental Engineering Department, Khon Kaen University
Khon Kaen, 40002, Thailand
e-mail: netnapid@kku.ac.th

Abstract

Pollution prevention was studied by applying cleaner technology on dairy mills in the Northeast of Thailand, with the aim to gain maximum productivity and minimize wastes. This research conducted details study of various elements within the dairy milk factory such as the raw material input, production process, and outputs, including the storage and handling of materials within the dairy mill. Moreover, measures were taken in conjunction with economic evaluation and proposals were made to minimize loss and environmental problems. From the study, water loss occurs in the homogenize units of the dairy mills and the cassava-washing process. The proposed measures include the reuse of the cooling water after from the homogenize unit and reuse of cooling water from the compressor with a capital investment of only 4,200 bahts per unit for a return of 5,460 bahts per month.

Keywords: clean technology, dairy mill, water reuse

1. Introduction

Industrial development is one of the major factors of economic growth. Currently, there are a total of 125,000 factories in Thailand. Approximately, 80 % of the total factories are small and medium enterprises (SMEs). SMEs play an important role to the economic status of the country as they generated about 50% of the Gross Domestic Product and stand for about 35% of Thailand's export. While SMEs are very important to economic growth, they unfortunately generate environmental pollution. When compared to large industries, small and medium industries might not generate a large quantity of wastes but instead generated a higher rate of waste quantity per unit of product. This is due to inappropriate and inefficient management of the environmental, technologies, capitals and operation cost of treatment as well as information updating and networks. The Ninth National Socio-Economic Development Plan (2002–2006) underscores the need to upgrade Thailand's international competitiveness where one of the goals is to increase productivity through more effective use of natural resources and through better environmental management in conjunction with the application of appropriate technologies.

Cleaner Technology (CT) or Green Productivity (GP) is the paradigm of sustainable concept by developing a manufacturing production coupled with environmental protection as well as socio-economic consideration. CT applies both technology and management to maximize production while minimizing waste generation, leading to lower operational and production costs. Currently, the Government sector and the Non-Government Organizations have put efforts into intensively promoting CT or GP to industries to make them shift from the end of pipe treatment towards pollution prevention and waste minimization.

The project on pollution prevention of agro-industry in the northeast of Thailand was conducted in eights factories aiming to assess the use of CT implemented in industry. Of which the project's specific objectives are: 1) To minimize wastes generation and/or prevent pollution, 2) To compare the outcome before and after implementing CT and 3)To comply with the ISO 14000. In this paper, the sucess of reducing water usage in a Dairy Mill factory by using CT approaches is presented.

2. Materials and Methods

This research is an industry-university collaboration project, which consisted of 4 main steps: preparation, problem analysis with recommendation of improvement measures, implementation and evaluation, and finally information-dissemination.

The research started with organizing a research team where members are experts from various fields of technology and governmental representatives in the environmental field. In the first step, a seminar was organized to teach the fundamentals of CT to owners of the selected factories, followed by a visit to each factory to make an initial assessment of the existing problems, production processes, the use of resources in production, and environmental pollution in order to prepare a detailed assessment for each factory.

In the problem-analysis step, information on the production process, production capacity, use of raw materials and energy in production, waste generation and others relevant data and processes were collected. The collected data were analyzed to identify problems and their causes, as well as to prepare the mass-balance calculations.

The problem analysis was followed by a detailed study of the possibilities for improvements in each factory. This included an assessment of the economic implications of the proposed measures, and the results were presented at a meeting between the research team and the factories' owners. Based on the researchers' lists and assessment of alternatives, the factory owners were able to select the measures that they felt could be best implemented in practice. During implementation, a follow-up evaluation was performed to monitor the progress. The situation before and after the implementation of CT was compared and then the success rate of the use of CT was evaluated.

In the final step of information dissemination, a seminar was organized for the owners of the factories, the academics players and the involved governmental officers to elaborate on the use of CT in terms of pollution prevention and economic benefit.

3. Results and Discussion

A Dairy Mill Factory located in Northeastern Thailand was studied using the method described above. The factory processes pasteurized milk with the production capacity of 3,600 tons/year.
The process of dairy mill production is shown in Figure 1 and can be described as follows:

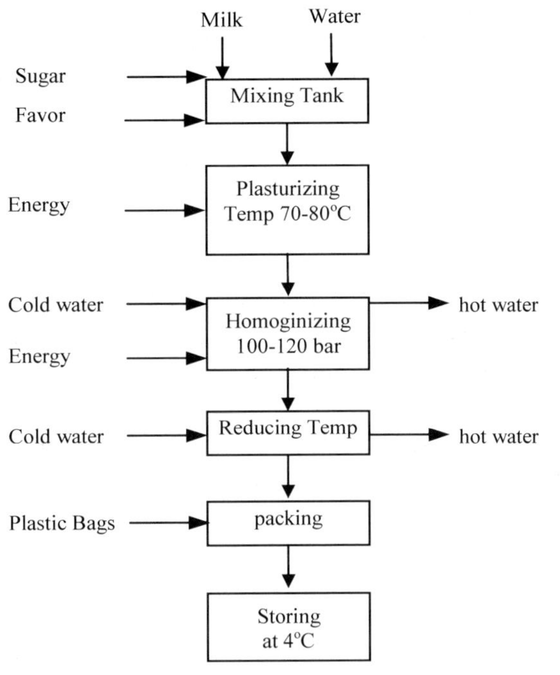

Figure 1. Dairy Mill Production Process

Milk is kept in a temperature control room where the temperature is set below 4^0C. The first process is where the milk is mixed with water, sugar and favor in a mixer tank before pasteurizing in order to prohibit growth of microorganism. Using a heat exchanger operating at temperature of 70-80^0C performs the milk pasteurized. Then it is transferred to the homogenized unit where the pressure is increased to 110-120 bars. Before going to the packaging and storing units, the milk temperature is reduced to 4^0C by the heat exchanger.

Based on a mass balance of the entire process especially utility usage, the environmental and production process related problems were identified for each step in the production process focusing on water usage. Aspects and CT alternatives for these steps are presented below.

Source of water for the dairy milk factory is mainly from the irrigation channal which is treated by coagulation, filtration and softening before going in to the process. Besides, deep well water is pumped to be used for cleaning. Water used daily is approximately 25-30 m3/day. The following chart represents the mass balance of the water used in the process.

Wastewater is produced during the above process from cooling the heat exchanger and machine and from floor washing which are approximately 10- 15 m3/day. Wastewater is treated by four septic tanks before going into the final filtration tank with carbon and sand and then discharged to community drainage system. The ratio of wastewater produced from different activities is shown as the below chart.

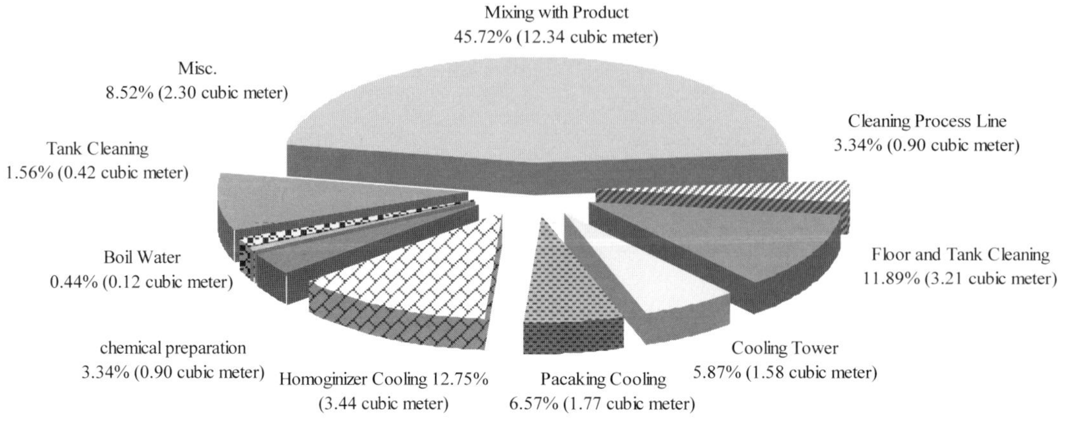

Figure 2. Dairy Mill Water Usage

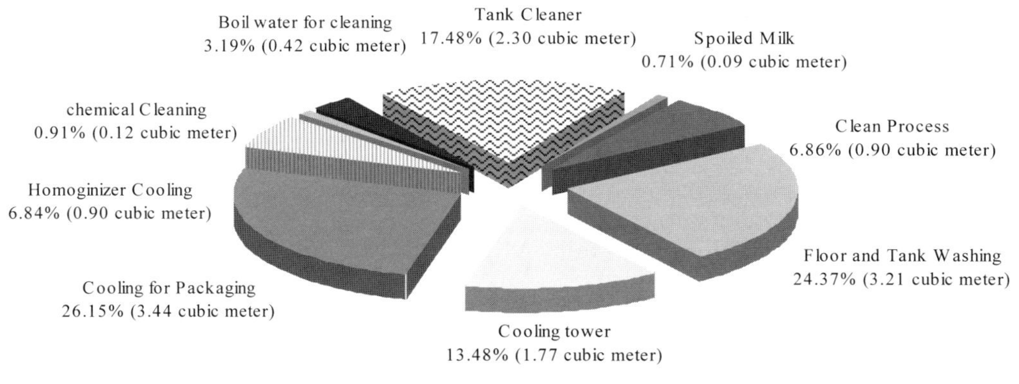

Figure 3. Dairy Mill Wastewater

The problem-analysis step was conducted to identify problems and their causes, as well as to prepare the mass-balance calculations. Problems relating to the water usage and waste waster is as the following

1) Homogenized machine cooling system: Homogenized machine pistons' temperature will rise during the operation. Therefore the cooling water for the pistons is needed to cool the system. In the Dairy Mill 1, water from water supply was used with high feeding rate. In order to reduce the water loss, a lower water-feeding rate is required. One CT alternative is to reduce the pump feed rate to be just enough to keep the pistons cool. Applying this option, the water loss will be reduced by 1 m3/day. Besides, the quality of water needed to cool down the piston is not strict. Another CT option is to use the cool water from the packaging in place of the clean water from the water supply system. This option will reduce the water loss 3.44 m3/day.

2) Reuse of condensed water: Approximately, 3.44 m3 per day of cooling water was discharged to a wastewater drain. The quality of the condensed water was checked and it was found that it is an acceptable quality. CT alternative comprises construction of a collection tank and pump to the homogenized machine cooling system. This option saves water and also reduces wastewater.

3) Re-circulate use of cooling water in cold water stream: it was found that the compressor that produces cold water to the storage room cannot lower the temperature to the required level. Thus, additional ice is purchased for the storage room everyday. The process was checked and found that it is possible to re-circulate the cold water in the system. CT alternative comprises construction of a collection tank and pump to the compressor cooling system. This option will saves a large amount of water, and also reduces energy usage. This would help reduce the cost of ice purchase by about 120 bath per day.

Table 1. The Economic Evaluation of CT Options

CT options	Investment (bath)	Pay back period (days)	Benefit (bath/month)	Equipment life span (year)
Reuse of cooling water	200	4	1,890	5
Reuse of cold water	4,000	45	3,600	4

It was found that water usage could be reduced from 2.59 m3 per ton to 1.74 m3 per ton (or by 33%). Which can make additional profit to the company estimated to be just about 5,490 baht per month.

4. Conclusions

CT alternatives proposed to the dairy milk industry could be divided into 3 categories, no-investment cost, low medium, and high investment cost. Obviously, the factories would prefer to implement the no-investment cost alternatives, followed by either low or medium investment practically. On the other hand, the high investment cost has to be carefully considered. However, the outcomes of the implemented CT alternatives in terms of economic benefit would convince them for further investments in CT alternatives. CT would therefore be beneficial not only in terms of economy, but also in terms of environmental health.

CT is the analytical tool for determining appropriate measures to maximize productivity and minimize waste generation. It is likely a technological measure to solve the problems of environmental pollution, prevent pollution and guide for continual improvement. The CT approach can lead towards environmental performance evaluation (ISO 14030) and environmental management systems (ISO 14001).

The research also shows that relative simple but systematic analysis can lead to substantial environmental improvements and more efficient use of resources, all through applications of well known low technology at low investment costs, which actually over a short period of time generates profit for the factories.

This is an important lesson not only for the industrial sector, but certainly also for the higher education system in order to prepare all relevant engineers and scientists for the challenge.

5. Acknowledgement

The author wishes to thank the Thailand Research Fund for financial and academic support to the research project. All successes are given to the research team of the Industry-University Collaboration Research Project on Green Productivity of Agro-Industry in Northeast, Thailand. Without our productive work, this project could not have been completed.

6. References

Allen, D. T. and Rosselot, K. S. (1997). *Pollution Prevention for Chemical Process*, John Wiky & Sons, inc.

Mycock, J. C., McKenna, J. D. and Theodore, L. *Handbook of Air Pollution Control Engineering and Technology.*

Theodore, L. and McGuinn, Y. C. (1992). *Pollution Prevention*, Van Nostrand Reinhold.

Bishop, P. L. (2000). *Pollution Prevention: Fundamentals and Practice*, McGraw-Hill.

Khon Kaen University. (2001). *Final Report of Waste Minimization of Agro-Industry in Northeast Thailand, submitted Thailand Research Fund.*

Use of Wastewater from Intensive Hybrid Catfish (*Clarias macrocephalus x C. gariepnus*) Pond Culture as Fertilizer for Rice Crop

Y. Yi[1)], L. M. Lan[2)] and C. K. Lin[3)]

1) Aquaculture and Aquatic Resources Management, School of Environment, Resources and Development
Asian Institute of Technology, P.O Box 4, Klong Luang, Pathumthani 12120, Thailand
e-mail: yangyi@ait.ac.th
2) College of Aquaculture and Fisheries, Can Tho University, Can Tho, Vietnam
e-mail: lmlan@ctu.edu.vn
3) Aquaculture and Aquatic Resources Management, School of Environment, Resources and Development
Asian Institute of Technology, P.O Box 4, Klong Luang, Pathumthani 12120, Thailand
e-mail: lin@ait.ac.th

Abstract

This study was aimed at investigating the feasibility of using wastewater from hybrid catfish *(Clarias macrocephalus x C. gariepnus)* pond culture as fertilizer inputs for rice crop. There were four treatments in triplicate: (A) rice plots were irrigated with canal water without fertilizer input; (B) rice plots were irrigated with canal water and inorganic fertilizer inputs (nitrogen, phosphorus and potassium); (C) rice plots were irrigated with water from the catfish pond as the only nutrient input; (D) irrigated with water from the catfish pond supplemented with phosphorus and potassium fertilizers.
Grain yields in treatments B, C, and D, ranging from 2.86 to 3.08 t/ha, were not significantly different ($P>0.05$). The highest filled grain yield was achieved in treatment D, intermediate in treatments C and D, and lowest in treatment A ($P<0.05$), while N and P recovery was higher in treatment C than that in treatments B and D ($P<0.05$). This experiment indicates that wastewater from hybrid catfish ponds can be reused for rice crop to reduce environmental impacts of intensive aquaculture effluents and to lower fertilizer costs for rice cultivation.

Keywords: aquaculture effluents, rice-fish integrated culture, wastewater reuse

1. Introduction

Scientific interests in the impact of nutrient pollution from aquaculture have increased in recent years. Many treatments have been used for aquaculture effluents in many countries (Metcafl and Eddy, 1991; Kristiansen and Cripps, 1996). However, setting up this kind of treatment systems is not suitable for most fish or shrimp farms because of the high cost of facilities. Wastewater from fish culture has been reused for agricultural activities, and integrated aquaculture-agriculture farming systems have led to technical and community sustainability and environmental enhancement (Huat and Tan, 1980; Edward, 1993).

Intensive catfish culture has become a serious environmental problem. Uneaten feeds and fish feces are major sources of suspended solids and inorganic nutrients which has led to the increase in biochemical oxygen demand (BOD) and eutrophication in receiving waters (Muir, 1982). Concentration of nutrients in wastes from intensive catfish culture is high (Veerina, 1989), and more than 64% of total N and 77% of total P from feed input were released to water (Udomkam, 1989). In order to obtain environmentally sustainable aquaculture development, nutrient-rich wastes from intensive catfish culture have been reused as a fertilizer input to semi-intensive tilapia culture (Udomkarn, 1989; Veerina, 1989; Lin *et al.,* 1990; Lin and Diana, 1995; Yi, *et al*., 2003).

The purpose of this study was to investigate the feasibility of using wastewater from hybrid catfish *(Clarias macrocephalus x C. gariepnus)* pond culture as fertilizer inputs for rice crop.

2. Materials and Methods

An earthen pond of 400 m^2 in surface area at Can Tho University, Vietnam, was used to culture hybrid catfish. A total of 3,600 hybrid catfish fingerlings with mean weight of 16.7±4.0 g/fish were stocked in the pond, and fed with homemade feeds, depending on the availability of by-products from a fish processing plant.

The experiment was conducted in a completed randomized block design in 1,450-m^2 rice field, which was divided into 12 plots of 50-m^2 (6.25 m x 8 m) each by using soil to compact dykes (0.2 m in depth and 0.2 m in width). Each plot had 2 small canals (6.25 m x 0.2m x 0.1m and 8m x 0.2m x 0.1 m). There were four treatments in triplicate: (A) rice cultivation without fertilizer input; (B) rice cultivation with inorganic fertilizer input (urea, super phosphate and potassium chloride); (C) rice cultivation using wastewater from the hybrid catfish pond as only nutrient input; (D) rice cultivation using wastewater from the hybrid catfish pond as nutrient input supplemented with super phosphate and potassium chloride fertilizers.

A high yielding rice variety OM 1490 was used in this experiment. Seeds were sowed directly at 1 kg/plot (20 kg/1,000 m^2). Rice seeds were soaked for 24 hours, husked by washing and incubating for 48 hours prior sowing. Rice plots were supplied with water from different sources starting from 5 days after direct sowing. Water from the hybrid catfish pond was pumped by an electric pump to 6 plots of treatments C and D, while treatments A and B received water supply from a irrigation canal. The level of water in rice plots was dependent on the growth stages of rice. Water supply in each plot was adjusted through the flow speed and depth gauge in rice plot, thus the rice plots could receive the same volume of water. Urea (46% N), super phosphate (16% P_2O_5) and potassium chloride (60% K_2O) were applied to the rice plots in treatments B and D for three times on day 9, 18 and 34 after sowing. Total amount of fertilizers applied during the cultivation period was 1,013 g urea, 663 g super phosphate and 497 g potassium chloride in the treatment B, and 663 g super phosphate and 497 g potassium chloride in the treatment D.

Dissolved oxygen (DO), pH and temperature of source water were measured biweekly in situ. Source water for rice plots were sampled during every irrigation time for analyses of total ammonium nitrogen (TAN), nitrite-N, nitrate-N, total Kjeldahl nitrogen (TKN), total nitrogen (TN), soluble reactive phosphorus (SRP), total phosphorus (TP), total suspended solids (TSS) and total volatile solids (TVS) (APHA, 1985).

At the beginning and the end of the experiment, fishpond sediments and soils at top 5 cm of rice plots were collected at different points and homogenized for measuring pH, soil moisture, bulk density, TP and TN. Moisture, TN and TP were analyzed for rice seeds and fish carcass at the beginning, and for rice grains, rice straw and fish carcass at the end of experiment.

Every batch of homemade feed were also analyzed for moisture, TN and TP. Feed conversion ratio (FCR), percentage nutrient recovery and nutrient distribution for fish culture were also determined. Grain yield, straw weight, percentage of filled grain, harvest index, grain-straw ratio, N and P uptake, apparent nutrient recovery were determined for rice cultivation.

Data were analyzed statistically by analysis of variance (ANOVA) (Steele and Torrie, 1980) using SPSS (version 7.0) statistical software package (SPSS Inc., Chicago, USA). Differences were considered significant at an alpha of 0.05.

3. Results and discussion

Total production of hybrid catfish was 368 kg/pond with mean individual weight of 105 g/fish and mean daily weight gain of 1.05 g during the 105-day culture period. Survival rate was high (97.1%). Total feed input (dry weight basis) was 980.5 kg, giving a FCR of 3.18.

Water quality of outflow water from the hybrid catfish pond to rice plots fluctuated during the fish grow-out period. Water temperature and pH ranged from 27.3-30.5°C and 6.90 - 7.25, respectively. DO decreased from the initial level of 8.96 mg/l to the final level of 2.40 mg/l. Concentrations of nitrite-N, nitrate-N, TAN, and SRP were quite stable, ranging from 0.003-0.068 mg/l, 0.087-0.261 mg/l, 0.022-0.560 mg/l, and 0.025-0.253 mg/l, respectively, while TKN, TN and TP concentrations increased towards the end of the culture period, from 1.282 to 7.142 mg/l, 1.531 to 7.395 mg/l, and 0.044 to 1.680 mg/l, respectively. TSS and TVS fluctuated between 74-193 mg/l and 22-82 mg/l, respectively, during the entire experimental period.

Nutrient budget showed the sources of nutrient inputs in the hybrid catfish pond fingerlings, feed, irrigation water and rainfall. The dominant source of nutrient inputs was feed, which accounted for 95.0% TN and 97.0% TP inputs. Most of the inputted nutrients (42.6% TN and 72.5% TP) accumulated in the pond sediment. Only 14.5% TN and 10.1% TP inputs from feed were harvested by hybrid catfish, while 35.5% TN and 10.4% TP were unaccounted for.

For normal rice cultivation, the major nutrients supplied to rice are from inorganic and organic fertilizers that are widely practiced by rice farmers worldwide to obtain high crop yields (De Datta, 1989). Grain yields (2.86-3.08 t/ha) in treatments with nutrient inputs (B, C and D) were similar to the average yields (3.05 t/ha) of this rice variety obtained by normal practice in that area. There was no significant difference ($P>0.05$) in grain yields among treatments B, C and D, which was significantly higher than that in treatment without nutrient inputs (A) ($P<0.05$, Table 1). The harvested grain included unfilled and filled grain. The filled grain yield in treatment D was the highest, intermediate in treatments B and C, and lowest in treatment A ($P<0.05$). However, there were no significant differences in the percentage of filled grain among all treatments ($P>0.05$). The highest straw weight was obtained in treatment B, which was not significantly different from that in treatment D ($P>0.05$), but was significantly higher than that in treatments A and C ($P<0.05$). Straw weight in treatment A was lowest, but was not significantly different from that in treatments C and D ($P>0.05$). There were no significant differences in mean harvest index and grain-straw ratio among all treatments ($P>0.05$).

Table 1. Grain and straw yields, harvest index and grain-straw ratio in the four treatments

Parameters	Treatments A	B	C	D
Grain yield				
Wet weight (kg/plot)	10.53 ±2.62 ^{b}	14.78 ±1.36 ^{a}	14.32 ±1.11 ^{a}	15.40 ±2.08 ^{a}
Dry weight (kg/plot)	8.96 ± 2.22 ^{b}	12.58 ± 1.18 ^{a}	12.22 ± 0.97 ^{a}	13.18 ± 1.82 ^{a}
Extrapolated wet weight (t/ha)	2.11±0.52^{b}	2.96± 0.27 ^{a}	2.86± 0.22 ^{a}	3.08± 0.42 ^{a}
Filled grain yield				
Wet weight (kg/plot)	7.87 ± 2.20^{c}	8.98 ± 2.83 ^{bc}	10.87 ± 1.86 ^{ab}	12.24 ± 1.84 ^{a}
Total grain yield (%)	74.39 ± 2.51	59.98 ± 14.81	75.61 ± 7.99	79.42 ± 5.18
Straw weight				
Wet weight (kg/plot)	22.83 ± 2.83 ^{b}	35.92±5.03 ^{a}	26.82±5.35 ^{b}	28.18 ± 4.01 ^{ab}
Dry weight (kg/plot)	14.85 ±1.66 ^{b}	23.81 ± 3.88 ^{a}	17.43 ±3.18 ^{b}	18.88 ±3.28 ^{ab}
Extrapolated wet weight (t/ha)	4.57 ± 0.57 ^{b}	7.81 ± 1.01 ^{a}	5.36 ± 1.07 ^{b}	5.64 ± 0.8 ^{ab}
Grain-straw ratio	0.46 ± 0.07	0.43 ± 0.09	0.55 ± 0.12	0.55 ± 0.06
Harvest index	0.31 ±0.03	0.29 ±0.04	0.35 ±0.05	0.35 ±0.02

Mean values with different superscripts in the same row were significantly different (P<0.05).

Nutrient budgets in the rice plots during the 95-day cultivation were summarized in Table 2. In treatment B, inorganic fertilizer was an important source of nutrient, whereas wastewater was an important nutrient source in treatments C and D. Wastewater from the hybrid catfish pond for treatments C and D contained much higher nutrients than that from the irrigation canal for treatments A and B. Nitrogen uptake by grains was highest in treatment B, intermediate in treatments C and D, and lowest in treatment A (P<0.05). Phosphorus uptake by grains was significantly lower in treatment A than that in treatments B, C, and D (P<0.05), among which there was no significant difference (P>0.05). Nitrogen uptake by rice (total of grain and straw) was not significantly different among all treatments (P>0.05), while phosphorus uptake by rice in treatment A was significantly lower than that in other three treatments (P<0.05), among which there were no significant differences (P>0.05). TN and TP contents in soil were lower at the end than at the beginning of the experiment in all treatments, and the nutrients released from soil accounted for large portions of total nutrient inputs in all treatments. There was no significant difference both TN and TP contained in soils at initial and harvest among all treatments (P>0.05). Unaccounted loss of nutrients was significantly higher in treatment B than that in other treatments (P<0.05).

Nutrients supplied to the rice plots of treatments C and D from the wastewater of the hybrid catfish pond were 1,106 g N and 182 g P, accounting for 32% N and 24% P in the pond water. The apparent nitrogen recovery by grains from inputed nitrogen was not significantly different among treatments B, C and D (P>0.05, Table 3). Apparent phosphorus recovery in treatment C was highest, which was significantly higher than that in treatments B and D (P<0.05, Table 3).

Table 2. Inputs and outputs of TN and TP in the rice fields during the 95-day cultured period

Items	TN (g/plot)				TP (g/plot)			
	Treatment A	Treatment B	Treatment C	Treatment D	Treatment A	Treatment B	Treatment C	Treatment D
Input								
Fertilizers	0	466	0	0	0	106	0	106
Water	48	48	184	184	14	14	30	30
Rainfall	12	12	12	12	6	6	6	6
Rice seeds	10	10	10	10	6	6	6	6
Soil	12,624±926	11,429±443	11,580±728	11,838±766	2,066±139	1,942±65	2,010±220	2,139±91
Total	12,694±926	11,965±423	11,787±728	12,044±766	2,091±139	2,074±65	2,052±220	2,287±91
Output								
Grain	90±23 [b]	171±22 [a]	140±30[ab]	134±18 [ab]	42±13 [b]	68±10 [a]	63±12 [a]	69±11 [a]
Straw	131±16	222±53	164±67	173±31	31±3	80±14	71±38	68±15
Subtotal	221±40	393±45	304±95	307±53	73±15 [b]	148±22 [a]	133±34 [a]	137±26 [a]
Soil	12,335±936	11,174±435	11,295± 691	11,531± 758	1,925± 117	1,817± 67	1,865± 223	2,029± 104
Total	12,556± 944	11,567± 478	11,598± 658	11,839± 734	1,997± 129	1,966± 59	1,999± 210	2,165± 118
Nutrient gain								
Grain	80±20	161±23	129±28	124±16	36±12	63±11	57±11	63±12
Straw	131± 16	222± 53	164± 67	173± 31	31± 3	80± 14	71± 38	68±15
Soil	-289± 10	-255± 22	-286± 42	-306± 17	-141± 47	-124± 11	-144± 4	-110± 16
Unaccounted loss	137± 39 [b]	398± 36 [a]	188± 100 [b]	205± 46 [b]	94± 35	108± 31	53± 30	122± 29

Mean values with different superscripts within each nutrient in the same row were significantly different (P<0.05).

Table 3. Apparent nutrient recovery (%) by grains in all treatments of the present experiment

Treatments	Apparent N recovery	Apparent P recovery
A	----	----
B	17.32 ± 4.81	25.35 ± 9.60 [b]
C	35.96 ± 22.21	128.71 ± 73.97[a]
D	32.11 ± 20.63	21.99 ± 8.97[b]

Mean value with different superscript letters in the same column were significantly different (P<0.05).

The applied N in the inorganic fertilization treatment (B) was higher than that in the wastewater treatments (C and D), however, the filled grain yield in the inorganic fertilization treatment was not higher than that in the wastewater treatments, because the percentage of filled grains in the inorganic fertilization treatment was lowest among all treatments. Guong *et al.* (1994) mentioned that lower yields were obtained at higher N rates because of a high percentage of unfilled and lodging. Thus, wastewater from intensive catfish culture system is suitable for rice cultivation, however, the use of the catfish pond water may require nutrient supplementation to improve the rice yield. The hybrid catfish culture in the present experiment as well as in Vietnam is not as intensive as that in some other countries such as Thailand (Lin and Diana, 1995). The wastewater from the more intensive hybrid culture contains much higher concentrations of nutrients, suspended particles and algae, which may cause clogging problem when the wastewater is used for irrigating crops such as wheat (Al-Jaloud *et al.*, 1993) and beans (Wood *et al.*, 2001).

The unaccounted nutrient losses in the present experiment were rather high. Besides nutrient loss from leaching by seepage, volatilization, nitrification and denitification, large amounts of TN and TP were taken up by the rice root systems and grass removal during the cultivation period. Loss of applied N by denitrification may vary from 0 to 70% (De Datta, 1989). Yoshida (1981) observed that apparent nitrogen recovery by grains was low at low level of fertilizer application. Guong *et al.* (1994) reported that apparent recovery of nitrogen and phosphorus ranged from 0.8 - 46.0% and from 15 - 36.5% at 60 kg N/ha and from 18.9 - 38.4% and from 25.9 - 101.6% at 120 kg N/ha, respectively. In the present experiment, apparent nitrogen recovery ranged from 17.32 - 35.96%. The efficiency of N application was better in the treatment with low TN input than that with high TN input. Apparent phosphorus recovery by grains in the treatment with wastewater input alone (C) exceeded 100% (128.71%) at a low level of TP input and thus extracted phosphorus from the soil, while low apparent phosphorus recovery (21.99% and 25.35%) was observed at high levels of phosphorus input in treatments D and B.

4. Conclusions

This study has demonstrated that it is feasible to use wastewater from intensive hybrid catfish ponds to irrigate rice crop, and to reuse nutrients in the wastewater as fertilizer for rice cultivation. Such an integrated system has a potential to reduce the environmental impacts of intensive aquaculture and also to reduce the fertilizer costs for rice production. However, more research should be conducted to develop appropriate application strategies for reusing aquaculture wastewater for rice cultivation.

5. Acknowledgement

The authors wish to thank the faculty and staff of the Department of Freshwater Aquaculture and Department of Soil Science, College of Agriculture, Cantho University, and Aquacultural Extension staff at Chauthanh and Omon Districts, Cantho Province, Vietnam for their encouragement, support, and field and lab assistance. This is part of Mrs. Lam My Lan's M.Sc. thesis research. Special thanks are extended to Swedish International Development Cooperation Agency (Sida) for providing her scholarship.

6. References

Al-Jaloud, A.A., Hussain, G., Alsadon, A.A., Siddiqui, A.Q. and Al-Najada, A. (1993). Use of aquacultgure effluent as a supplemental source of nitrogen fertilizer to a wheat crop, *Arid Soil Res. Rehab.*, Vol. 7 (3), pp. 233-241.

APHA AWWA and WPCF (1985). *Standard Methods for the Examination of Water and Wastewater*, 16th ed., American Public Health Association, American Water Works Association and Water Pollution Control Federation, Washington DC.

Cripps, S.J. and Kelly, L.A. (1996). Reductions in wastes from aquaculture. In: *Aquaculture and Water Resource Management*, D.J. Baird, M.C.M. Berveridge, J.F. Muir (Eds), Blackwell Science Publisher, pp. 166-201.

De Datta, S.K. (1989). Integrated nutrient management in relation to soil fertility in low land rice-based cropping systems. In: *Rice Farming Systems: New Direction*, Proceedings of an International Symposium, 31 January-3 February 1987, International Rice Research Institute, Philippines, pp. 141-160.

Edwards, P. (1993). Environmental issues in integrated agriculture-aquaculture and wastewater-fed fish culture system. In: *Environment and Aquaculture in Developing Countries*, R. Pullin, H. Rosenthal, J. Maclean (Eds), International Center for Living Aquatic Resources Management, Manila, Philippines, pp. 139-170.

Guong, V.T., Lap, T.T., Hoa, N.M., Castillo, E.G., Padilla, J.L. and Singh, U. (1994). Nitrogen use efficiency in direct-seeded rice in the Mekong river delta: varieties and phosphorus response. In: *Vietnam and IRRI: A Partnership in Rice Research*, G.L. Denning, V.T. Xuan (Eds), pp. 151 - 158.

Huat, K.K. and Tan, E.S.P. (1980). Review of rice-fish culture in Southeast Asia. In: *Integrated Agriculture Aquaculture Farming System*, R.S.V. Pullin, Z.H. Shehadeh (Eds), International Center for Living Aquatic Resources Management, Manila, Philippines, pp. 1-14.

Kristiansen, R. and Crispps, S.J. (1996). Treatment of fish farm wastewater using sand filtration. *Journal of Environmental Quality*, Vol. 25, pp. 545-551.

Lin, C.K. and Diana, J. S. (1995). Co-culture of catfish *(Clarias macrocephalus x C. gariepinus)* and tilapia *(Oreochromis niloticus)* in ponds. *Aquat. Living Resour.*, vol. 8, pp. 449-454.

Lin, C.K., Jaiyein, K. and Muthuwan, V. (1990). Integration of intensive and semi-intensive aquaculture: concept and example. *Thai Fisheries Gazette*, Vol. 43, pp. 425 - 430.

Metcalf and Eddy (1991). *Wastewater Engineering: Treatment, Disposal and Reuse*, 3rd ed. (revised by G. Tchobanoglous, F. Burton), McGraw-Hill, Inc., Boston, MA.

Muir, J.F. (1982). Economic aspects of waste treatment in fish culture. In: *Report of the EIFAC Workshop on Fish Farm Effluents*, J.S. Alabaster (Ed), EIFAC Technical Paper 41, pp. 123-135.

Steele, R.G.D. and Torrie, J.H. (1980). Principles and Procedures of Statistics, 2nd ed., McGraw-Hill, New York.

Udomkarn, C. (1989). *Integrated of Catfish (Clarias macrocephalus) Cage Culture with Tilapia in Earthen Pond.* M. Sc. thesis, Asian Institute of Technology, Bangkok, Thailand.

Veerina, S. S. (1989). *Intensive Culture of Walking Catfish (Clarias macrocephalus) in an Outdoor Recirculating Systems Integrated with Tilapia.* M. Sc. thesis, Asian Institute of Technology, Bangkok, Thailand.

Wood, C.W., Meso, M.B., Veverica, K.L. and Karanja, N. (2001). Use of pond effluent for irrigation in an integrated crop/aquaculture system. In: Eighteenth Annual Technical Report, A. Gupta, K. McElwee, D. Burke, J. Burright, X. Cummings, H. Egna (Eds). Pond Dynamics/Aquaculture CRSP, Oregon State University, Corvallis, Oregon, pp. 69-78.

Yi, Y., Lin, C. K. and Diana, J. S. (2003). Hybrid catfish (*Clarias macrocephalus* x *C. gariepinus*) and Nile tilapia (*Oreochromis niloticus*) culture in an integrated pen-cum-pond system: growth performance and nutrient budgets. *Aquaculture*, Vol. 217, pp. 395-408.

Yoshida, S. (1981). *Fundamentals of Rice Crop Science*. International Rice Research Institute, Philippines.

Trihalomethane Species and NOM Parameters in the Water Supply of a Small Rural Waterwork

Suraphong Wattanachira[1)],Orathai Permsuk[2)] and Taha F. Marhaba[3)]

1) Department of Environmental Engineering/NRC-EHWM, Chiang Mai University,
Chiang Mai, 50000, Thailand
e-mail: suraphong@eng.cmu.ac.th
2) National Research Center for Environmental and Hazardous waste Management (NRC-EHWM),
Chulalongkorn University, Bangkok, 10330, Thailand
e-mail: tooktanet @yahoo.com
3) Department of Civil & Environmental Engineering, New Jersy Institute of Technology, University Heights
Newark, NJ 07102, USA
email:marhaba@adm.njit.edu

Abstract

This study was to investigate trihalomethane (THMs) species and natural organic matter (NOM) parameters in the water supply of a selected small rural waterwork in Chiang Mai, Thailand. Chloroform and Bromodichloromethane were the only two species of THMs observed in this study. About 96.3% of the main THMs species found in the produced water supply was Chloroform while as little as about 3.7% was Bromodichloromethane. Total trihalomethane (TTHMs) concentration in raw water was not detected but averaged at 62.5 μg/l in the water supply. Total Organic Carbon (TOC), Dissolved Organic Carbon (DOC) and Ultraviolet absorbtion at wavelenght 254 nanometer (UV-254) were considerably used as NOM surrogate parameters. In raw water, the average value of TOC was 2.05 mg/l, DOC was 0.41 mg/l and UV-254 was 0.1447 1/cm, whereas TOC, DOC and UV-254 in the produced water supply were 1.35 mg/l, 0.31 mg/l and 0.0745 1/cm, respectively. Based on the investigated data and the Maximum Control Level (MCL) of 80 μg/l TTHMs in drinking water, as proposed by the United States Environmental Protection Agency (US.EPA), it could be stated that there is a 79% probability of finding TTHMs in the produced water supply at a level of lower than US.EPA MCL.

Key words: DOC, NOM, small rural waterwork, THMs, TOC, UV-254

1. Introduction

In general, small rural waterworks may not be operated under the optimal dosage of coagulant in a coagulation process. In addition, the objective of treatment process in small waterworks mostly focus on turbidity removal and by maintaining free chlorine residual for disinfection in the produced water so as to make the consumers safe from bacteria, viruses and parasites. Chlorine is commonly used as a disinfectant because it is relatively cheap and has a prolonged action. Due to an improper coagulation process prior to chlorination, NOM may not be reduced efficiently, thus chlorine can react with NOM to form THMs. Under this situation, THMs may directly affect the consumer.

In 1974, the Dutch scientist Johannes Rook was the first to identify chloroform and bromotrihalomethanes, the first class of halogenated disinfection by products (DBPs) discovered in chlorinated drinking water (Rook, 1974). A year later, US.EPA conducted a survey which identified chloroform as being dominant in most chlorinated drinking water. Chloroform was identified in 1976 as a suspected human carcinogen, which led the US.EPA to set the Maximum Control Level (MCL) for TTHMs. Because of wide occurrence and potential health risks of DBPs, the US.EPA proposed the Disinfectants /Disinfection By-Products (D/DBP) Rule in two stages. Stage 1 of the D/DBP Rule proposed MCL of 0.080 mg/l. Stage 2 of the D/DBP Rule proposed MCL of 0.040 mg/l which has even lower MCL for DBPs than that of Stage 1 (US.EPA, 1994 and US.EPA, 1997).

THMs are organohalogen compounds that are named as derivatives of methane when three out of the four hydrogen atoms are substitued by three halogen atoms such as chloride (Cl), bromine (Br), iodine (I) or a combination thereof. THMs have a general formula, CHX_3, where X is the halogen atom. THMs are formed during the reaction between a chemical disinfectant (such as chlorine) and NOM in water. The primary biochemical ancestors of THMs identified by many researchers are humic substances including humic acid and fulvic acid (Trussell and Umphes,1978), algal materials mainly the chlorophyll-a, algal biomass and their extra-cellular products (Hoenh *et al.*, 1984) occur naturally in the water. These materials also contribute to the natural color of the water.

2. Materials and methods

2.1 The selected small rural waterwork

A small waterwork using raw water from surface water in Mae-Hia reservoir located in Chiang Mai province, Thailand was selected as the plant to be studied. Water samples were collected between October 2002 and February 2003. The waterwork has a capacity of producing the water supply of approximately 500-800 m^3/day. The water supply from this plant is distributed for drinking, bathing and household use for all communities, faculties, offices and dormitories located in the area of Chiang Mai University at Mae-Hia campus, Chiang-Mai, Thailand

2.2 Water Treatment process of the selected small rural waterwork

Figure 1 shows the diagram of the water treatment process of the selected small rural waterwork including the sampling points and the analytical parameters. The diagram consists of a coagulation stage using polyaluminum chloride (PACl) as a coagulant, flocculation, sedimentation, chlorination and filtration processes. The properties of PACl used in this plant were briefly described as follows; Code name: PACl, Apperance: Liquid, Al_2O_3: 10-11%, pH: 3.5-5.0 (at 1% wt/V) and Specific gravity: 1.256.

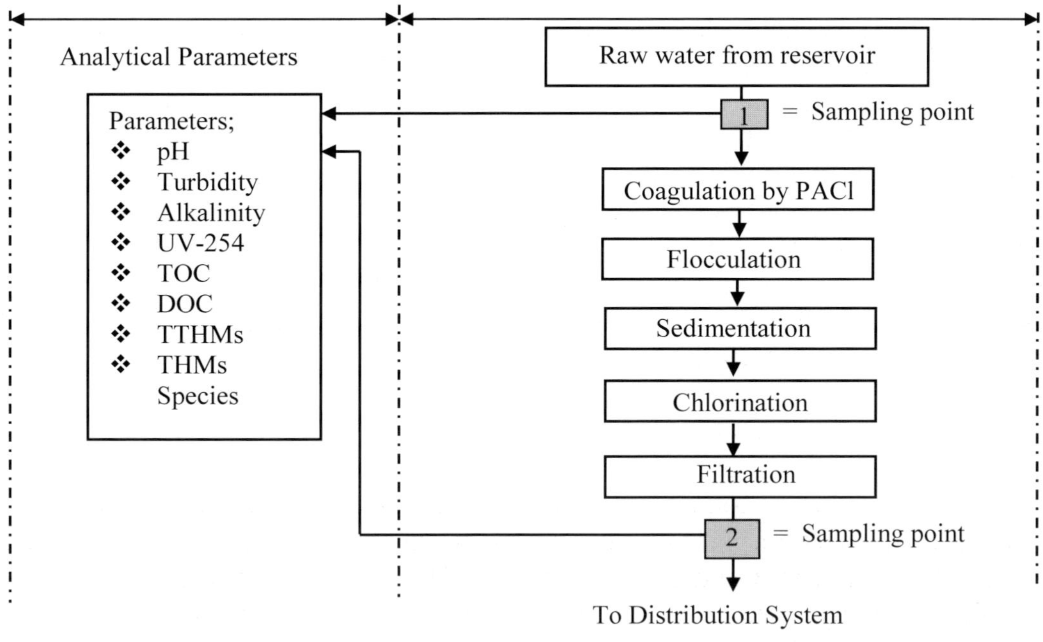

Figure 1. Water treatment process of the selected small waterwork, sampling points and analytical parameters

2.3 Analytical method, standard and instrument

TTHMs, THMs species and NOM parameters in terms of TOC, DOC and UV-254 along with pH, turbidity and alkalinity were determined in both raw water and the produced water supply. The analytical methods used for analyzing each parameter are depicted in Table 1.

Table 1. Analytical method, standard and instrument

Parameter	Analytical Method	Standard	Analytical Instrument
pH	Direct measurement	-	Horiba pH-meter, Model D-13E
Turbidity	Direct measurement	-	HACH, 2100 Turbidity Meter
Alkalinity	Tritation Method	Standard method 2320B	-
UV-254	Ultraviolet Absorption Method	Standard method 5910B	Perkin-Elmer Model Lambda 25, UV/VIS spectrometer
TOC	Persulfate-Ultraviolet Oxidation Method	Standard method 5310C	O.I. analytical 1010 TOC analyzer
DOC	Persulfate-Ultraviolet Oxidation Method	Standard method 5310C	O.I. analytical 1010 TOC analyzer
$TTHM_s$ and THMs species	Liquid-Liquid Exraction Gas Chromatography method	Standard method 5710 and 6232B	Agilent 6890 Series Gas Chromatography with ECD detector

3. Results and discussion

3.1 TTHMs and NOM parameters in raw water

In order to describe the characteristics of raw water from the selected small waterwork, as shown in Table 2, pH, turbidity, alkalinity, UV-254, TOC, DOC, TTHMs were determined for understanding the physical and chemical characteristics as well as the quantity of NOM parameters in raw water.

Table 2. TTHMs and NOM parameters in raw water

Parameter	Raw water			
	No. of sampling	Range of values	Average value	Standard deviation
pH	15	6.3-6.9	6.5	± 0.15
Turbidity (NTU)	15	8.10-45.80	24.80	± 12.52
Alkalinity (mg/L as $CaCO_3$)	15	25.7-42.9	34.0	± 4.55
UV-254 (1/cm)	15	0.1097-0.2056	0.1447	± 0.03
TOC (mg/L)	10	1.74-3.60	2.05	± 0.25
DOC (mg/L)	10	0.18-0.65	0.41	± 0.16
TTHMs (μg/L)	15	0.00	0.00	-

As shown in Table 2, the average pH value was 6.5 which was suitable for a water treatment process by coagulation. A relatively low alkalinity of about 34 mg/l as $CaCO_3$ in raw water was observed. As a result, sufficient alkalinity may be added in water during coagulation since alkalinity must be destroyed in the reaction with coagulant to produce floc.

The average value of 24.80 NTU turbidity in raw water was not clear enough to allow direct utilization as potable water. However, turbidity or the cloudiness of raw water caused by multiple factors such as clay, silt, fine organic and inorganic matter, soluble colored organic compounds, and others, is expected to be removed by coagulation.

In general, TOC and DOC could be used as NOM surrogate parameters. The average value of TOC was 2.05 mg/l, whereas that of DOC was 0.41 mg/l. With regard to UV-254, UV is represented as the organic compounds that are aromatic or that have conjugated double which bonds absorb light in the ultraviolet at wave length 254 nm. Thus the UV-254 absorbance is also a well known NOM surrogate parameter for creating THMs. The average of UV-254 at 0.1447 1/cm was also measured in raw water. As the results of such NOM parameters were determined in raw water as shown in Table 2, it could be expected that natural organic compound, which is THMs precursor in raw water, may create THMs if it is not removed efficiently in a coagulation stage prior to chlorination. However, THMs were not detected in raw water; this may be due to the fact that available chlorine and halogen compounds may not be present in raw water which leads to the absence of reaction with NOM to form THMs.

3.2 TTHMs and NOM parameters in the produced water supply

TTHMs and NOM parameters including pH, turbidity and alkalinity in the produced water supply were also determined as shown in Table 3. Based on the results in Table 3, it was found that acceptable quality of the water supply in terms of pH, turbidity and alkalinity was produced in accordance with Thai Drinking Water Standard. NOM parameters namely TOC, DOC and UV-254 still remained in the produced water supply, but were relatively lower than those found in raw water. These could be explained by the fact that water treatment process using PACl coagulation could reduce some natural organic matter but not all.

Table 3. TTHMs and NOM parameters in the produced water supply

Parameter	The produced water supply			
	No. of sampling	Range of values	Average value	Standard deviation
pH	15	6.5-7.1	6.8	± 0.16
Turbidity (NTU)	15	0.02-1.65	0.58	± 0.50
Alkalinity (mg/l as $CaCO_3$)	15	10.7-34.8	23.7	± 7.01
UV - 254 (1/cm)	15	0.0585-0.0957	0.0745	± 0.01
TOC (mg/l)	10	0.51-2.60	1.35	± 0.87
DOC (mg/l)	10	0.14-0.54	0.31	± 0.16
TTHMs (μg/l)	15	35.2-106.4	62.5	± 18.57

As depicted in Table 3, an average of 62.5 μg/l TTHMs in the produced water supply was obtained while TTHMs in raw water was not detected. TTHMs occurred in the produced water supply due to the use of chlorine in the disinfection process where chlorine can react with the remaining NOM in water to form THMs. Since the recommended concentration or permissible level of TTHMs in drinking water by Thailand's agency was not currently available, the MCL of TTHMs of 80 μg/l proposed by US.EPA was considerably utilized as the standard criteria for TTHMs discussion. The probability plot of TTHMs concentration data in the produced water supply is established as shown in Figure 2.

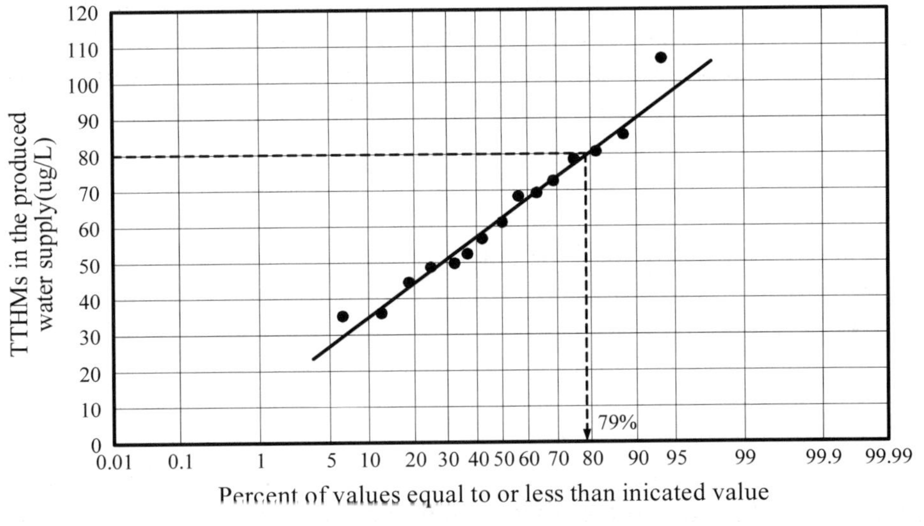

Figure 2. Probability plot of TTHMs concentration data in the produced water supply

From Figure 2, there is a 79 % probability of finding TTHMs in the produced water supply at a level of lower than the US.EPA MCL of 80 μg/l. This could be concluded by saying that the selected small waterwork could not produce potable water 100% safe from problematic THMs.

3.3 Correlation between TTHMs and NOM parameters in the produced water supply

The correlation plots between TTHMs and NOM parameters in the produced water supply were established as shown in Figure 3.

As illustrated in Figure 3, which compares the relationships between TTHMs and TOC, TTHMs and DOC as well as TTHMs and UV-254, it was found that the relationship between TTHMs and TOC has the best correlation with R^2 of 0.6445 and could be represented by the equation of TTHMs = 19.781TOC + 36.472, whereas the correlation between TTHMs and

DOC was considerably categorized as poor with R^2 of 0.4839 and that of TTHMs and UV-254 was not acceptable with R^2 of 0.0934.

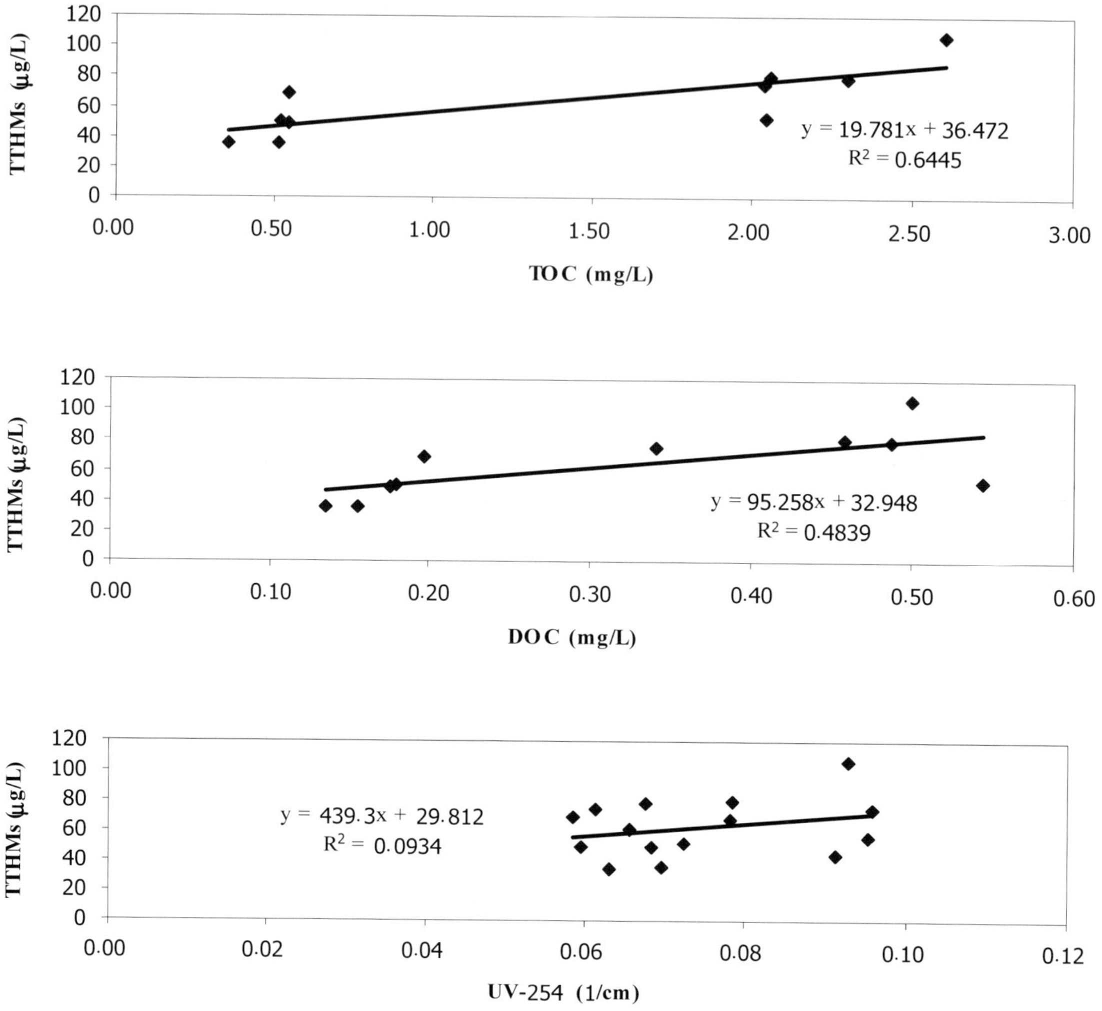

Figure 3. Correlation between TTHMs and NOM parameters in the produced water supply

3.4 TTHMs species in the produced water supply

Overall, four trihalomethane species of concern in drinking water are chloroform ($CHCl_3$), bromodichloromethane ($CHCl_2Br$), dibromochloromethane ($CHClBr_2$) and bromomethane ($CHBr_3$). Those usually occurred in the water containing NOM during disinfection, which is accomplished by chlorination. Therefore, the above-mentioned four forms of THMs species were expected to be formed in the produced water supply from small rural waterworks.

Table 4. Average concentration and percentage of THMs species in the produced water supply

Water Sample	THMs Species								TTHMs
	Chloroform		Bromodichloromethane		Dibromochloromethane		Bromoform		
	Conc. (µg/l)	% of TTHMs	Conc. (µg/l)	% of TTHMs	Conc. (µg/l)	% of TTHMs	Conc. (µg/l)	% of TTHMs	(µg/l)
The produced water supply	60.2	96.3 %	2.3	3.7 %	not detected	-	not detected	-	62.5 (100 %)

From the results in Table 4, it appeared that Chloroform and Bromodichloromethane were the only two species of THMs found in the produced water supply. Chloroform of approximately 96.3 % was the major THMs species while only 3.7% of Bromodichloromethane was observed. The reason used to explain this phenomenon is that the halogen atom in the disinfectant compound used in this selected small rural waterwork was chlorine and it may be assumed that only small amount of bromine was present in raw water. Hence, Chloroform with a small amount of Bromodichloromethane were the only THMs species found in the produced water supply during the course of this research.

4. Conclusions

1. Chloroform and Bromodichloromethane were the only two THMs species observed in the produced water supply. About 96.3% of Chloroform was found as the main THMs species and as little as approximately 3.7% was Bromodichloromethane.

2. In raw water, the average value of TOC was 2.05 mg/l, DOC was 0.41 mg/l and UV-254 was 0.1447 1/cm, whereas TOC, DOC and UV-254 in the produced water supply were 1.35 mg/l, 0.31 mg/l and 0.0745 1/cm, respectively.

3. The relationship between TTHMs and TOC has the best correlation with R^2 of 0.6445 whereas the correlation between TTHMs and DOC was poor with R^2 of 0.4839 and that of TTHMs and UV-254 was not acceptable with R^2 of 0.0934.

4. Average concentration of TTHMs in the produced water supply was 62.5 µg/l. The 79% probability of finding TTHMs in the produced water supply was lower than 80 µg/l, which is the MCL proposed by US.EPA.

5. Acknowledgement

This research was funded by the National Research Center for Environmental and Hazardous Waste Management (NRC-EHMW) program of Thailand. The authors would like to thank the NRC-EHMW at Chulalongkorn University and Chiang Mai University for their supports.

6. References

Hoehn, R.C., Dixon,K.L. and Randall, C.W.(1984). Biologically Induced Variations In the Nature and Removability of THM Precursors by Alum Treatment, *J.AWWA.*, Vol. 76 (4), pp. 134-140.

Rook, J.J. (1974). Formation of Haloforms During Chlorination of Natural Water, *Wtr.Trmt & Exam.*, Vol. 23, pp. 234.

Trussell, R.R., and Umphres, M.D. (1978). The Formation of Trihalomethanes, *J.AWWA.*, Vol. 70 (11), pp. 604-611.

US.EPA (1994). National Primary Drinking Water Regulation: Disinfectants And Disinfection Byproducts (D/DBP), Notice of Data Availability; Proposed Rule. *Fed.Reg.*, 59:145:38668.

US.EPA (1997). National Primary Drinking Water Regulation: Disinfectants And Disinfection Byproducts (D/DBP), Notice of Data Availability; Proposed Rule. *Fed.Reg.*, 62:212:59388.

Combination of Chemical Precipitation and Upflow Anaerobic Floating Filter for Piggery Wastewater Treatment

S. Lee[1)] and J. Iamchaturapatr[2)]

1) Dept. of Environmental Engineering, Kumoh National Institute of Technology, Gumi 730-701, Korea
e-mail: dlee@kumoh.ac.kr
2) School of Environmental, Resources and Development, Asian Institute of Technology, P.O. Box 4, Klong Luang, Pathumthani 12120, Thailand
e-mail: janjit@ait.ac.th

Abstract

This study was conducted to investigate the applicability of chemical precipitation (CP) combined with upflow anaerobic floating filter (UAFF) for the treatment of piggery wastewater. Several series of experiments employing CP and UAFF separately, and its combination have been carried out in the laboratory-scale unit. Combination of chemical precipitation (CP) with upflow anaerobic floating filter (UAFF) shows 90-95% of COD, 95-98% of SS, 75-80% of Color and 91-95% of TP removal efficiencies. There was 10-25% of increase in the removals of COD, SS and Color compared to independently operated modes of process, namely CP and UAFF only. Combination of UAFF with CP provides the highest performance of treating piggery wastewater among all other types of operational arrangement. There was 3-10% of increase in the removals of COD, SS, Color and TP compared to the combination of CP with UAFF, while nutrients removal was insignificant. Organic loading rate (OLR) and chemical constitutes caused by acid coagulants in CP have shown a significant effect on biogas production rate.

Keywords: anaerobic reactor, biogas generation, chemical precipitation, piggery wastewater

1. Introduction

Animal production is growing rapidly due to the increase of population and related rise in demand for animal protein in most Asian countries. The change from traditional small-scale farming to large-scale industrial farming has led to environmental pollution problems causing environmental degradation and deterioration to public health in Thailand (EPA, 1998 and 2001). Pork production is a major agricultural enterprise where approximately seven million pigs are farmed in Thailand (TDA, 1997). A major concern arising from the expansion of pig farming not only in Thailand but also in many other countries where pigs are intensively farmed, is the problem of disposing of the pig manure to the waterways without proper treatment. Several innovative technologies of wastewater treatment system of pig wastewater that are potentially appropriate for urban environments have been continually explored and developed. Chemical precipitation (CP) has the advantage of requiring a smaller land area for unit operations compared to biological treatment processes. The unit operations are completely enclosed and often require low maintenance (Bratby, 1980). Disadvantages include high capital cost and heavy metal leakage. Chemicals can be expensive and easily worked with corrosive chemicals (Metcarf and Eddy, 2003; Eilbeck and Mattock, 1987). Several literatures also show that anaerobic treatment was found to be an effective option for the treatment of piggery wastewater due to its low investment and maintenance costs (Kongsricharoern, 1991; Jayamanne, 2002; Reynolds and Richards, 1996). Few studies have attempted, however, to look at the treatment of piggery wastewater by combined system of anaerobic treatment and CP. This study has been conducted to study the effect of CP in upflow anaerobic floating filter (UAFF) system and to evaluate the performance of a combined system, and to investigate the operating conditions of this combined process for piggery wastewater treatment.

2. Materials and experimental methods

A laboratory-scale experimental unit was designed to investigate the performance of CP and UAFF separately, and their combination for treating piggery wastewater. Figure 1 shows the experimental framework. Initially the raw piggery wastewater was treated by CP and UAFF independently. The appropriate operational condition was monitored by analyzing the removal efficiencies of major wastewater quality parameters such as pH, COD, SS, Color, TKN, TP and NH_3-N (APHA, AWWA and WEF, 1998). Secondly a combined system was employed by two different approaches; (1) UAFF followed by CP, and (2) CP followed by UAFF.

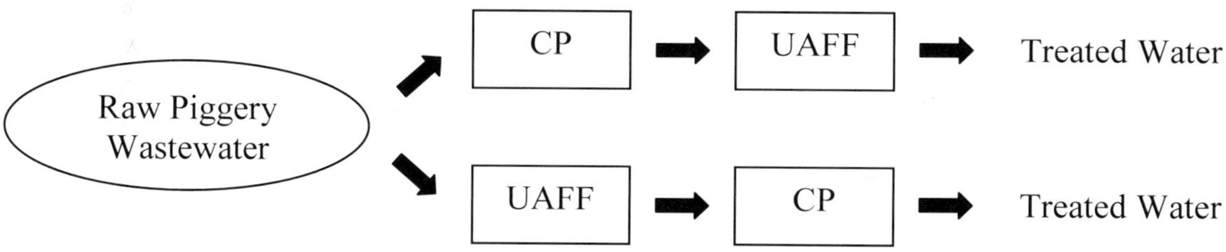

Figure 1. Experimental Framework

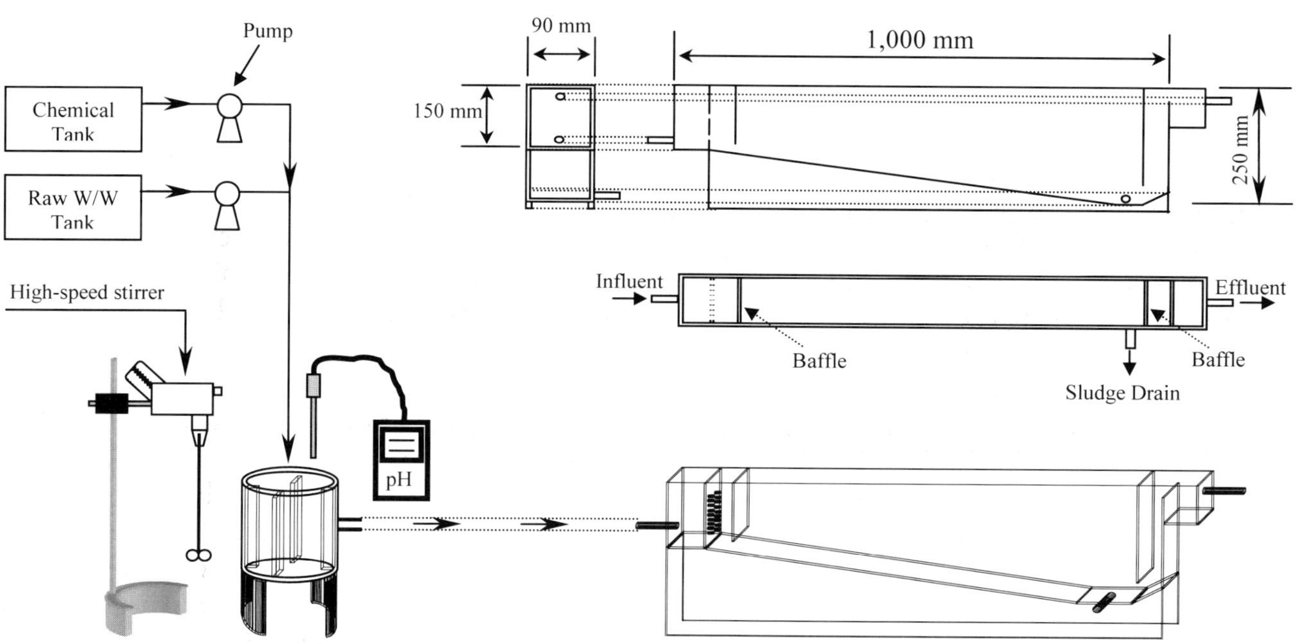

Figure 2. Experimental Set-up of CP

Experimental units used in this study are separately operated. Details of the experimental set-up for CP and UAFF systems are shown in Figures 2 and 3 respectively.

2.1. Chemical Precipitation (CP)

In the CP unit, continuous reactors consisting of rapid mixing and sedimentation tanks was interconnected as shown in Figure 2. Aluminium sulfate ($Al_2(SO_4)_3$ was used as an acid coagulant for all experiments. Rapid mixing reactor was made of acrylic tube with internal dimensions of ∅150mm x 200mm x 4mm (diameter x height x thickness), accommodating a volume of between 0.3 to 1.5 l. For sludge sedimentation, a long shaped rectangular tank was assembled by acrylic plate with total volume of 18 l. Optimum acid coagulant dosage obtained from the batch experiment was applied in continuous reactor. Optimal operating conditions applied in the CP process are shown in Table 1.

Table 1. Operating Conditions of CP

Parameters	Unit	Value
1. Rapid mixing tank		
- Detention time, DT	s	60
- Rotational speed, n	r/min	1,000
2. Sedimentation tank		
- Detention time, DT	min	60
- Average overflow rate	$m^3/m^2.d$	4.8
3. Chemical feeding tank		
- Aluminium sulfate, $Al_2(SO_4)_3$	mg/l	1.25

2.2. Upflow Anaerobic Floating Filter (UAFF)

Each UAFF reactor was made of PVC tube with dimension of ∅150mm x 350mm (diameter x height). Entire UAFF experimental set-up is shown in Figure 3. Approximately 20% upper part of each UAFF reactor was occupied by poly propylene fibre, which functioned as an anaerobic bacterial attached media and filter. Anaerobic bacteria in UAFF reactor was cultured and practiced by the feeding of pig sludge obtained from anaerobic pond treatment in the pig farm for the period of one week. Afterwards fresh piggery wastewater was continuously fed to the three set of UAFF reactor (Figure 3) with an average daily COD loading rate of 1.6-1.8 kg COD/m^3 of wastewater. During the operation time, percentage removal of COD was recorded until the system reaches the stabilizing stage (up to 52 days). Treated piggery wastewater from CP was continuously fed to the UAFF reactors with an average daily COD loading rate of 1.1-1.2 kg COD/m^3 of wastewater (53-81 days). Effluents from UAFF were collected and analyzed periodically and the biogas generated at each reactor was kept in gas collector. Table 2 shows the characteristics of raw piggery wastewater and CP treated water which was adopted as a raw wastewater in UAFF process.

Table 2. Characteristics of Raw Piggery Wastewater and CP Treated Water

Parameters	Unit	Raw Piggery Wastewater	CP Treated Water
COD	mg/l	3,500-4,700	1,140-1,150
SS	mg/l	1,000-1,500	85-120
Color	ADMI	600-700	180-230
TKN	mg/l	700-800	620-730
NH_3-N	mg/l	500-550	490-520
TP	mg/l	50-60	3-5
pH	-	7.8-8.2	7.3-7.5

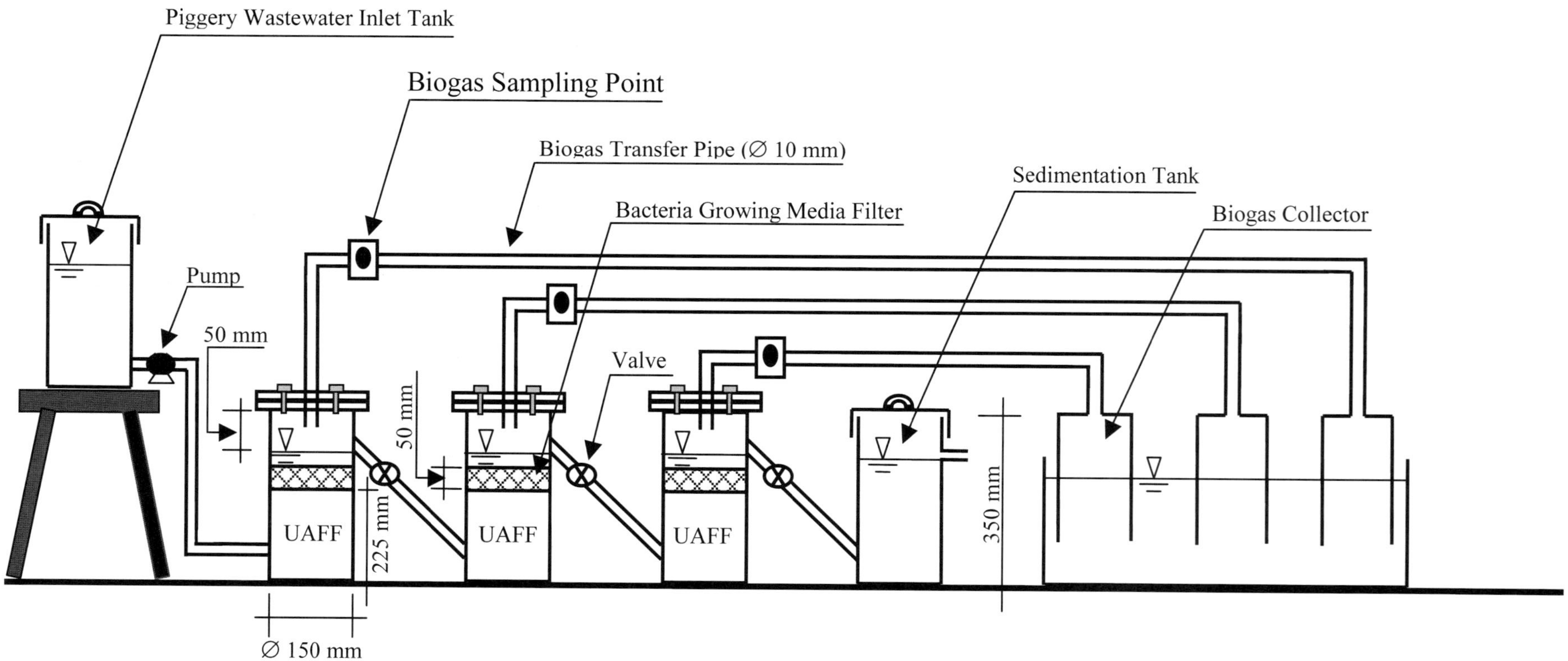

Figure 3. Experimental Setup of UAFF

3. Results and discussion

3.1. Organics and Nutrients Removal

Figure 4 shows the removal efficiencies of COD, SS, Color, TKN, NH_3-N and TP at the UAFF system. Here the raw wastewater was collected from CP treated water and the operation time was counted from the CP process. Average removal efficiencies of COD started from 30% at the initial stage of operation and increased to 65%. More than 55% of removal efficiency was maintained after 65 days of operation. Removal efficiency of SS was in the range of 40-55%. Effluent SS concentration was kept below 60 mg/l throughout the operation. Color removal was insignificant (25-40%). Nutrients (TKN and NH_3-N) removal was also in the range of 15-25%.

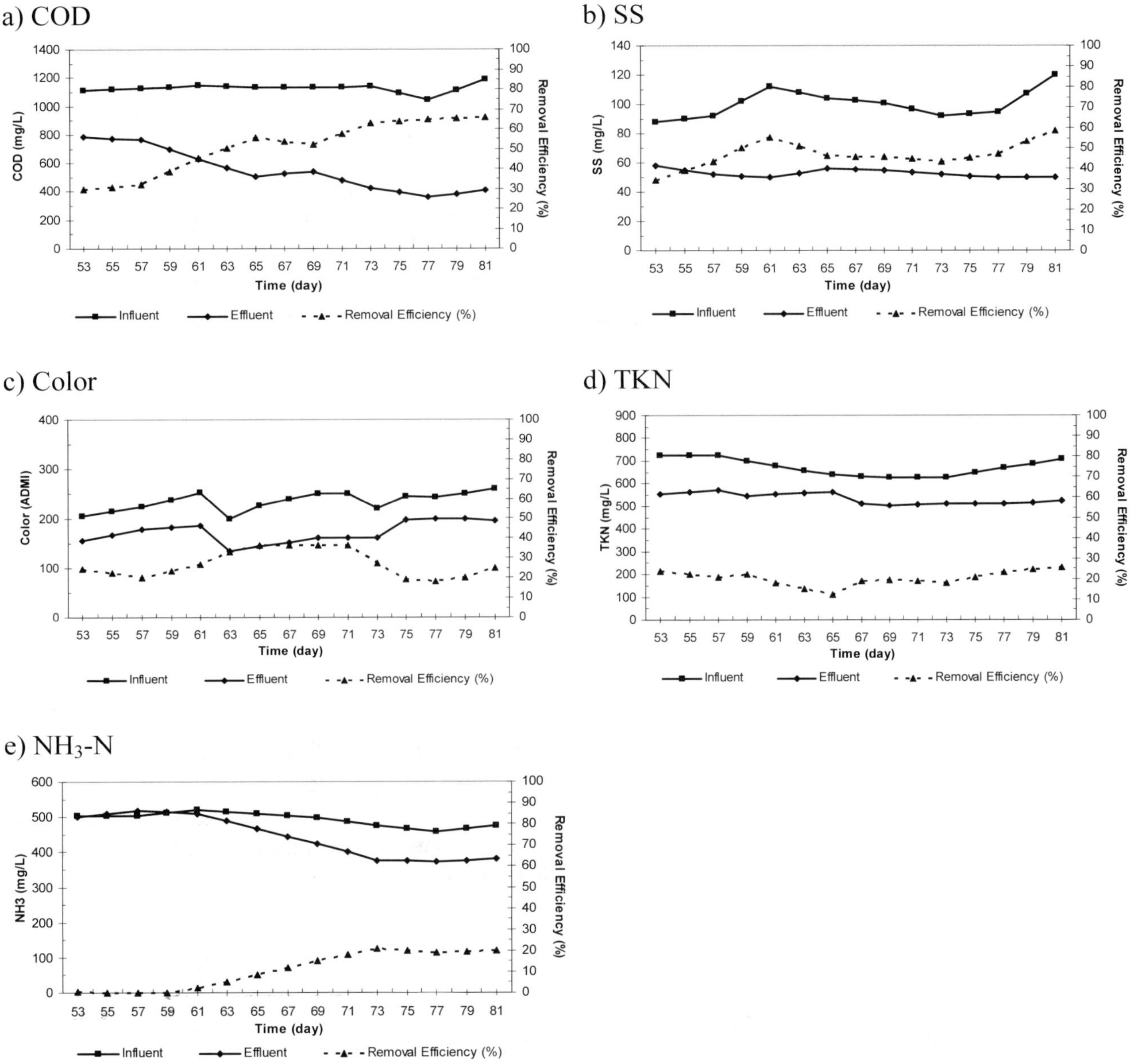

Figure 4. Performance of UAFF Using CP Treated Water

Generally anaerobic filter achieves high SS removal (80-90%) in wastewater (Metcalf and Eddy, 2003). In this UAFF process, SS removal was relatively low (up to 65%), as more than 90% of SS removal was achieved by CP. Color removal was also minimized in this process. Nitrogen removal was also found to be limited. Approximately 70-80% of TKN in piggery wastewater was in the form of NH_3-N, which can not be eliminated by CP. Small reduction of ammonia nitrogen may occur from air bubble of rapid mixing from acid coagulant process that stirs ammonia to release out to atmosphere (Adin *et al.*, 1998). Effluent TP concentration has slightly increased compared to the influent concentration. Phosphorus can leak from the sludge under anaerobic condition for the reactions of ATP to ADP resulting in the increase of phosphorus concentration in effluent from UAFF (Metcalf and Eddy, 2003; Bratby, 1980).

3.2. Variation of pH and Volatile Fatty Acid (VFA)

Figure 5 shows the variation of pH and volatile fatty acid (VFA) in UAFF reactor using CP treated water. With the increase of VFA, pH of the influent gradually decreased. The pH of the third reactor, however, has kept constant (pH=8.0). Although the initial pH of influent was adjusted by acetic acid between 6.0 to 7.5, gradual increase of pH was observed with the increase of anaerobic reactor. Reaction activities of anaerobe bacteria give rise to alkalinity in wastewater, which results in the pH incremental phenomena (Bratby, 1980).

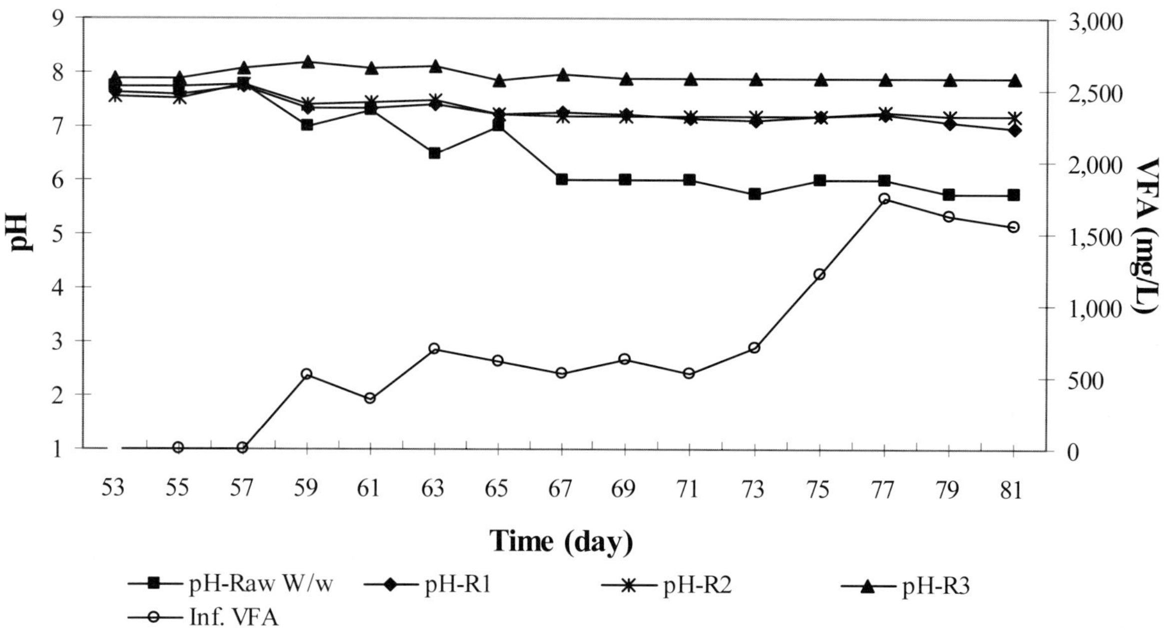

Figure 5. Variation of pH in Each UAFF Reactor and VFA Concentration

3.3. Biogas Generation

Figure 6 shows the variation of biogas generation and organic loading rate at two different modes of operation: UAFF alone, and combination of UAFF and CP. This figure clearly shows that the application of CP followed by UAFF process resulted in the remarkable decrease of biogas production rate. Decrease of organic loading rate (OLR) from 1,400-1,800 to 1,100-1,200 g COD/m^3.d after 53 days of operation has brought the decrease of biogas production rate. Chemical constitutes in CP treated water such as SO_4^{2-} generated from $Al_2(SO_4)_3$ disturbs the methane producing bacteria resulting in lower biogas generation rate (Bratby, 1980; Clair *et al.*, 1994). Increase of anaerobic reactor in UAFF has brought the increase of biogas production. Biogas generation rate in the third reactor, however, was negligible after employing CP and UAFF process. Table 3 summarizes the range of removal efficiencies at different modes of operation; CP alone, UAFF alone, Combination of CP with UAFF and Combination of UAFF with CP. Combination of UAFF with CP has shown the best performance of COD, SS, Color and TP removals for the treatment of piggery wastewater. In every mode of operation, Ammonia (NH_3-N) and TKN removals were very low (maximum 30%) compared to other water quality parameters. Combination of UAFF with CP gives better performance than the combination of CP with UAFF to all the water quality parameters monitored in this experiment.

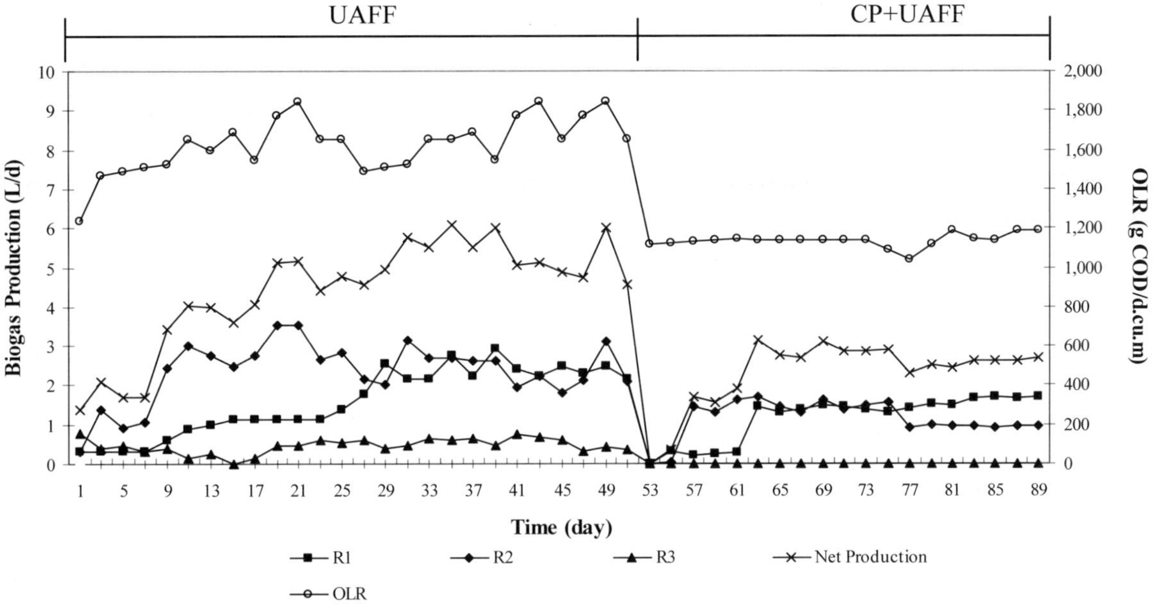

Figure 6. Variations of OLR and Biogas Production

Table 3. Comparison of Removal Efficiency at Different Modes of Operation

Parameters	CP	UAFF	CP+UAFF	UAFF+CP
COD	70-72	82-87	90-95	96-98
SS	90-95	84-92	95-98	96-99
Color	65-73	60-68	75 80	88-91
TKN	5-17	20-27	25-32	27-33
NH_3-N	1-5	15-24	20-30	15-30
TP	91-95	64-73	91-95	>98

4. Conclusions

A combination of chemical precipitation (CP) with upflow anaerobic floating filter (UAFF) shows 90-95% of COD, 95-98% of SS, 75-80% of Color and 91-95% of TP removal efficiencies. There was 10-25% increase in the removal of COD, SS and Color compared to independently operated modes of process, namely CP alone and UAFF alone. A combination of UAFF with CP provides the highest performance of treating piggery wastewater among all other types of operational arrangement, namely UAFF alone, CP alone and combination of CP with UAFF. There was 3-10% increase in the removal of COD, SS, Color and TP compared to the combination of CP with UAFF, while nutrient removal was insignificant at a different mode of operation. With the increase of VFA from the CP treated water, pH of the influent gradually decreased although the pH of the third reactor maintained constant (pH=8.0). Gradual increase of pH was observed with the increase of anaerobic reactor. A combination of CP with UAFF resulted in the remarkable decrease of biogas production rate due to the lower organic loading rate (OLR) and chemical constitutes generated in CP treated water

5. References

Adin, A., Soffer, Y. and Ben Aim, R. (1998). Effluent Pretreatment by Iron Coagulation Applying Various Dose-pH Combinations for Particle Separation, *Wat. Sci. Tech.*, Vol. 38, pp. 27-34.

APHA/AWWA/WEF (1998). *Standard Methods for the Examination of Water and Wastewater,* 20th ed., Amer. Public Health Assoc., Amer. Water Works Assoc. and Water Envir. Fed., USA.

Bratby, J. (1980). *Coagulation and Flocculation with an Emphasis on Water and Wastewater Treatment*, Uplands Press Ltd., Croydon CR91LB.

Clair, N., Sawyer, L., McCarty, P. and Parkin, G. (1994). *Chemistry for Environmental Engineering*, 4th ed., McGraw-Hill Inc., ISBN: 0-07-113908-7.

Eilbeck, W.J. and Mattock, G. (1987). *Chemical Processes in Waste Water Treatment*, Ellis Harwood Limited., ISBN: 0-85312-791-3.

EPA (1998). *Environmental Impacts of Animal feeding Operations*, U.S. Environmental Protection Agency, Office of Water, Standards and Applied Sciences Division, Washington D.C, December.

EPA (2001). *Development Document for the Proposed Revisions to the National Pollutant Discharge Elimination System Regulation and the Effluent Guidelines for Concentrated Animal Feeding Operations*, U.S. Environmental Protection Agency, Washington D.C., January.

Jayamanne, A. (2002). Application of Upflow Anaerobic Floating Filter (UAFF) System in Series for High Strength and Low Degradable Wastewater Treatment, Thesis, AIT, Bangkok, Thailand.

Kongsricharoern, N. (1991). Two – Stage Anaerobic Treatment of Pig Manure with Total Effluent Recycle, Thesis, AIT, Bangkok, Thailand.

Metcalf and Eddy (2003), *Wastewater Engineering: Treatment Disposal and Reuse*, International Edition, McGraw Hill Inc., New York.

Reynolds, T.D. & Richards, P.A. (1996). *Unit Operations and Processes in Environmental Engineering*, 2nd ed., pp. 414.

Thailand Department of Agriculture (TDA) (1997). *Strategies and Working Plans for Livestock Development during the Eighth National Development Plan,* Bangkok, Thailand.

Arsenic Contamination of Groundwater and Selected Removal Technologies for Rural Isolated Communities in Bangladesh

M. H. Rahman

Department of Civil Engineering, Bangladesh University of Engineering & Technology (BUET)
Dhaka-1000, Bangladesh
e-mail: habibr@ce.buet.ac.bd

Abstract

Groundwater contamination by arsenic is a major concern in Bangladesh where about 97% of the total rural population depend on tube-wells for drinking water, which is bacteriologically safe to drink. Arsenic in groundwater above 50 μg/l (Bangladesh standard for Drinking Water) was found in 61 districts out of total 64 districts and 433 out of the 496 thanas in Bangladesh. But this dimension of the arsenic occurrence problem in groundwater in Bangladesh is yet to be fully identified. The total population exposed to arsenic contaminated water above 50 μg/l and 10 μg/l are in the range of 28.1 - 32.5 million and 46.4 - 56.7 million respectively. The incidences of excess skin cancer are 0.29%, 0.043% and 0.012% from drinking arsenic contaminated water at the present level of arsenic contamination, satisfying the Bangladesh Standard (50 μg/l) and satisfying WHO guideline value (10 μg/l) respectively. In the context of the prevalence of high concentrations of arsenic in groundwater in Bangladesh, several methods/ technologies of treating water for arsenic reduction from drinking water have been tried. In many cases conventional technologies have been scaled down to suit the rural isolated households and communities to provide safe sources of water for drinking purpose. Recently a number of research studies have been conducted in Bangladesh to identify such novel technologies to remove arsenic from groundwater in rural isolated communities. This study presents a brief overview of some of these processes and technologies.

Keywords: arsenic, arsenic contamination, arsenic removal technologies in Bangladesh, groundwater contamination

1. Introduction

Groundwater contamination by arsenic (As) is a major concern in Bangladesh, where about 97% of the total rural population depends on tubewells for drinking water that is bacteriologically safe to drink. Arsenic was first detected in groundwater in West Bengal, India in the late 1980s and in the west of Bangladesh in 1993 following reports of extensive contamination of water supplies in the adjoining areas of India. At present several thousands of people are suffering from arsenic related diseases and millions are at risk of arsenic poisoning from drinking groundwater with arsenic in excess of the acceptable limit. The human consequence of this pollution has been described as the worst mass poisoning of a population in history (Smith et al, 2000).

Several technologies are currently being promoted by several organizations for field application in Bangladesh. In many cases conventional technologies have been scaled down to suit the rural isolated households and communities to provide safe sources of water for drinking purposes. The effectiveness, viability and sustainability of these technologies under field conditions in Bangladesh are yet to be authenticated.

This paper is aimed at identifying the arsenic contamination problems in Bangladesh and then an overview of technologies used for arsenic removal in the rural isolated communities will be presented.

2. Arsenic Contamination in Groundwater

Arsenic in groundwater above 50 μg/l (Bangladesh Standard for Drinking Water) was found in 61 districts out of the total 64 districts and in 433 police stations (thanas, the lowest level of administrative units of the country) out of the 496 thanas in Bangladesh during the time of the study period. This is based on the studies conducted by British Geological Survey (BGS), Department of Public Health Engineering (DPHE), Mott MacDonald Limited (MML) and Bangladesh Arsenic Mitigation Water Supply Project (BAMWSP). BGS, DPHE and MML conducted the study in two phases and examined 3,534 distributed water samples from 61 districts (except 3 hill districts) in an approximate grid of 6km x 6km (DPHE-BGS-MML, 1999 and BAMWSP, 2001). These include an average of 58 samples per district and 8 samples per thana, and 25% tested samples exceed the concentration of 50 μg/l, Bangladesh Standard for drinking water and 42% tested samples exceed the concentration of 10 μg/l, provisional World Health Organization (WHO) guideline value for arsenic in drinking water. This percentage of contaminated tubewells increases if only shallow tube-wells are considered. Then 27% tested samples exceed the concentration of 50 μg/l and 46% tested samples exceed the concentration of 10 μg/l. In the case of tested water samples collected from deep-tube-wells (strainer depth > 150m), only 1% and 5% samples exceed the allowable limits of 50 μg/l and 10 μg/l respectively. Figures 1 and 2 show the severity of the problems of arsenic contamination of the study areas in Bangladesh. These include the information obtained from arsenic analysis conducted by BGS and DPHE (BGS and DPHE, 2001), DPHE, BGS and MML (DPHE-BGS-MML, 1999), School of Environmental Studies, Jadavpur University (SOES, JU) and Dhaka Community Hospital (DCH) (SOES and DCH, 2000) and Bangladesh University of Engineering and Technology (BUET) but do not include test data obtained by test kits.

BAMWSP has been conducting a national screening survey to identify arsenic contaminated tube-wells in two phases mostly in arsenic contaminated areas (BAMWSP, 2001). In the first phase, the survey conducted by BAMWSP includes 80,390 tube-wells in 6 upazilas, and the number of contaminated tube-wells is 38,739 (48.19%). BAMWSP examined 5,44,975 tube-wells in the second phase (still ongoing) and out of which 2,52,214 (46.28%) tube-well waters have been found contaminated with arsenic.

The estimates of total population exposed to arsenic contaminated water by the DPHE, BGS and MML in phase-I is in the range of 18.5-22.7 million. But in phase-II, BGS and DPHE furnishes two estimates of population exposure based on project population of 125.5 million in 1999. The total population exposed to arsenic contaminated water above 50 μg/l and 10 μg/l are estimated, using Kriging method, to be 32.5 million and 56.7 million respectively. Based on these statistics the total population exposed to arsenic contaminated water above 50 μg/l and 10 μg/l are estimated to be 28.1 million and 46.4 million respectively. According to the SOES,JU and DCH (SOES and DCH, 2000) the population exposed to 50 μg/l and 10 μg/l in 43 districts in Bangladesh are 25 million and 51 million respectively.

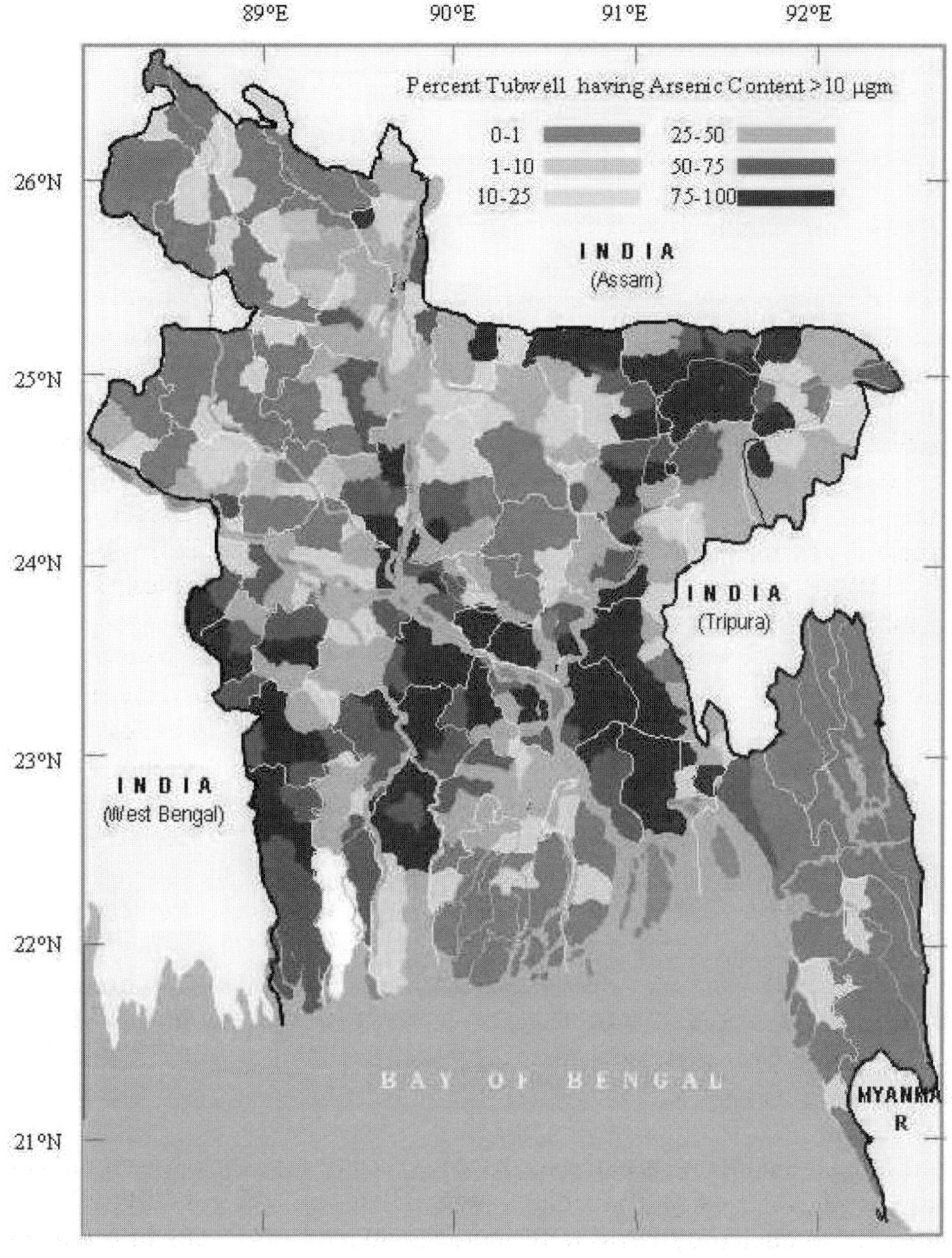

Figure 1. Intensity of Arsenic (As > 10) in Bangladesh

3. Health Effect

Arsenic is assumed to be the 12th most abundant element in the biosphere, and may be found in atmospheric air at concentrations of around 0.4 to 30 ng/m^3. The concentrations in food and soil may be in the ranges of around 0.4 to 120 μg/kg and 5 to 6 mg/kg respectively. The human being may therefore, naturally be exposed to some amount of arsenic.

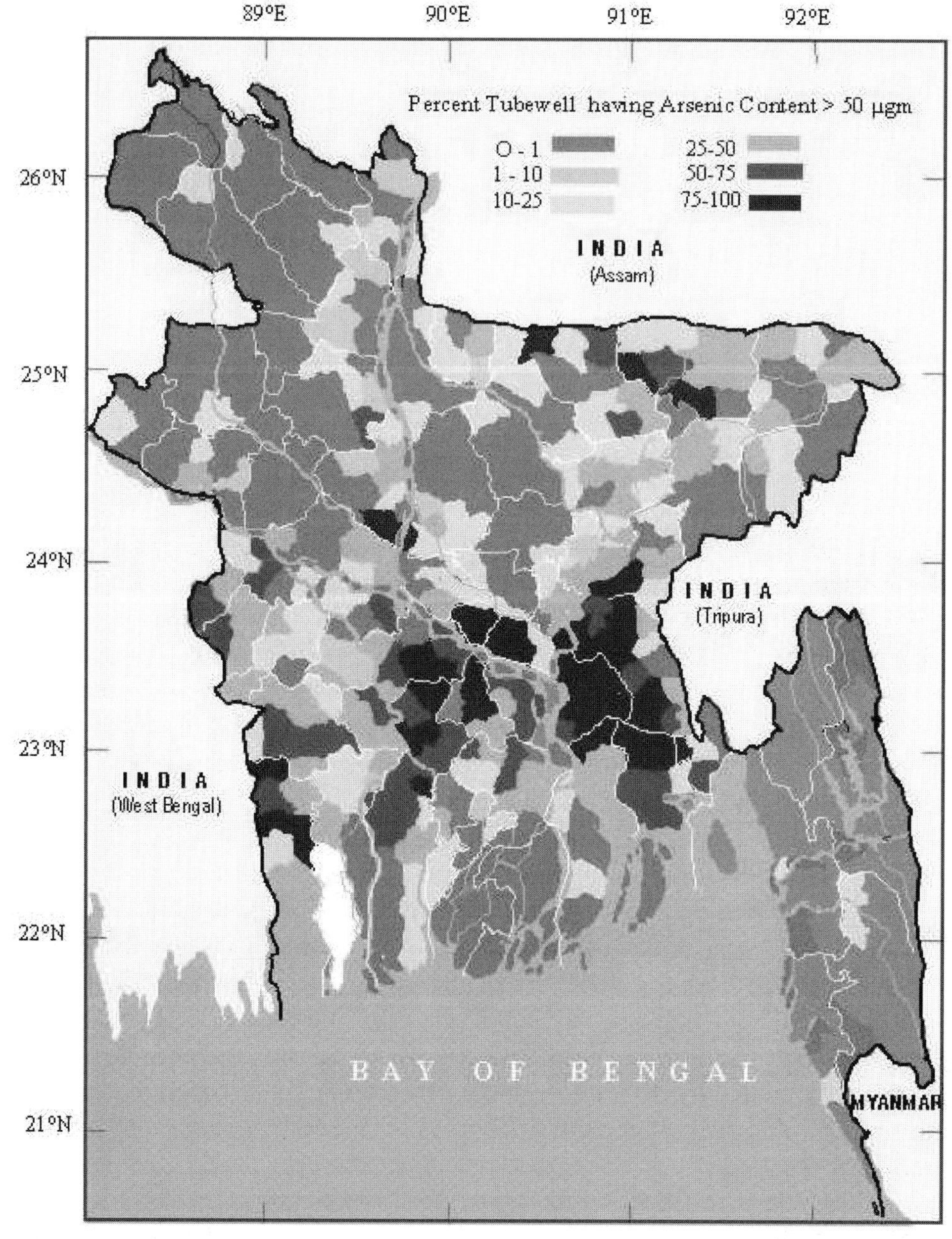

Figure 2. Intensity of Arsenic (As > 50) in Bangladesh

Though arsenic is said to be essential for some animal species, but any form of arsenic compound is not desirable for human consumption. The toxic effect of arsenic species mainly depends on their chemical form, route of entry, age, sex, dose and duration of exposure. The organic form of arsenic is several orders of magnitude less toxic than the inorganic form. The trivalent arsenic (As-III, arsenite) species are approximately an order of magnitude more toxic compared to the pentavalent arsenic (As-V, arsenate) species. The consumption of arsenic contaminated food and drinking water is the main concern of arsenic toxicity. Skin diseases are the common effects of arsenic poisoning. Long-term exposure to excessive arsenic generally causes changes in skin pigmentation and hyperkeratosis, and promotes development of ulceration of skin, skin cancer and a number of internal (liver, bladder, kidney, etc.) cancers. In BAMWSP (2001) study, the total number of identified arsenicosis patients are

1,139 (0.067% of the total population surveyed) and 9,647(0.112% of the total population surveyed) in phase-I and phase-II respectively.

There is no well-established relationship between consumption of arsenic through drinking water. Using USEPA model and distribution of population exposed to different levels of arsenic, LGRD (2002) estimated the incidence of excess lifetime skin cancer for different level arsenic contamination of drinking water for the present total population of 129.25 million in Bangladesh. The incidences of excess skin cancer are 0.29%, 0.043% and 0.012% for drinking arsenic contaminated water at present level arsenic contamination, satisfying the Bangladesh Standard (50 μg/l) and satisfying WHO guideline value (10 μg/l) respectively.

4. Arsenic Removal Technologies Used in Bangladesh

Many technologies have been developed for the removal of arsenic from drinking water as the occurrence of arsenic contamination in groundwater was reported in various parts of the world. They mostly utilize the principles of oxidation, precipitation/co-precipitation, adsorption onto sorptive media, ion exchange and physical separation by synthetic membranes (Rahman *et al*, 2003). In consideration of lowering drinking water standards by United State Environmental Protection Agency (USEPA), a review of arsenic removal technologies was made to consider the economic factors involved in implementing lower drinking water standards for arsenic (Chen *et al*, 1999). A comprehensive review of low-cost well-water treatment technologies for arsenic removal with the list of companies and organizations involved in arsenic removal technologies is available in Murcott (2000). Most of the documented experiences of arsenic removal technologies include large-scale application. In the context of prevalence of high concentration of arsenic in groundwater in Bangladesh, several methods/ technologies of treating water for arsenic reduction from drinking water have been tried. In many cases conventional technologies have been scaled down to suit them for the rural isolated households and communities to choose safe sources of water for drinking purposes. Recently a number of research studies have been conducted in Bangladesh to identify such novel technologies to remove arsenic from groundwater in rural isolated communities. A brief overview of some of these processes and technologies is presented here.

4.1 Passive sedimentation

Since a large part of Bangladesh (about 65% areas) contains iron in excess of 2 mg/l and in many acute iron problem areas, the concentration of iron is as high as 15 mg/l, therefore, arsenic has been found to co-exists with iron in many situations. In such situations arsenic can be removed by both co-precipitation and adsorption onto the precipitated $Fe(OH)_3$ by oxidation of this water during collection and subsequently storing them in household level. The study conducted by the author revealed that this process removed more than 60% of arsenic, where raw groundwater arsenic and iron concentrations were in the range of 150 μg/l to 713 μg/l and 8 mg/l to 14 mg/l respectively (Rahman *et al*, 2003). But the removal rate is largely controlled by the arsenic concentration, the iron/arsenic ratio, and pH.

4.2 Solar oxidation

In this process, transparent bottles containing water are exposed to sunlight for solar oxidation of arsenic in the presence of other oxidants like oxygen, followed by precipitation of arsenic with naturally occurring iron. Experiments in Bangladesh showed the removal efficiency in the range of 50% to 70%, when 50 μM citrate or 100-200 μl (4-8 drops) of lemon juice/l is added. The cost of this system is around US$2.07/person/year. The major limitation of this system is that it does not work on cloudy days and the addition of chemicals is cumbersome.

4.3 Arsenic and iron removal plants (AIRP)

The conventional small-community type iron removal plants, which operate on the principles of aeration of ferrous iron to convert them to ferric iron to co-precipitate arsenic. Groundwater drawn by hand tube-well drops into storage (aeration/ sedimentation) chamber for oxidation of iron and arsenic with air to co-precipitate. Water from storage chambers passes through filtration chambers, due to the pressure head of aeration/ sedimentation chamber and is subsequently collected into a storage tank for public use. Filtration media comprise brick chips, charcoal and sands. Filtration media is periodically (3 to 4 times a year) back washed, and sludge is collected in a holding pond. Iron and arsenic removal efficiencies of conventional AIRP plants operating in small communities are in the range of 60% to 90%. The average installation, operation and maintenance cost of a typical plant is around US$ 0.25/person/year if the design life of this system is considered as 25 years.

Most of the AIRP plants shown here have good arsenic removal performance and have been treating water to the satisfactory level except in those areas, where arsenic concentrations are high. It is evident from field surveys that these AIRP plants are well accepted by the community. Thus, some of the above described AIRPs have good potential for small isolated communities where arsenic co-exists with iron at suitable concentrations.

4.4 Bucket treatment unit

Bucket treatment unit developed by DPHE-Danida a project consists of two buckets (about 20 l capacity) placed one above the other. Chemicals (200 mg/l aluminum sulfate and 2 mg/l potassium permanganate) are mixed manually with arsenic contaminated water in upper bucket by vigorous stirring with a wooden sticks for about 30-60 seconds and then flocculated by gentle stirring for about 90 seconds. Mixed water is then allowed to settle for about 1-2 hours and top supernatant is allowed to flow into the lower bucket via a plastic pipe and sand filter installed in the lower bucket. Although this technology perform well (with arsenic removal efficiencies in the range of 65% to 90%) in many situations but in some cases under rural operating conditions it fails to remove to the desired Bangladesh standard level and produces undesirable sludge, and the addition of chemicals is cumbersome. The cost of this system is around US$0.6 /person/year.

4.5 Pitcher treatment

Pitcher filters are constructed by stacking a number of pitchers (Kolshis), one above the other, containing different filter media. Raw water is poured in the top pitcher and filtered water is collected from the bottom one. 3-pitchers filter uses zero valent iron filling and coarse sand in the top pitcher, charcoal and fine sand in the middle pitcher. The bottom pitcher is used to store treated water. The arsenic removal efficiencies of this system are in the range of 55% to 95% but it depends on the maintenance of the system and the quality of water. But if batches are left for too long, dissolved iron concentrations becomes unacceptably high and the system may be quickly clogged if groundwater contains excessive iron. Hardening of the filter media and breaking of pitchers increases the cost of treatment. The cost of this system is in the range of US$0.65/ person/year to US$0.85 /person/year.

4.6 Activated alumia

Water passes through the packed column of activated alumina, impurities present in the water including arsenic are adsorbed on the surface of activated alumina grains. At some stage, the alumina becomes saturated and then exposing the medium to 4% caustic soda regenerates these. The residual caustic soda is then washed out and the medium is neutralized with a 2% solution of sulfuric acid rinse. A number of activated alumina based sorptive media are used in Bangladesh and some of the alumina based sorptive media are more effective in removing arsenic. This process produces undesirable sludge and the addition of chemicals is cumbersome The performance and applicability of these arsenic removal units are yet to be determined through closed field monitoring.

4.7 Membrane technique

A nanofiltration process coupled with bicycle pumping system was examined by Oh *et al.* (2000) using arsenic contaminated tube-well water in rural areas of Bangladesh. Arsenite was found to have lower rejection than arsenate in ionized form. Therefore low-pressure nanofiltration with pre-oxidation or reverse osmosis with a bicycle pump device can be used for the treatment of arsenic contaminated groundwater where an electricity supply is not efficient or feasible. The capital and operational cost of this system is relatively high.

4.8 Other arsenic removal technologies

To mitigate arsenic contamination problems in Bangladesh, a number of organizations and industries have been trying to develop arsenic removal systems and chemicals. As discussed earlier, they mostly utilize the principles of oxidation, precipitation/co-precipitation, adsorption onto sorptive media, ion exchange and physical separation by synthetic membranes. The performance and applicability of these arsenic removal units and chemicals are yet to be determined through close field monitoring.

5. Conclusions

On a critical analysis of the existing arsenic contamination problems and available water supply options that have been tried to mitigate this problem in Bangladesh, the following observations and conclusions are made.

- Arsenic in groundwater above 50 µg/l (Bangladesh standard for Drinking Water) was found in 61 districts out of total 64 districts and 433 out of the 496 thanas in Bangladesh. It is evident from the test results of 3,534 distributed water samples from 61 districts (except 3 hill districts) in an approximate grid of 6km x 6km that 27% tested samples collected from sallow tube-wells exceed the concentration of 50 µg/l and 46% tested samples exceed the concentration of 10 µg/l (WHO guideline value for arsenic in drinking water). In the case of tested water samples collected from deep-tube-wells (strainer depth

> 150m), only 1% and 5% samples exceeded the allowable limits of 50 μg/l and 10 μg/l respectively. But this dimension of the arsenic occurrence problem in groundwater in Bangladesh is yet to be fully identified.

- The total population exposed to arsenic contaminated water above 50 μg/l and 10 μg/l are in the range of 28.1 - 32.5 million and 46.4 - 56.7 million respectively.
- The incidences of excess skin cancer are 0.29%, 0.043% and 0.012% from drinking arsenic contaminated water at the present level of arsenic contamination, satisfying the Bangladesh Standard (50 μg/l) and satisfying WHO guideline value (10 μg/l) respectively.
- The total number of identified arsenicosis patients, as indicated in BAMWSP study, are 1,139 (0.067% of the totals population surveyed) and 9,647(0.112% of the totals population surveyed) in phase-I and phase-II respectively.
- It is apparent from this study that a remarkable development in arsenic removal technologies has taken place during the last few years. These arsenic removal technologies mostly utilize principles of oxidation, precipitation/co-precipitation, adsorption onto sorptive media, ion exchange and physical separation by synthetic membranes. In many cases conventional technologies have been scaled down to suit the rural isolated households and communities to provide safe sources of water for drinking purposes.

6. References

BAMWSP (2001). *Status Report of Bangladesh Arsenic Mitigation Water Supply Project (BAMWSP)*, December.

BGS and DPHE (2001). *Arsenic Contamination of Groundwater in Bangladesh*, D. G. Kinniburg and P.L. Smedley (Eds), vol.2, Final Rep.British Geol. Survey, WC/00/19.

Cheng, H.W., Frey, M.M., Clifford, D., McNeill, L.S. and Edwards, M. (1999). Arsenic Treatment Consideration, *J.AWWA*, Vol. 91 (3), pp. 74-85.

DPHE-BGS-MML (1999). *Groundwater Studies for Arsenic Contamination in Bangladesh*, Final Report, Phase-I, Department of Public Health Engineering (Bangladesh), Department for International Development (UK) and British Geological Survey, January.

LGRD (2002). *Arsenic Mitigation in Bangladesh*, M. F. Ahmed, and C. M. Ahmed (Eds), Local Government Division, Ministry of Local Government, Rural Development (LGRD) and Cooperatives, Govt. of the Peoples Republic of Bangladesh.

Murcott, S. (2000). A Comprehensive Review of Low-cost, Well-water Treatment Technologies for Arsenic Removal, http://phys4.harvard.edu/_wilson./murcott2.html.

Oh, J. I., Yamamoto, K., Kitawaki, H., Nakao, S., Sugawara, T. Rahman, M. M., Rahman, M. H. (2000). Application of low-pressure nanofiltration coupled with a bicycle pump for the treatment of arsenic-contaminated groundwater, *J. Desalination*, Vol. 132, pp. 307-314.

Rahman, M. H., Rahman, M.M., Watanabe, C. and Yamamoto, K. (2003). Arsenic Contamination of Groundwater in Bangladesh and Its Remedial Measures, *Proc. UNU-NIES International Workshop on Arsenic Contamination in Groundwater – Technical and Policy Dimensions*, United Nations University, Japan, pp. 9-21.

Smith, A.H., Lingas, E.O. and Rahman, M (2000), Contamination of Drinking-water by arsenic in Bangladesh: a public health emergency, *Bull. World Health Org*., Vol. 78 (9), pp. 1093-1103.

SOES and DCH (2000). *Groundwater Arsenic Contamination in Bangladesh*, A. Summary of Field Survey from August 1995 to February 2000: B. Twenty-seven Days Detailed Field Survey from April 1999 to February 2000.

Various Treatment Technologies to Remove Arsenic and Mercury from Contaminated Groundwater: An Overview

S. Babel[1)] and T. A. Kurniawan[2)]

1) Environmental Technology Program, Sirindhorn International Institute of Technology (SIIT), Thammasat University, P.O. BOX 22, Pathumthani 12121, Thailand
e-mail: sandhya@siit.tu.ac.th
2) Environmental Technology Program, Sirindhorn International Institute of Technology (SIIT), Thammasat University, P.O. BOX 22, Pathumthani 12121, Thailand
e-mail: tonni@siit.tu.ac.th

Abstract

Contamination of groundwater due to toxic heavy metals, such as As and Hg, is of international concern. In this paper, the technical feasibility of some technologies to treat metal-contaminated groundwater is reviewed. All data presented are quoted from the original sources. It is conclusively evident from the literature survey that low cost technology, like adsorption, can demonstrate excellent metal removal capabilities comparable to reverse osmosis. Due to chemical modification, adsorption using inexpensive materials, such as coconut husk carbon, can remove As(III) with metal adsorption capacity of 159 mg/g. This high adsorption capacity suggests that adsorption treatment technique is capable of reducing a residual arsenic concentration to 10 μg/l, which is the maximum discharge limit set by the U.S.EPA (Environmental Protection Agency). In spite of the availability of various technologies to treat metal-contaminated groundwater, the suitability of a selected technology for such a purpose depends on the initial metal concentration, the residual metal concentration required, the removal performance on heavy metal compared to other technologies, and the overall treatment cost.

Keywords: developing country, groundwater purification, heavy metal, removal performance

1. Introduction

Contamination of groundwater due to toxic pollutants, such as As and Hg, is one of the most outstanding environmental problems in some developing countries, like Korea, India, and Bangladesh (Woo and Choi, 2001; Nickson *et al.*, 2000; Yokota *et al.*, 2001; Pandey *et al.*, 2002). Studies have linked the long-term exposure to As in the drinking water to cancer of the skin, liver, lung, and kidney. Non-cancer effects of ingesting As and Hg included cardiovascular, pulmonary, and neurological disorders as well as tremor and behavior disturbances (Ning, 2002). In order to neutralize the toxic effect of arsenic on living organisms, the context and approach taken to treat the metal-contaminated groundwater are country-specific, as the ideal treatment technologies for such a purpose not only should be appropriate, suitable, and applicable to the local condition of each country, but they should be able to meet the maximum contaminant level (MCL) established in their respective countries (Table 1).
Various physicochemical treatment methods, such as chemical precipitation, coagulation, oxidation followed by filtration, reverse osmosis, ion exchange, and adsorption have been

employed to treat metal-contaminated groundwater. All of the treatment technologies mentioned previously have inherent advantages and limitations in their application.

Table 1. Guideline MCL for arsenic in some countries

Country/Institution	MCL (µg/l)	References
Australia	7	Thirunavukkarasu *et al.*, 2002
Europe	10	
U.S.A	10	
WHO	10	Ning, 2002
India	50	Manju *et al.*, 1998
Thailand	250	Pollution Control Department (PCD), 2003

Due to the simplicity of the process, chemical precipitation is the most widely used method in developing countries for groundwater decontamination. During precipitation, the dissolved ions in the solution are converted to insoluble solid via a chemical reaction.

Coagulation has also been frequently used for the removal of arsenic from contaminated groundwater. Principally, it converts soluble heavy metals into insoluble reaction products, resulting in a separation by sedimentation. The general approach involves the addition of ferric or alum salts as the coagulant and also includes pH adjustment.

Due to its convenient operation, filtration is another conventional method commonly used in developing countries for groundwater purification from toxic metals. Common particulates removed in the filtration include As(V), Hg(II), and Cr(VI).

Although reverse osmosis (RO) is an ideal alternative for the production of drinking water from groundwater, its usage is limited due to high operational cost. Basically, RO is a pressure-driven membrane process. Presently, the application of reverse osmosis for water treatment is expanding rapidly due to stringent environmental regulations. Depending on the characteristic of its membrane, reverse osmosis retains all ions and passes water.

Ion exchange has also gained a considerable interest among environmental technologists to treat metal-contaminated groundwater. It is a reversible interchange of ions between the solid and the liquid phase where there is no permanent change in the structure of the solid.

Among the treatment technologies mentioned previously, adsorption is the most intensively studied treatment method for such a purpose. Principally, adsorption is a mass transfer where a solute substance (adsorbate) is transferred from the liquid phase to the surface of a solid (adsorbent) and becomes bound by chemical or physical forces. Materials that are available in large quantities, like waste from agricultural or industrial operation, can be used as a low-cost adsorbent. Conversion of waste, representing an unused resource, into activated carbon, which can be used as an adsorbent for water treatment, would add its economic value, reduce the cost of waste disposal, and provide a good alternative to the costly activated carbon. Although many research works have been done for treatment of contaminated groundwater, so far no effort has been made to obtain a comprehensive overview of all the treatment methods mentioned previously in terms of their removal performance. An overview of some treatment technologies based on recent publications is presented here and their removal performance is compared to each other.

2. Various Technologies for Heavy Metal Removal from Contaminated Groundwater

2.1 Chemical precipitation

Various research studies on the use of the precipitation method to treat metal-contaminated groundwater have been done in recent years. In 1991, the use of lime softening to reduce arsenic concentration was performed by Dutta and Chaudhuri. It was reported that a lime dosage of 1250 mg/l at pH 11.5 could achieve 90% of arsenic removal. It was also found that the presence of $Mg(OH)_2$ floc played a significant role in the removal of As(III). In 1997, McNeill and Edwards used precipitative softening to remove arsenic from contaminated groundwater by using $Mg(OH)_2$ and $CaCO_3$. It was reported that the former is preferable to the latter for arsenate, as it is capable of removing arsenic higher than 95% for an initial As concentration ranging from 15 to 150 μg/l at a pH higher than 10.

Treatment of As from the well water in Bangladesh using co-precipitation (ferric hydroxide and sodium hypochlorite) was also studied (Meng *et al.*, 2001). It was reported that a Fe/As mass ratio higher than 40 was required to reduce As(V) concentration to less than 50 μg/l. For As concentration of 500 μg/l, 20 mg/l of Fe(III) was required to treat the well water.

Despite chemical precipitation is a popular treatment method for As removal in developing countries, major drawbacks with this method are excessive sludge production, the increasing cost of landfill disposal, and the long-term consequences for the local environment where the sludge is disposed. The cost effectiveness of precipitation is also diminished when disposing of large quantities of low concentration arsenic waste.

Table 2. Arsenic removal using chemical precipitation method

Species	Precipitant	Dose (mg/l)	Initial concentration (μg/l)	Removal efficiency (%)	pH	Reference
As	$CaCO_3$	1250	56	90	11	Dutta and Chaudhuri, 1991
As(V)	$Mg(OH)_2$	20	25	>95	10-12	McNeill and Edward, 1997
	$CaCO_3$	100	20	42.5	10-12	
As(V)	$Fe(OH)_3$	20	500	85	-	Meng *et al.*, 2001

2.2 Coagulation

In 1994, the removal of As(III) and As(V) using $FeCl_3$ and alum as coagulants was studied by Edward. It was reported that arsenic removal efficiency was solely dependent on pH, instead of initial As(V) concentration and coagulant dose. It was also suggested that both adsorption and co-precipitation had major roles for the arsenic removal by $FeCl_3$ and alum.

A further study on the removal of arsenic by enhanced coagulation was conducted by Chang *et al.* (1994). It was reported that $FeCl_3$ was more effective than alum for removing arsenic from groundwater and that there was settleable floc formation for $FeCl_3$ and alum dosage higher than 40 and 80 mg/l, respectively.

In 1997, Hering *et al.* studied arsenic removal from drinking water using coagulation. It was reported that coagulation is capable of achieving final dissolved As(V) concentration less than 2 μg/l for the influent As(V) concentration ranging from 50 to 100 μg/l. It was also found that both alum and ferric chloride could be used to achieve an efficient As(V) removal, as presented in Table 3. Like chemical precipitation, the need to dispose of large amounts of arsenic-contaminated iron sludge makes coagulation an unfavorable method for treatment of contaminated ground water, as it requires further treatment to convert the toxic sludge into a stabilized product, which would not reveal arsenic in the environment.

Table 3. Arsenic removal using coagulation treatment method

Species	Coagulant	Dose (mg/l)	Initial concentration (μg/l)	Removal efficiency (%)	pH	Reference
As(III)	$FeCl_3$	5	50	100	7.0	Edward, 1994
As(V)		6	5	72	8.0	
As(V)	$FeCl_3$	40	43	97	8.0	Chang *et al.*, 1994
	Alum	60	43	97	8.0	
As(V)	Alum	40	20	80	6-8	Hering *et al.*, 1997
	$FeCl_3$	4.9	20	70	4-8	

2.3 Filtration

In 1990, Cooper and Thompson studied the removal of arsenic using a mixed media filtration process. It was reported that an iron to arsenic ratio of 10:1 resulted in 98% of arsenic removal at pH 7.0. It was also found that a higher iron to arsenic ratio improved the removal efficiency of arsenic.

In 1994, a further investigation on the removal of arsenic using greensand regenerated with $KMnO_4$ was performed by Lauf. It was reported that an initial As(III) concentration of 200 μg/l could be removed to a threshold level of 5 μg/l over 400 bed volumes.

In addition to the co-precipitation method, Meng *et al.* (2001) also used a filtration system to treat As-contaminated well water in Bangladesh. It was reported that the removal efficiency of As(V) was about 97% for an initial concentration of 485 μg/l (Table 4). In fact, the residual arsenic concentration decreased to 5 μg/l when 200 l of water was treated. It is important to note that a pre-oxidation step is usually required to achieve an efficient removal of As(III).

Table 4. Arsenic removal using oxidation process followed by filtration treatment method

Species	Oxidant	Oxidant concentration (mg/l)	Initial concentration (μg/l)	Removal efficiency (%)	pH	Reference
As	Cl_2	4.5	260	98	7	Copper and Thompson, 1990
As(III)	$KMnO_4$	-	200	97.5	-	Lauf, 1994
As	Cl_2	2	485	97	-	Meng *et al.*, 2001

2.4 Reverse osmosis (RO)

Reverse osmosis is the most commonly used method for potable water treatment in the U.S. RO is capable of rejecting contaminants with diameters as small as 10^{-4} μm. Success in desalination has encouraged further consideration of RO for the production of drinking water totally free from toxic heavy metals, like As, Hg, and Cr.

In 1981, Huxtep investigated the removal of arsenic using high pressure (400 psi) and low pressure (200 psi). It was reported that the high pressure RO system achieved 91-98% rejection of As(V), while the rejection of a low pressure system ranged from 77-87%; thus, indicating that both high and low pressure systems were effective for As(V) removal.

In 1986, Clifford *et al.* studied the removal of dissolved As from water using some types of RO membranes. It was reported that the removal efficiency of As(V) ranged from 98-99% for an initial As concentration of 0.11-1.9 mg/l, while a low As(III) removal efficiency of 46-75% was achieved at an initial As concentration ranging between 0.04-1.3 mg/l (Table 5).
In 2002, the arsenic removal by reverse osmosis was investigated by Ning. It was reported that As(V) was effectively removed by RO. One pilot study reported As(V) removal at 96-99% and As(III) removal at 46-84%. To improve the As removal efficiency, it was suggested that As(III) needs to be oxidized to As(V) using potassium permanganate prior to treatment.
Although RO is very effective in removing As(V) with a rejection percentage over 95%, it was found that it is not equally efficient at removing As(III). Moreover, the use of RO for such a purpose presents some major problems, rendering it unfavorable for specific cases. The small pores of membrane make it more prone to fouling. If cations, such as Fe and Mn, are present in groundwater, they can also promote membrane fouling due to precipitation, which might be irreversible. Other drawbacks of using RO are the increased operational cost of membrane processes and the need of experienced personnel to run the overall process.

Table 5. Arsenic removal by reverse osmosis treatment

Species	Pressure (psig)	Initial concentration (mg/l)	Removal efficiency (%)	pH	Reference
As(V)	400	-	91-98	-	Huxtep, 1981
	200	-	77-87	-	
As(V)	-	0.11-1.9	98-99	-	Clifford *et al.*, 1986
As(III)	-	0.04-1.3	46-75	-	

2.5 Ion Exchange

In 1983, Ficklin studied the removal of As(III) and As(V) in groundwater using a strong anion-exchange resins (Dowex 1 x 8, 100-200 mesh, acetate form). It was found that the removal of both ions was accomplished by the gravity flow of the samples through anion exchange resins in a small column with NaCl as an eluent. Removal of As(V) by chloride from a strong base anion-exchange resin can be represented by the following reactions:

$$2RCl + HAsO_4^{2-} \leftrightarrow R_2HAsO_4 + 2Cl^- \qquad (1)$$
$$R_2HAsO_4 + 2NaCl \leftrightarrow 2RCl + Na_2HAsO_4 \qquad (2)$$

A further study on the As species removal using ion exchange was conducted by Miller *et al.* (2000). It was reported that the Ficklin anion exchange method for such a purpose was a good solution to the problem for the removal of arsenic species. This method was recommended, as it was very rapid and simple in terms of the nature of removal. Unlike chemical precipitation, ion exchange does not present sludge disposal problems.
A similar study on the separation of inorganic arsenic species in groundwater using ion exchange method was done by Kim (2001). It was reported that the optimum pH for the removal of As(III) and As(V) ranged from 4 to 7 and that ion exchange was good for fieldwork, as the speciation results were reliable, the required equipment was portable, and the experiment could be done quickly. However, in spite of its advantages, high operational costs for ion exchange makes it unsuitable for further use in developing countries.

2.6 Adsorption

Due to the high operational cost of some technologies, like RO and ion exchange, to treat metal-contaminated groundwater, adsorption using commercial activated carbon (CAC) has been developed to be the most effective and feasible option for such a purpose. Since the use of CAC may not be suitable for developing countries due to its high cost, various adsorbents, originated from inexpensive and readily available materials, have been investigated to achieve the economically feasible and effective treatment of metal-contaminated groundwater.

In 1982, Pandey and Chaudhuri studied the removal of inorganic mercury from water using bituminous coal. It was reported that sulfonation of coal increased the sorption capacity of coal, as it formed chelating groups on the coal surface, such as xanthates and dithiocarboxylates. It was also found that these groups improved Hg binding.

A similar study on Hg(II) sorption using bituminous coal was also conducted by Karthikeyan and Chaudhuri (1985). It was found that chemical treatment of bituminous coal with HNO_3 significantly enhanced Hg sorption to a level higher than that exhibited by activated carbon in terms of adsorption kinetics and sorption capacity. It was also reported that an adsorption capacity of 10 mg of Hg(II)/g was observed at a pH range of 7.0-8.5 by bituminous coal.

In 1996, Singh *et al.* used hematite and feldspar as adsorbents to treat As(V)-rich effluent. It was reported that both unconventional adsorbents are effective for the removal of As(V) from aqueous solution. However, hematite had a better arsenic removal capacity than feldspar, as can be seen in Table 6.

Chemical modification of coconut coir using amine functional group was also conducted to improve its adsorption capacity for As(V) and Cr(VI) (Baes *et al.*, 1997). It was reported that the maximum adsorption capacities of As(V) and Cr(VI) were found to be 6.45 and 17 mg/g, respectively, and that an amine functional group could increase the positive charge of adsorbent, which favored more adsorption of arsenate and chromates anions due to columbic forces, resulting in a higher uptake of both As(V) and Cr(VI) ions.

The performance of coconut husk carbon (CHC) for removing arsenic from groundwater was evaluated by Manju *et al.* (1998). It was reported that surface modification of CHC with impregnating Cu have significantly improved the sorption capacity of As(III) at the optimum pH of 12.0. It was also found that for removing 50 mg/l of arsenic (IIII), an optimum dose of adsorbent of 4 g of Cu-impregnated CHC or 20 g of CHC was required, respectively.

Table 6. Heavy metal removal by adsorption treatment method

Species	Type of adsorbent	Dose of adsorbent (g/l)	Initial concentration (mg/l)	Optimum pH	Adsorption capacity (mg/g)	Reference
Hg(II)	coal	1.0	0.10	6.2	-	Pandey, 1982
Hg(II)	coal	0.50	0.20	7.0-8.5	10	Karthikeyan, 1985
As(V)	hematite	40	1.00	4.2	0.204	Singh *et al.*, 1996
	feldspar	40	1.00	6.2	0.190	
As(V)	Coconut	4.0	-	4-6	6.45	Baes *et al.*, 1997
Cr(VI)	coir	4.0	-	4-6	17	
As(III)	CHC	20	50	8.5	2.5	Manju *et al.*, 1998
	Cu-impregnated CHC	4.0	50	12	158.65	
As(III)	FMO	2.0	0.18	2-8	0.04	Chakravarty *et al.*, 2002
As(V)	FMO	2.0	0.18	2-8	0.07	
As(III)	polymeric	-	0.20	7	67	Katsoyianis and
As(V)	materials	-	0.20	4	161	Zouboulis, 2002

Recently, the use of ferruginous manganese ore (FMO) as an adsorbent, (cost about US$ 0.05/kg), for removing As from groundwater was studied by Chakravarty *et al.* (2002). It was reported that both As(III) and As(V) were adsorbed by the adsorbent in the pH range of 2.0-8.0. It was also found that the adsorbed As did not become desorbed from the filter sludge even at pH 2.0.
A further investigation on the removal of arsenic from contaminated water using iron-oxide-coated polymeric materials, such as polystyrene and polyHIPE (high internal phase emulsion) was performed by Katsoyianis and Zouboulis (2002). It was reported that arsenic removal was dependent on the pH and contact time and that As(V) was removed more efficiently than As(III); thus, suggesting the need of pre-oxidation of As(III) to As(V) for better treatment.
Overall, the use of adsorption for water purification offers environmental technologists a lot of promising benefits, as it uses locally available unused resources with simple and safe operation. The easy availability of low-cost adsorbent, fast adsorption rate, reasonable adsorption capacity, and sludge-free operation are expected to cut down the operating cost and render this adsorption method attractive for treatment of metal-contaminated groundwater in developing countries.

3. Concluding remarks

It is evident that many treatment schemes can be effective for removing As and Hg from contaminated groundwater. However, no single treatment has been found to be effective for all of them. Due to this reason, a pilot test needs to be conducted before selecting a specific treatment process, as the suitability of a technology to treat contaminated groundwater depends on the initial metal concentration, the residual metal concentration required, the metal removal performance compared to other technologies, and the overall treatment cost. It is important to note that simplicity and cost effectiveness are the two major factors that should influence the selection of a treatment system for heavy metal removal in developing countries. Despite reverse osmosis system being suitable for use in developed countries, such a treatment method may not be appropriate for use in developing countries due to its high operational cost. Other simple and economical treatment systems, such as manganese greensand filtration and adsorption, may offer a better alternative treatment system for groundwater purification.
So far, cost information of all technologies available for treatment of contaminated ground-water is not reported in any of the publications, as the overall cost of a treatment system varies, depending on the processing required and the local condition of each country. This situation makes a comprehensive comparison among the available treatment technologies difficult to materialize due to inconsistencies in data presentation.
In spite of this fact, the use of adsorption treatment method for groundwater decontamination in developing countries is recommended due to its technical feasibility, engineering applicability, and cost effectiveness. A modified low-cost adsorbent, like Cu-impregnating CHC can perform well in removing As(III) with metal adsorption capacity of 159 mg/g compared to as-received CHC with adsorption capacity of 2.5 mg/g. Surface modification of coconut coir using amine functional group can also improve its metal removal performance with adsorption capacities of 6.4 and 17 mg/g for As(V) and Cr(VI), respectively. Thus, the use of various low-cost adsorbents for heavy metal removal may significantly contribute to the sustainability of surrounding environment.

4. References

Baes, A.U., Okuda, T., Nishijima, W., Shoto, E. and Okada, M. (1997). Adsorption and ion exchange of some groundwater anion contaminants in an amine modified coconut coir, *Water Sci. Technol.*, Vol. 35 (7), pp. 89-95.

Chang, S.D., Ruiz, H. and Clark, D.L. (1994). Removal of As by enhanced coagulation and membrane technology. Critical Issues in Wastewater Treatment, In *Proceeding of the 1994 National Conference on Environmental Engineering*, July 11-13 1994, Colorado, USA.

Chakravarty, A., Dureja, V., Bhattacharyya, G., Maity, S. and Bhattacharjee, S. (2002). Removal of As from groundwater using low cost ferruginous manganese ore, *Water Res.*, Vol. 36, pp. 625-632.

Clifford, D., Subramanian, S. and Sorg, T.J. (1986). Removing dissolved inorganic contaminants from water, *Environmental Science Technology*, Vol. 20 (11), pp. 1072-1080.

Cooper, K.H. and Thompson, R.E. (1990). Iron, manganese, and arsenic removal using high-rate adsorption clarification-mixed media filtration process, In *Proceeding of AWWA Annual Conference*, Ohio, USA, pp. 1147-1166.

Dutta, A. and Chaudhuri, M. (1991). Removal of arsenic from ground water by lime softening with powdered coal additive, *J.Water SRT Aqua*, Vol. 40, pp. 25-29.

Edwards, M. (1994). Chemistry of arsenic removal during coagulation and Fe-Mn oxidation, *Journal American Water Works Association*, Vol. 86 (9), pp. 64-78.

Ficklin, W.H. (1983). Separation of arsenic(III) and arsenic(V) in groundwater by ion exchange, *Talanta*, Vol. 30 (5), pp. 371-373.

Hering, J.G., Chen, P., Wilkie, J.A. and Elimelech, M. (1997). Arsenic removal from drinking water during coagulation, *J. Env. Eng.*, Vol. 123 (8), pp. 800-807.

Huxtep, M.R. (1981). Inorganic contaminant removal from drinking water by reverse osmosis, USEPA report No. EPA-600/2-81-11, USEPA, Florida, USA.

Karthikeyan, J. and Chaudhuri, M. (1985). Enhancement of mercury(II) sorption from water by coal through chemical pretreatment, *Water Res.*, Vol. 20 (4), pp. 449-452.

Katsoyiannis, I.A. and Zouboulis, A.I. (2002). Removal of arsenic from contaminated water sources by sorption onto iron-oxide-coated polymeric materials, *Water Res.*, Vol. 36, pp. 5141-5155.

Kim, M.J. (2001). Separation of inorganic arsenic species in groundwater using ion exchange method, *Bull. Environ. Contam. Toxicol.*, Vol. 67, pp. 46-51.

Lauf, G.F. (1994). Arsenic removal using potassium permanganate generated greensand, In *Proceeding of 1994 AWWA-WQTC Conference*, San Fransisco, USA.

Manju, G.N., Raji, C. and Anirudhan, T.S. (1998). Evaluation of coconut husk carbon for the removal of arsenic from water, *Water Res.*, Vol. 32 (10), pp. 3062-3070.

McNeill, L.S. and Edwards, M. (1997). Arsenic removal during precipitative softening, *J. Env. Eng.*, Vol. 123 (5), pp. 453-460.

Meng, X.G., Korfiatis, G.P. and Bang, S.B. (2001). Treatment of As in Bangladesh well water using a household co-precipitation and filtration system, *Water Res.*, Vol. 35 (12), pp. 2805-2810.

Miller, G.P., Norman, D.I. and Frisch, P.L. (2000). A comment on arsenic species separation using ion exchange, *Water Res.*, Vol. 34 (4), pp. 1397-1400.

Nickson, R.T., McArthur, J.M., Burgess, W.G. and Ahmed, K.M. (2000). Mechanism of arsenic release to groundwater, Bangladesh and West Bengal, *Applied Geochemistry*, Vol. 15, pp. 403-413.

Ning, R.Y. (2002). Arsenic removal by reverse osmosis, *Desalination*, Vol. 143, pp. 237-241.

Pandey, M.P. and Chaudhuri, M. (1982). Removal of inorganic mercury from water by bituminous coal, *Water Res.*, Vol. 16, pp. 1111-1118.

Pandey, P.K., Yadav, S., Nair, S. and Bhui, A. (2002). Arsenic contamination of the environment: a new perspective from central-east India, *Environment International*, Vol. 28, pp. 235-245.

PCD, "Water Quality Standards", Thai Ministry of Natural Resource and Environment, July 2003. (URL:ftp://pcd.go.th/information/Regulations/WaterQuality/Effluents.htm#Industry)

Singh, D.B., Prasad, G. and Rupainwar, D.C. (1996). Adsorption technique for the treatment of As(V)-rich effluents, *Coll. & Surf. A: Physicochemical and Engineering Aspects*, Vol. 111, pp. 49-56.

Thirunavukkarasu, O.S., Viraraghavan, T., Subramanian, K.S. and Tanjore, S. (2002). Organic arsenic removal from drinking water, *Urban Water*, Vol. 4, pp. 415-421.

Woo, N.C. and Choi, M.J. (2001). Arsenic and metal contamination of water resources from mining wastes in Korea, *Environmental Geology*, Vol. 40 (3), pp. 305-311.

Yokota, H., Tanabe, K., Sezaki, M., Akiyoshi, Y., Miyata, T., Kawahara, K., Tsushima, S., Hironaka, H., Takafuji, H., Rahman, M., Ahmad, SkA, Sayed, M.H.S.U. and Faruquee, M.H. (2001). Arsenic contamination of ground and pond water and water purification system using pond water in Bangladesh, *Engineering Geology*, Vol. 60, pp. 323-331.

Investigations on the Status of Arsenic Contamination in Southern Thailand

R. Jindal[1)] and P. Ratanamalaya

1) School of Environmental Engineering, Institute of Engineering, Suranaree University of Technology
111 University Avenue, Nakhon Ratchsima 30000, Thailand
e-mail: rjindal@ccs.sut.ac.th

Abstract

The status of arsenic contamination in Southern Thailand was investigated through review of past studies as well as field surveys to have insight into source identification, mechanism of its transportation, and the severity of toxicity. Sources of arsenic contamination in the Ronpiboon district were identified to be the previous tin mining activities around the area. Oxidation of arsenic in the soil caused contamination in groundwater and flooding spread out the contamination around the area. Villagers in Ronpiboon are suffering from chronic arsenic poisoning with skin cancer or "black fever." However, there is no clear answer about how arsenic was distributed in this region. Government agencies and research groups have had many studies conducted over the last few yeas, but no conclusion of the arsenic contamination mitigation in the area had been reported. Thus, any knowledge about the simple methods of arsenic removal from drinking water would be very helpful to the general public in Ronpiboon area.

Keywords: arsenic poisoning, contamination, groundwater, mining, Southern Thailand

1. Introduction

The environmental impact of arsenic contamination is causing serious concerns in some countries such as Bangladesh, India, Taiwan, Thailand, Mexico, and Ghana in recent years. According to a recent report, the Mekong Delta groundwater may not escape from an arsenic contamination problem. Apparently, this has not yet turned into a known major public health disaster such as in Bangladesh. No outbreak of arsenic related disease has been reported to the international health community, although that does not mean there is none, or they were not mistakenly dismissed as something else. The Mekong Delta may have a little more time than Bangladesh but certainly no time to waste either. The pressure of population growth and local economic development has led to rapid extraction and so receding levels of ground water, an identical path to Bangladesh and West Bengal. More importantly, the proposed diversion of 8.8 BCM (billion of cubic meters) of water from the Mekong River in Thailand and the planned development of 37 MW of hydropower reservoirs in China and Laos would exacerbate the arsenic problem into a catastrophe (Than and Long, 2000). Severe problems of arsenic pollution have occurred in Ronpiboon District, Nakhon Si Thammarat Province in southern part of Thailand, where arsenic in groundwater was found as $FeAsS$ and FeS_2 in the tin mining belt within that region.

Arsenic is a naturally occurring element present in the environment in both inorganic and organic forms and it is toxic and carcinogenic. Inorganic arsenic is considered to be the most toxic form of the element and is found in groundwater and surface water, as well as in many foods. A wide variety of skin and internal cancers, and cardiovascular and neurological effects, have been attributed to chronic arsenic exposure, primarily from drinking water (Smedley and Kinniburgh, 2001). Human exposure to inorganic arsenic occurs through inhalation, ingestion, or skin absorption. Occupational exposure may occur among smelter workers and the workers exposed in production and use of arsenic containing pesticides. Ingestion occurs mainly by drinking contaminated water. Most ingested arsenic is absorbed through the gastrointestinal tract and lungs and then into blood stream. It is distributed in lungs, liver, kidney, spleen, and intestine within 24 hours of ingestion and to skin, hair, and bone within two weeks. Harmful effects of arsenic on human health include acute and chronic toxicity. Acute toxicity occurs only from the ingestion of arsenic compounds. The symptoms of acute toxicity include severe vomiting and diarrhea, muscular cramps, facial edema, and cardiac abnormalities. An ingested dose of 70-180 mg of arsenic (III) oxide has been reported to be fatal in people. Symptoms may occur within a few minutes of exposure if the arsenic compound is in solution but may be delayed for several hours if it is in solid form or taken with a meal. Chronic toxicity is manifested in the organ systems affecting the skin, nervous system, liver, cardiovascular system, and respiratory tract (Kamal, 1999).

The WHO guideline for As in drinking water was provisionally reduced in 1993 from 50 μg/l to 10 μg/l. The new recommended value is based largely on analytical capability. If the standard basis for risk assessment applied to industrial chemicals were applied to arsenic, the maximum permissible concentration would be lower still. The USEPA limit was also reduced from 50 μg/l to 10 μg/l in January 2001 following prolonged debate over the most appropriate limit. The European Community's maximum admissible concentration (MAC) for As in drinking water is also to be reduced to 10 μg/l. The Japanese limit for drinking water is 10 μg/l and the interim maximum acceptable concentration for Canadian drinking water is 25 μg/l.

Whilst many national authorities are seeking to reduce theirs limits in line with the WHO guideline value, many countries and indeed all affected developing countries, still operate at present to the 50 μg/l standard, in part because of lack of adequate testing facilities for lower concentrations (Smedley and Kinniburgh, 2001). Assessment of arsenic exposure via drinking water is often based on the measurements of arsenic concentration in drinking water and assumptions regarding the amount of water consumed. Such data are estimates, the uncertainty of which will depend on the method used.

2. Background Information

After recognizing the alarming situation of arsenic contamination in southern Thailand, intensive surveillance and studies have been carried out by various organizations, such as the Environmental Research and Training Center (ERTC), the Department of Environmental Quality Promotion (DEQP), the Pollution Control Department (PCD), the office of Environmental Policy and Planning (OEPP), the Department of Health (DOH), the Department of Mineral Resources (DMR), and the British Geological Survey (BGS). All such studies indicated that water resources in Ronpiboon were contaminated with arsenic, at concentrations exceeding the World Health Organization's standards (the standard specifies a maximum value in potable water of 10 μg/l). The water in some shallow wells had arsenic levels 50-100 times higher than the standard value. In addition, it was found that several sites in Ronpiboon were heavily contaminated. Old tin mines, and primary and secondary tin ore deposits were suspected to be the source of arsenic contamination. Run-off from some areas of ore deposits, and tin dressing processes, were suspected to have caused soil contamination, which then passed through to shallow groundwater (JICA, 2000).

Although, studies of arsenic contamination of groundwater in the area have been carried out by various agencies, the source of the arsenic, and the contamination mechanisms, have not been yet studied in detail. In an early stage of tackling this problem, the Thai Government, through concerned agencies, have transported water from outside the area, and also provided water jars to local residents for collecting rainwater. In addition, recently, the government has constructed simple water pipelines; however these did not cover the entire area affected, and the water quality was not controlled according to Water Quality Standards set by the Ministry of Public Health.

Arsenic contamination in Ronpiboon has caused sickness, and seriously endangered public health. Contaminated water has been spread through groundwater, and residents drinking the groundwater have been exposed to arsenic contamination. An urgent action is therefore needed in order to clarify the precise causes and contamination mechanisms and to devise and implement appropriate mitigation measures to remedy the contaminated areas. The overall aim of this research was to study the status of arsenic contamination in Southern Thailand through source identification of arsenic contamination; and the assessment of arsenic concentration levels in the study area.

3. Methods

Scope and methodology of this research included: a) identification of the study area of arsenic contamination in southern Thailand as Ronpiboon district in Nakhon Si Thammarat province; b) review of the various past studies on identification of sources of arsenic contamination in the study area; c) field testing on samples collected from shallow wells, groundwater pumps, as well as surface water and analyzed for arsenic concentrations for assessment of contamination.

Field Test Kit

Asia Arsenic Network (ANN) Field Test Kit modified by National Institute of Preventive and Social Medicine of Bangladesh (NIPSOM) was utilized in Ronpiboon field trip. Modified Arsenic Field Test Kit is a cheap and easy device for detection of arsenic in water. This improved field kit technique offers a user friendly and simple way of testing a large number of samples at the community level to screen out the contamination of the tubewell water. The results can be obtained within 5 minutes and it can detect arsenic from 0.01 mg/l to 0.5 mg/l or higher. The field kit method is based on the principle of Gutzeit reaction (Eq.1), which is a qualitative or roughly quantitative method; it is sensitive to as little as 0.001 mg As and provides an accuracy of ± 0.005 mg (Ahmad et al., 1997).

Gutzeit Reaction:

$$\underset{}{AsH_3} + \underset{\text{(mercuric bromide)}}{2HgBr_2} \rightarrow \underset{\text{(arseno mercuric bromide)}}{AsH(HgBr)_2} + 2HBr \qquad (1)$$

In this method, arsenic gas is liberated from an arsenic containing water sample; arsenic gas produces color on filter paper treated with mercuric bromide. A range of colours from faint yellow to dark brown may appear on the filter paper, which can be translated into the corresponding arsenic concentration from a precalibrated chart provided with the kit.

In ground water, arsenic usually occurs as arsenite (As(III)) and arsenate (As(V)). To determine the arsenic in water, arsenate is reduced to arsenite by the reducing agents (Potassium iodide and Stannous chloride). As(III) is then reacted with Zinc and HCl to produce arsine gas, as shown in Eqs. (2)-(4).

Reduction:

$$As^{5+} \xrightarrow[KI, SnCl_2]{} As^{3+} \quad (2)$$

$$Zn + HCl \rightarrow ZnCl_2 + H_2 \quad (3)$$

$$As^{3+} + H_2 \rightarrow AsH_3 \quad (4)$$

Color change (yellow to reddish brown) produced on the disc paper by reaction of arsine gas with modified mercury bromide indicates the presence of arsenic in water. When no color appears on the disc paper, it indicates absence of arsenic or the concentration to be below detectable limit. To avoid interference with hydrogen sulfate, cotton ball impregnated with lead acetate solution may be used.

4. Description of the Study Area

Location: In Thailand, one of the areas heavily contaminated with arsenic is at Ronpiboon sub-district, Nakhon Si Thammarat province. Ronpiboon area is at the center of a tin mineralized zone, about 800-km south of Bangkok on the Malay Peninsula, as shown in Figure 1. It is situated at the foothills of Khao Suan Chan Mountain (318-m in elevation) with Khao Mak Mountain (265-m in elevation) north of the district, and the Ban Trok Bon Mountain (over 260-m in elevation) in the south. Khao Ron Na and Khao Suan Chan Mountains are a part of the Khao Laung Mountains. The main rivers that pass through the study area are Huai Hua Mueng and Huai Ron Na. The water route is connected to Pak Panang River and finally discharge to the open sea at Pak Panang Gulf (JICA, 2000). Ronpiboon sub-district is divided into 16 villages, starting from Moo 1 - Moo 16.

Mining Sites: Primary deposits in Khao Suan Chan and Khao Ron Na Mountains have been mined since one hundred years ago. As the mining activities, which are associated with arsenic contamination in the area, were banned by the Department of Mineral Resources (DMR), those site were closed down 15 years ago. However, arsenic contamination caused by past mining activity remains in the area. In the vicinity, downstream from the foothills, contaminated waste, was found to have been transported and spread by flooding. The waste from the concentrator of previous mining sites may have been dumped into the dredging pond located west of the town. Concentrator waste of the town was considered a major cause of arsenic contamination. Along the Ronpiboon route, near Num Khun canal, a layer of clay one meter thick had been brought to cover the laterite soil. This resulted in a noticeable increase in measured contamination; (laterite under such reducing conditions causes arsenic to be released into the groundwater). The site along the Ronpiboon – Khao Chum Thong route was covered by a two-meter thick layer of landfill material, which might have been contaminated by arsenic. Here, under chemically reducing conditions created by the landfill, and some domestic sewage, laterine may have released arsenic (JICA, 2000).

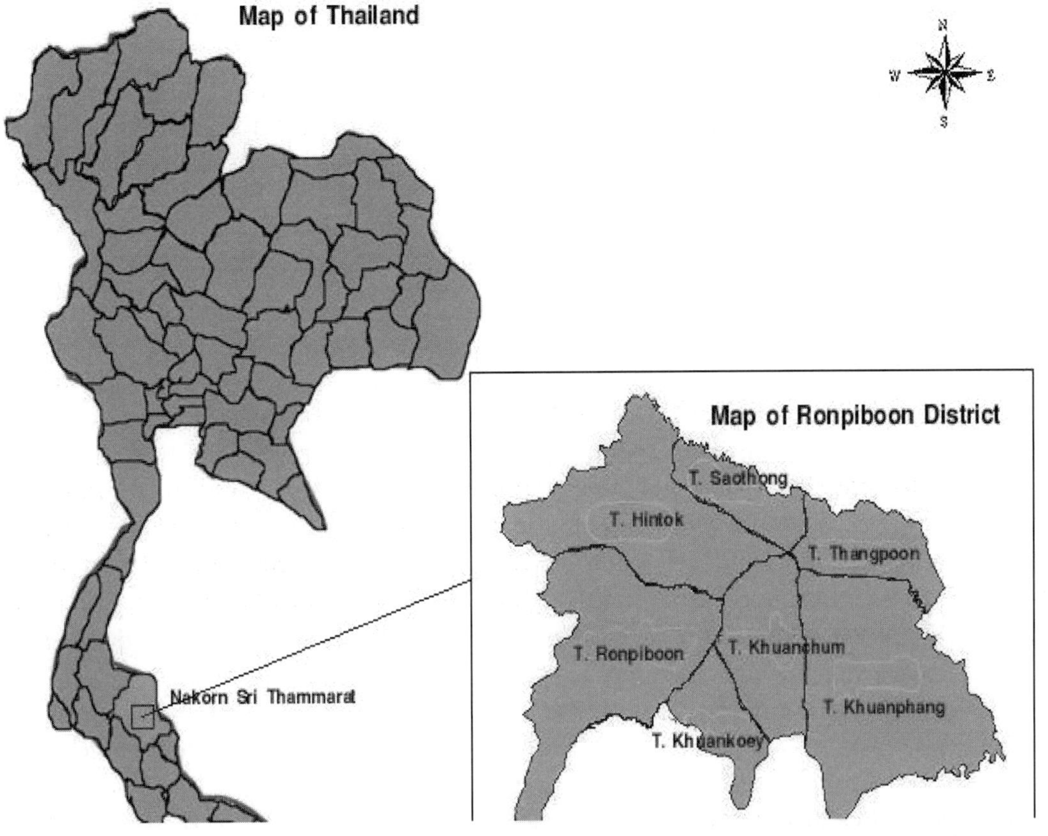

Figure 1. Regional Setting of Ronpiboon District (Source: JICA, 2000, http://www.medipe.psu.ac.th/ars_map.php).

Water Supply in Ronpiboon: Currently, water supply in Ronpiboon District is mainly provided by 2 agencies, the Provincial Waterworks Authority (PWA) and the Ministry of Health (MOH). Water is also supplied partly through the use of deep wells. Villages 2, 3, 8, 10 and 11 draw water from streams, using the flowing surface water directly without any prior treatment (JICA, 2000).

Water Use Conditions: Water supplied by the waterworks is mostly used for laundry, bathing and other miscellaneous purposes, hardly for drinking and cooking. Rainwater is used for drinking and cooking. The results of water use survey carried out by the Pollution Control Department (PCD) by interviewing 307 households in Ronpiboon subdistrict are shown in Table 1.

Table 1. Water use in Ronpiboon Subdistrct

Water source	**No. of households**	**% usage**
Rainwater	262	85
Waterworks	144	47
Shallow wells	129	42
Surface water	14	5
Bottled water	15	4

Source: JICA, 2000.

5. Results and Discussions

Identification of Contamination Sources

From Khao Suan Chan Mountain and Khao Ron Na Mountains to Pak Panung River, there have been tin-mining activities in Ronpiboon District, Nakhon Si Thammarat Province for more than 100 years. Considering the arsenic load to the Ronpiboon sub-district, a primary ore deposit or mineralized zone, in nature, contributes very little to the arsenic contamination of area. On the other hand, it is recognized that roasting process, in which arsenic was removed from tin crude concentrates, existed in the concentrator ruins in the study area. Ore dressing waste containing arsenic in sulfide form (arsenopyrites), oxide form (as As_2O_3), and sulfate form, was generated during their operation. Some part of these wastes have remained inside and/or along circumference of old concentrator. Large amounts of them might have been dumped at some places within Ronpiboon district (JICA, 2000; Praijitprapaon, *et al.*, 1995; Vitayavirasuk and Thongboriboon, 1997; Williams, *et al.*, 1996; and Wittayawarawat, 1994). Five identified arsenic contaminated areas were reported by a recent study as indicated below (JICA, 2000):

i. **Around the foothill concentrator and its downstream**: The concentrator waste was transported by flood and spread in the area.
ii. **Around the town concentrator**: The concentrator waste might have been dumped to the dredging pond west of the town concentrator
iii. **The area of laterite soil**: Clay of one meter thick was brought to cover the area and laterite soil was made to be in a reduction condition. Laterite under a reduction condition released arsenic into groundwater.
iv. **Landfill site**: The site is covered with two-meter thick landfill material. The landfill material might be contaminated by arsenic, like the concentrator waste. Laterite may release some arsenic because of reduction condition created by the landfill and some domestic waste.
v. **New and old waste dump**: Contamination might have been caused by reduction condition created by organic domestic waste and some concentrator waste might be mixed with domestic waste.

Mechanism of Arsenic Release from a Contaminant Source

a) Geochemical Phase of Arsenic When arsenic is concentrated in the sedimentary strata, generally, it is in the form of sulfide such as arsenopyrite or co-precipitated with or adsorbed onto oxide/hydroxide of Fe and Mn. Conventional elution test of soil can detect As_2O_3 included in the roasting waste but it cannot easily detect arsenic in sulfide or oxide/hydroxide of Fe and Mn. That means the elution test alone may not be able to identify the contamination source and even if it could, it is difficult to clarify the mechanism of arsenic release. From this viewpoint, sequential extraction of soil samples should be carried out to find the geochemical phase of arsenic to identify the contamination source and clarify the arsenic release mechanism.
b) Physical and Chemical Conditions Release of arsenic from Fe oxide or hydroxide was controlled by the oxidation-reduction potential. Fe (II) ion co-precipitates with arsenic in water in oxidizing condition and forms hydroxide. Fe oxide and hydroxide also adsorbs arsenic very well in the oxidation condition. However arsenic co-precipitated with Fe, and arsenic adsorbed onto Fe oxide will be released into the environment from the Fe (II)

ion under the reducing condition. On the other hand, the possibility of arsenic release from sulfide is considered small because arsenopyrite is relatively stable under normal conditions.

c) Arsenic Species Analysis and Behavior Arsenic is the element whose toxicity is changed by its chemical speciation. There are over several hundreds to thousands differences in toxicity between inorganic arsenic in water and organic arsenic in an aquatic organism. Even within inorganic arsenic, trivalent arsenic and pentavalent arsenic have several times difference in toxicity. As (III) is more toxic than As (V). Furthermore, behavior in soil and groundwater changes by its speciation. As (III) is near neutral in terms of its electric charge and hardly adsorbed, while As (V) is positively charged and easily adsorbed. Therefore, arsenic speciation is important in study of arsenic pollution (JICA, 2000).

Results of the On-site Measurement for Arsenic Contamination

Based on the results of the on-site measurements in Ronpiboon town at 6 sites, on 2nd May 2001, arsenic concentrations in the shallow stream, shallow groundwater, and shallow well pumps in the study area were exceeding the USEPA and WHO's drinking water standards. Table 2 shows the overall results of the on-site measurement.

Table 2. Results of the On-site Measurement for Arsenic Contamination

Location No.	Samples	Concentration (μg/l)	
		Sample 1	Sample 2
1	Surface water – shallow stream (near the bridge)	20	150
	Shallow groundwater (near the bridge)	> 700	> 700
	2 soil samples collected	-	-
2, 3, 4	Mud and a few soil samples collection only	-	-
5	Shallow well, house # 245, Village 12	10	< 10
	2 soil samples collected	-	-
6	Shallow well pump	10	< 10

6. Conclusions

Sources of arsenic contamination in the Ronpiboon district were identified to be the previous tin mining activities around the area. Oxidation of arsenic in the soil caused contamination in groundwater and flooding spread out the contamination around the area. Villagers in Ronpiboon are suffering from chronic arsenic poisoning with skin cancer or "black fever." However, there is no clear answer about how arsenic was distributed in this region. Government agencies and research groups have had many studies conducted over the last few years, but no conclusion of the arsenic contamination mitigation in the area had been reported. Thus, any knowledge about the simple methods of arsenic removal from drinking water would be very helpful to the general public in Ronpiboon area. The two low-cost arsenic removal techniques for drinking water could be kalshi (pitcher) filtration used in Bangladesh and/or small-size water-filtering system enhanced with coagulation used by some researchers in Prince of Songkla University in Thailand.

7. Acknowledgements

The principal investigator of this study wishes to express her special thanks to the Suranaree University of Technology for funding this research. She is also greatly indebted to Mr. Abu Sayed Md. Kamal of Asian Institute of Technology for lending the field test kit and to Dr. Udomphon Puetpaiboon, and his students and staff at Prince of Songkhla University for their help and assistance provided during the field visit to Ronpiboon district. The Environmental Research and Training Center (ERTC) is also acknowledged for providing some data and information.

8. References

Ahmad, S.A., Sayed, M.H.S.U., Hadi S.A. and Khan A.W. (1997). Modified Arsenic Filed Test Kit: A Cheap and Easy Device for Detection of Arsenic in Water, *Journal of Preventive and Social Medicine*, Vol. 16 (2), pp.143-150.

Boriboon, P. (2000). Emerging Chemical Food Problems, *Journal of Environmental Medicine*, Vol. 2 (1), pp. 78-86.

Japan International Cooperation Agengy (JICA), The Kingdom of Thailand, and The Ministry of Science, Technology and Environment (2000). Final Report (Summary) on the Environmental Management Planning Survey for Arsenic Contaminated Area of the Nakhon Si Thammarat Province in the Kingdom of Thailnad, Misui Mineral Develoment Engineering Co., Ltd. And Kokusai Kogyo Co., Ltd.

Kamal, A.S.Md. (1999). A Preliminary Study on Source Identification of Arsenic Contamination in Hizla Bangladesh, Master of Science Thesis, Asian Institute of Technology, Thailand.

Paijitprapaon, A., Charoenchaisri, P. and Williams, M. (1995). Chronic arsenic poisoning in Ronpibool District, Nakorn Srithammarat Province, Southern Thailand: how should we balance between economics and health?, Second International Conference on arsenic exposure and health effects, June 12-14, 1995, San Diego, CA.

Smedley, P.L. and Kinniburgh, D.G. (2001). Source and behaviour of arsenic in natural waters, United Nations Synthesis Report on Arsenic in Drinking Water (http://www.who.int/water_sanitation_health/Arsenic/ArsenicUNRep1.htm).

Thanh, M.T. and Long, P.P.P.E. (2000). Groundwater Arsenic Contamination: Can it happen in the Mekong Delta? A Vietnamese Perspective (http://www.mekongforum.org/long4e.html).

Vitayavirasuk, B. and Thongboriboon, T. (1997). Arsenic contamination in shallow groundwater and surface water environment in Ron Phibun Subdistrict, Ron Phibundistrict, Nakhon Si Thammarat Province, Southern Thailand, Toxic metal studies of Pak Panang and Pattani river basins, Oct, 1997, Research and Development Office, Prince of Songkla University (http://www.psu.ac.th/epidemiology).

Williams, M., Fordyce, F., Paijitprapapon, A. and Charoenchaisri, P. (1996). Arsenic contamination in surface drainage and groundwater in part of the southeast Asian tin belt, Nakhon Si Thammarat Province, Southern Thailand, Environmental Geology, Vol. 27, pp. 16-33 (http://www.psu.ac.th/epidemiology/inpub/william.htm).

Wittayawarawat, W. (1994). Contamination of arsenic, cadmium and lead in Pak Phanang river basin, Changwat Nakorn Si Thammrat Environmental Management, Prince of Songkla University (http://www.psu.ac.th/epidemiology).

Assessment of Combined Sewer Overflow impacts on receiving water: A case study of Bangkok

P. Bipin Kumar[1)] and H. Satoh[2)]

1) Environmental Engineering and Management Program, Asian Institute of Technology
Pathumthani, 12120, Thailand,
e-mail: bipin@ait.ac.th
2) Environmental Engineering and Management Program, Asian Institute of Technology
Pathumthani, 12120, Thailand,
e-mail: satohi@ait.ac.th

Abstract

The effect of combined sewer overflow on receiving water body in tropical rain climate was investigated in Rattanakosin, Bangkok. Rattanakosin is historic and tourism area, and is surrounded by Chao Phraya River and canals. The area is serviced by a combined wastewater collection and treatment system. Most of the combined sewer overflow (CSO) in rainfall event is discharged to the canals. A CSO chamber located in the study area, with catchment area of 0.50 ha was selected for the monitoring of the inflow, overflow of sewage and quality of sewage in rain events. The monitoring was conducted on three rain events, dated on June 4, July 9, and July 12, 2003 and the average intensity of rainfall were 52 mm/hr, 13 mm/hr and 37 mm/hr respectively. The overflow from CSO chambers on June 4 and July 9 were measured to be 132 m^3/ha and 26 m^3/ha respectively and no overflow on July 12. The behavior of soluble COD variation was almost similar in all cases that initial runoff had higher concentrations due to wash-off of contaminants from the ground surface and accumulated matters in sewer pipes. It was found that the peak of organic load per unit area came several minutes after the peak of soluble COD load concentration. The amount of organic pollutants discharged to the environment on the first and second events were calculated to be 13.4 kg/ha and 0.69 kg/ha respectively. Based on the monitoring results, the estimated COD load to the canals from CSO was 29.2 ton/month while COD treated by the treatment plant was estimated to be 160 ton/month. While wastewater collection and treatment system is contributing a lot to reduce the COD load to the canals, there could be room for improvement about CSO.

Keywords: tropical climate, urban runoff, combined sewer overflow, Bangkok, environmental management.

1. Introduction

The focus of urban drainage in the 1990s has shifted from not only the need to provide structurally sound drainage systems with an adequate hydraulic capacity but also to meet environmental quality objectives for the receiving water course. There has therefore been a need to develop the appropriate technical procedures and tools for the control of urban sewer flow quality. In combined sewer systems, the sewer flow quality is critical mainly during the first flush. Thornton and Soul, (1986) defined the first flush as the initial period of storm flow during which the concentration of pollutants is significantly higher than those observed at a later stage

of the storm event. Gupta and Soul (1996) developed the regressional relationships and found that the maximum rainfall intensity, rainfall duration and the antecedent dry weather period were found to be most important parameters influencing the first flush load of suspended solids. On the other hand, urban wet-weather runoff from impervious surfaces is routinely delivered into surface waters without further treatment, having significant impact on the quality of the receiving water, sediments and biota (Pitt *et al.*, 1995). Urban water management's aim is to reduce the amount of runoff into the surface waters (Marsalek *et al.,* 1993). The adverse effects of discharging wastewater from CSO to receiving waters are even stronger in terms of oxygen depletion and eutrophication than those of pure urban runoff discharge (Marsalek *et al.*, 1993).
In Bangkok, CSO discharging in wet weather, directly into the receiving water is frequent. In spite of the installation of sewerage systems, discharging of such untreated wastewater into the environment in Bangkok may still be a significant source of pollutants. The causes of CSO are due to intense rainfall and insufficient capacity of existing runoff collection and treatment systems. The average annual rainfall in Bangkok in the last eleven years was 1503.1mm and maximum daily rainfall recorded in the last five year was 165 mm on September 02, 1998 (TMD, 2002). It is necessary to assess how much organic pollutants come through first flush and overflow from the CSO chambers into the environment without being treated.

2. Study Area and Monitoring Methods

Study area overview

The study area is located in Rattanakosin, western part of Bangkok. It is the centre of culture, art, architecture and a popular tourist destination in Bangkok. The total population of Rattanakosin city in 2001 was 73,541. The area of Rattanakosin is 414 ha and separated from other areas by Chao Phraya River in the west and South, Banglumpoo canal in the north, Ong-Ang canal in the East, as shown in Figure 1. It also includes Khoomuang Derm canal, Wat Thepthida canal and Wat Rachabopit canals which are the main network for wastewater discharge. The study area consists of 35% residential area, 20 % commercial area, 30 % institutional area, 10 % open parks and 5% others. Bangkok Metropolitan Administration (BMA) constructed the sewage collection network and a wastewater treatment plant for the area which is in operation since 2000.
Figure 2 shows an overview of the present schematic of wastewater and rain water collection and treatment in the study area. Human excreta is collected in septic tanks and the supernatant overflow from the septic tank is either discharged to the collection systems or infiltrated to ground water. Grey wastewater and urban runoff are collected by combined sewer line. The capacity of the wastewater treatment plant and the main pipelines has capacity to handle 5 times of dry weather flow. As shown in Figure 2, the excess wastewater in rain event is discharged to the environment through CSO chambers.

Sewage flow monitoring

A CSO chamber located in the study area, with catchment area of 0.50 ha as shown in Figure 1 was selected for the monitoring of flow rate and quality of sewage during the rain event. The velocity of flow and discharge at the inlet and overflow of CSO chambers were measured at 0, 1, 3, 5, 7, 10, 15, 20, 25, 30, 40, 45 minutes after the start of rainfall for three rain events on June 4, July 9 and July 12, 2003 using current meter and staff gauge respectively. The quantities of inflow and overflow of sewage were calculated by using continuity equation. Rainfall was measured by a rain gauge installed at the site at same time

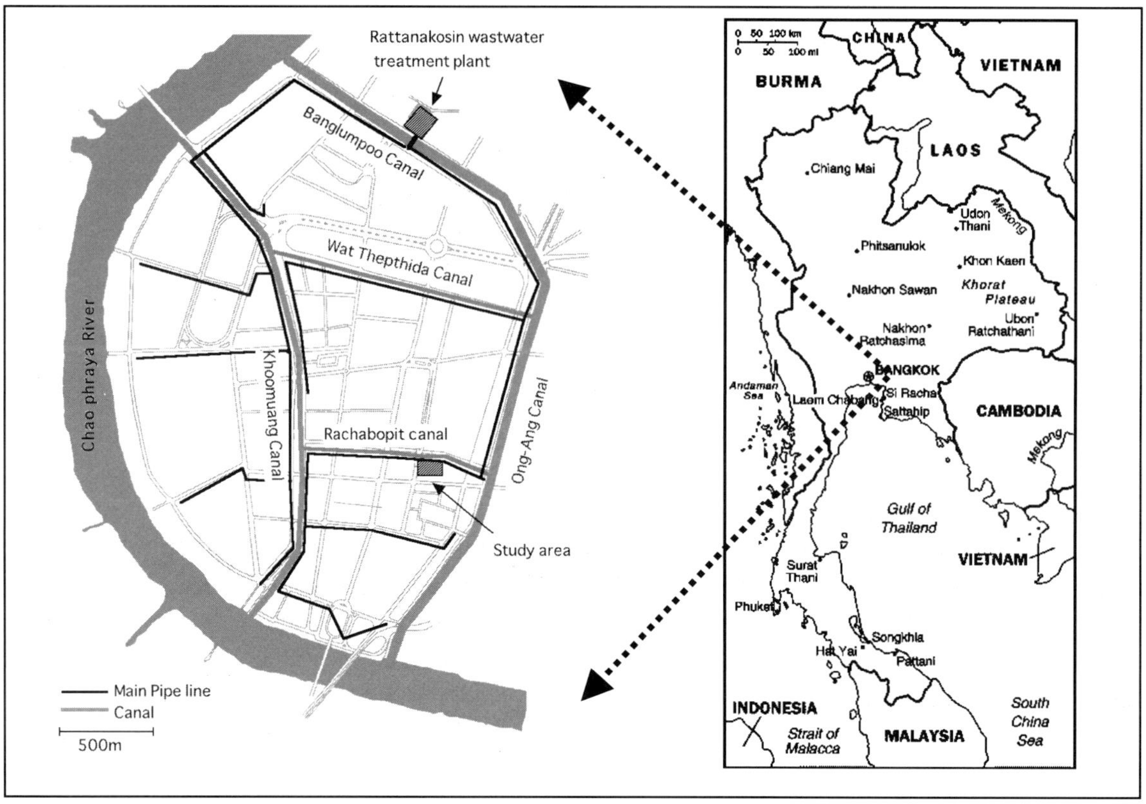

Figure 1. Overview of wastewater collection and treatment systems

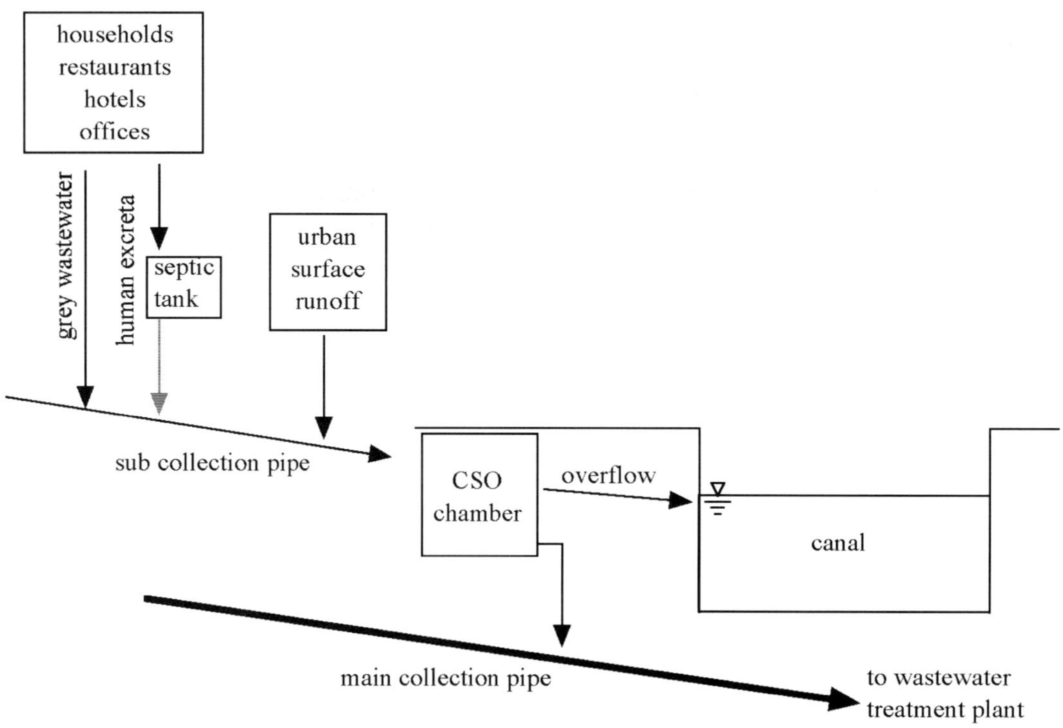

Figure 2. Schematic of the combined sewer system in Bangkok

interval as in flow measurement. Water samples were collected at the inlet of CSO chambers at the same time interval as in flow measurement. At the end of monitoring, the water of the receiving water body Rachabopit canal was sampled also. The water samples were taken to the laboratory, filtered by glass fiber filters and subjected to the analysis of chemical oxygen demand of soluble (filtrate) part (SCOD). The COD measurement was conducted by the closed tube digestion followed by colorimetric determination method (APHA, 1998)

3. Results and Discussion

Sewage quantity and quality

Figure 3 shows the intensity of rainfall and variation of flow rate observed on three sampling days. On June 4, the principal characteristics were: (1) the intensity of rainfall ranged from 62 mm/hr to 105 mm/hr with the duration of 30 min (2) the CSO was observed after 7 minute of start of rainfall and the total discharged volume was estimated to be 132m^3/ha. Rainfall on July 9 had intensity of 20mm/hr to 40mm/hr in the first 5 minute, and then reduced to less than 10mm/hr with total duration of 45 minutes. The overflow was observed after 15 min of the rain and the amount of CSO was 26m^3/ha. However the rainfall intensity was lower than that of June 4 and July 12, CSO was observed due to the proceeding heavy rain on July 8. Furthermore, there were two medium rain events on July 5 and July 6. The rain on July 12 had intensity of 60mm/hr to 120mm/hr for about 15 min and then gradually decreased and ceased in about 45 min. But all combined sewage was collected to the sewer pipe, and no CSO was observed.

Figure 4 shows the soluble COD concentration at CSO chambers. On June 4, soluble COD at the beginning was 50 mg/l, jumped up to around 479 mg/l at the 7th min as the peak, and then was reduced to between 40mg/l to 140mg/l during the rain. The COD concentration was very high because of dry weather since May 21. Observation showed that soluble COD concentration was remarkably increased at last quarter of first flush. The increment in COD concentration might be due to the discharge of wastewater from mobile restaurant at once and erosion of accumulated matters in sewers. The soluble COD concentration at the beginning was 247 mg/l and gradually decreased to 24 mg/l. In contrary to June 4, the soluble COD concentration on July 9 was highest at the beginning.

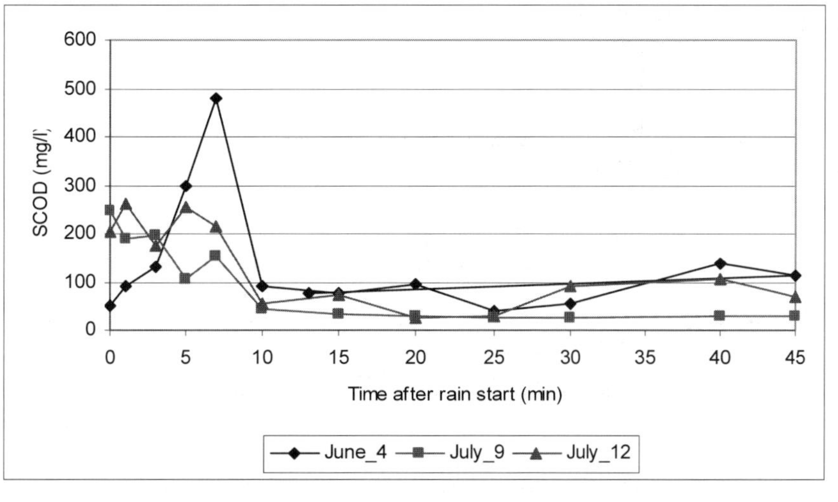

Figure 3. Variation of soluble COD concentrations (mg/l)

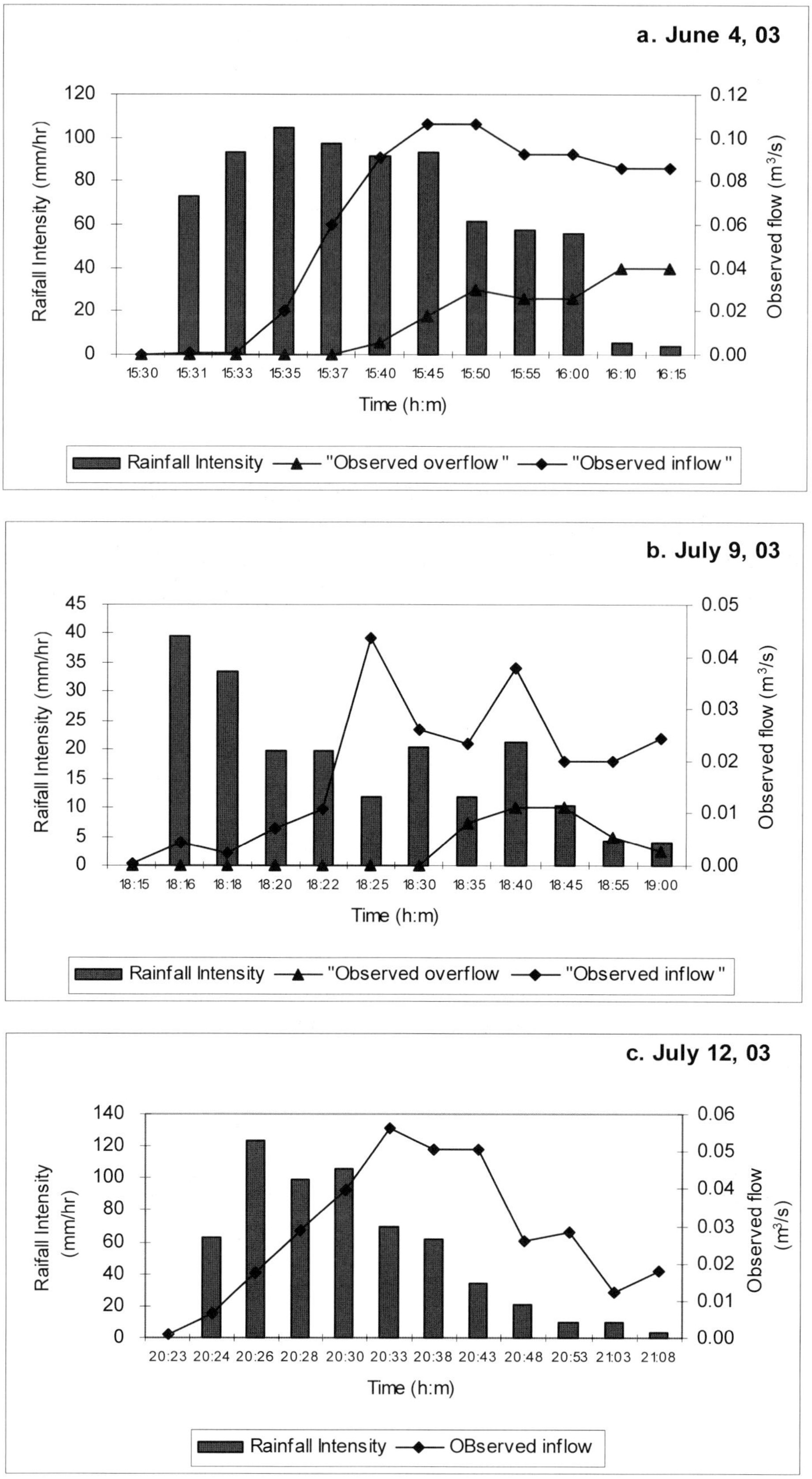

Figure 4. Temporal variation of Rainfall observed inflow and overflow

The cause of high initial COD might be due to remnant of soluble organic matter inside CSO chambers which resulted from the choking of CSO chambers caused by heavy rain on July 8. On July 12, soluble COD was around 200 mg/l in the initial 8 min, and then decreased to around 30 mg/l. Like the first event, the COD concentration was increased at the last quarter of first flush which indicates that the source of pollution was the same. From Figure 3, it is apparent that the pattern of SCOD concentration variation was almost similar in all cases indicating that initial phases of the hydrograph were extremely polluted due to wash-off contaminants from the ground surface and flushing of accumulated matters in sewers.

Figure 5 shows the inflow and overflow load of soluble COD at CSO chambers per unit area. On June 4, the inflow COD load was not a significant amount for initial five minutes and it was significant when the overflow stared to occur. The inflow values ranged from 2 kg/ha to 14 kg/ha and overflow ranged from 1 kg/ha to 7 kg/ha. Figure 5, depicts that COD overflow was about the 50% of total COD inflow and it continued till the end of first flush. It indicates that only 50% of COD load was received to the treatment plant. COD overflow load on July 9, was ranged from 0.1 kg/ha to 0.6 kg/ha and no overflow on July 12. It is concluded that the peak of the organic load per unit area was high at the end of first flush and overflow starts at the first quarter of first flush. It indicates that the flow which brings high organic load goes to the surface water during the rain event unless detention facilities

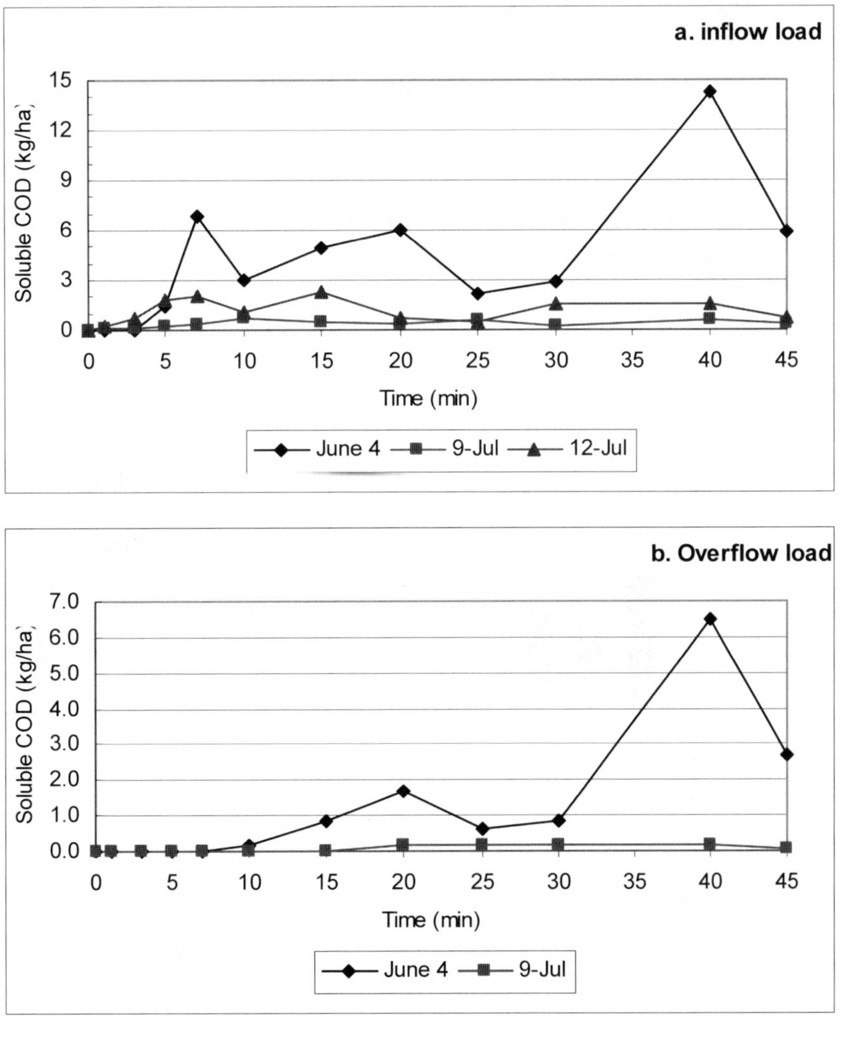

Figure 5. Variation of soluble COD load

are available. The variation of soluble COD loads can be justified by considering, for instance, the local variability of precipitation itself, as well as the duration of dry weather period preceding to the storm event.

Based on the monitoring results, the organic load to the canals caused by CSO was estimated. Statistical data showed that there were 14 heavy rainfall events per month with CSO in rainy season. It was supposed that there were 10 heavy rain events that made the overflow from the CSO chambers. The average COD load 7.05 kg/ha is adopted for the estimation of organic load to the canals caused by CSO. The estimated COD load was observed to be 29.2 ton/month.

The amount of COD that is collected by the collection pipe and treated by the wastewater treatment plant is estimated to be around 160 ton/month. When the above value is compared with this, it is expected that the wastewater collection and treatment system is contributing a lot to improve the canal water quality. On the other hand, the average soluble COD concentration in the canal was 42 mg/l and organic load in the canal itself was estimated to be in the order of 13.7 tons. An assessment showed that the velocity of water in the canal is less than 1 cm/s and polluted water could remain longer than one week. That is, weekly COD load deposited to the canal could account for about 50 % of total COD in the canal. It is indicated that while the wastewater collection and treatment system is significantly reducing the COD load to the canals, further study is needed to clarify the effect of CSO to the water quality in canals.

4. Conclusions and Recommendations

At Rattanakosin area, Bangkok, the impact of combined sewer overflow to receiving canals was investigated. As has been suggested by many researchers, it was found that the sewage soluble COD was the highest at the beginning or shortly after the start of rain. It was also found that in one of the rains monitored, higher COD was also observed at the latter stage of rain, suggesting that strong wastewater was discharged in the water collection area. The COD load from CSO could be affecting the water quality of the receiving canals, and further study is needed to more accurately assess the effect.

The best solution would be to construct in-line or off-line storage of combined sewage in rain events, but this would be very costly. Still, counter measures could be taken such as; (1) to better manage water levels in the collection pipes so that it can absorb initial runoff more effectively, and (2) to advise people in the area to discharge strong wastewater in dry weather but not in rain. It should be noted that the wastewater treatment plant is significantly contributing to the reduction of the organic load to the canals. But there might be a chance to further increase the contribution of the wastewater collection and treatment system of the area by the control of CSO.

5. Acknowledgement

The authors would like to express their sincere thanks to Ms. S. Kessomboon and Dr. K.-R. Klangkrong in Bangkok Metropolitan Administration and the staffs in Rattanakosin Wastewater Treatment Plant for their kind support for the project. The authors would also like to note that the study was financially supported by Japanese Government through JICA and GMS-JG fellowship. The study was also partially supported by SACWET.

6. References

American Public Health Association (APHA) (1998). Standard methods for water and waste water examination, 20th Ed., Washington, DC.

Gupta, K. and Saul, A. (1996). Specific relationships for the first flush load in combined sewers flows, *Water Res.*, Vol. 30 (5), pp. 1244-1252.

Marsalek, J., Barnwell, T. O., Geiger, W., Groukert M., Huber, W. C., Saul, A. J., Schilling, W. and Torno H. C. (1993). Urban drainage systems: design and operation, *Water Science and Technology*, Vol. 27 (12), pp. 31-70.

Pitt, R., Field, R., Lalor, M. and Brown, M. (1995). Urban stormwater toxic pollutants: assessment, sources, and treatability, *Water and Environmental Research*, Vol. 67, pp. 260-275.

Thai Metrological Department (TMD) (2002). Weather report (www.tmd.go.th).

Thornton, R. C. and Saul, A. (1986). Some quality characteristics of combined sewer flows. *Publ. Hith Engr*, Vol. 14, pp. 35-38.

The Distribution of Endocrine Disrupting Chemicals Bisphenol A and Alkylphenols in Water Environments around Bangkok

K. Zang[1)] and H. Satoh[2)]

1) Environmental Technology & Management, Asian Institute of Technology
zangkaisai@hotmail.com
2) Environmental Technology & Management, Asian Institute of Technology
SERD/UEEM, P.O. Box 4 Klong Luang, Pathumthani 12120, Thailand
satohi@ait.ac.th

Abstract

The concentration levels of endocrine disrupting chemicals (EDCs), bisphenol-A (BPA), 4-n-nonylphenol (NP) and 4-tert-butylphenol (BP) was determined in three polluted canals and treated and untreated wastewater from two wastewater treatment plants in and around Bangkok, Thailand to discuss the potential risk of their environmental estrogenic effects on aquatic organisms.
Target EDCs in water samples were analyzed using liquid-liquid extraction (LLE) followed by GC/MS analysis. High BPA contamination levels (~1 μg/l) were detected in all three selected canals. The highest concentration of BPA (8.2 μg/l) was found in the secondary effluent of an industrial wastewater treatment plant (Bang Pa-In). NP was detected at the concentration of 0.5 μg/l in Klong Dan and 0.095 μg/l in Klong San Sap. 0.78 μg/l of BP was detected in Klong Dan and 0.11 μg/l in Klong San Sap.
At the existing concentration levels of target EDCs in three examined canals, the risk of estrogenic effects on local aquatic organisms could exist. The findings in this study also imply the contributions of both domestic and industrial wastewaters to the BPA contamination and the significant contribution of industrial wastewater to both BP and NP contaminations.

Keywords: alkylphenols, Bangkok, bisphenol A, endocrine disrupting chemicals, GC/MS

1. Introduction

Since the 1950's, there has been an increasing number of reports in wild life on abnormal reproductive functions and behaviors, demasculinization of males, and decreased hatching success (Colborn, 1996). Later, evidence showed that endocrine disrupting chemicals are the contributing factors for these observations.

Bisphenol A (BPA) was reported to affect the sexual differentiation of frogs even at very low exposure levels, (Kloas, 1999). 4-Nonylphenol was observed to stimulate vitellogenin gene expression in trout hepatocytes (White, 1994). 4-tert-butylphenol and nonlyphenol, as well as bisphenol A, were reported to cause proliferation of human estrogen-sensitive MCF-7 breast tumor cells (Soto *et al.*, 1995 Colborn, 1996).

Despite the anticipation pointed out by many researchers that EDCs might be affecting ecosystems and even human health, different kinds of suspected EDCs are still being used in industrial, agricultural and household activities. BPA is the raw material in the production of polycarbonate, epoxy resins, unsaturated polyester-styrene resins and flame retardants (Jason, 2002). Nonionic surfactants alkylphenol polyethoxylates are used widely in heavy-duty laundry powders, personal care products and liquid detergents (SPEED'98/JEA, 1998). After discharged into environment, they degrade into more persistent products, alkylphenols, such as nonylphenol (NP) and butylphenol (BP).

Water is one of the environmental media where pollution by EDCs is anticipated. A report estimated that 37% of total nonylphenol ethoxylate industrial wastes in U.K. enter the water environment as wastewater. An estimated 60 to 65% of all nonylphenolic compounds entering the sewage treatment plant are finally discharged into surface water with the effluent (Jason, 2002). The concentration of nonylphenol varies greatly in different surface water systems. The concentration of nonylphenol in six rivers in United Kingdom was reported in the range from 20 to180,000 ng/l (Blackburn, 1995). Though BPA is often detected in contaminated running surface water, the concentrations of them are not high. In the national wide survey in U.S.A., only 5% of the concentrations for these compounds exceeded 1.0μg/l (Kolpin, 2002).

In this study, six water bodies with different pollution backgrounds in the Bangkok area were selected to exam the contamination levels of bisphenol A and alkylphenols (BP and NP). Simple assessment on the risk of estrogenic effects by the EDCs on local aquatic living things was performed. The selected water bodies include three canals (mainly polluted by domestic wastewater, industrial wastewater or urban pollutants) and treated and untreated wastewater from two wastewater treatment plants in office-tourism area and in industry area.

2. Materials and methods

The following chemicals were used in the study. As standards, 4,4'-isopropylidenediphenol (bisphenol A) (Fluka), 4-tert-butylphenol(Fluka), 4-n-nonylphenol (Riedel-dehaen), as the internal standard, 2,4,6-tribomophenol (Fluka), dichloromethane for organic analysis of traces (Merck), and as the derivatization reagent, N, O-bis (trimethylsilyl) trifluoroacetamid with 10% trimethyl-chlorosilane (BSTFA) (Fluka).

Standard solutions with concentrations higher than 100mg/l were prepared in acetone.

Site selection

Samples were collected at the 5 locations shown in Figure 1.

Bang Pa-In wastewater treatment plant is located in an industrial zone, which involves polyester, plastic, electricity and electric industries. BPA, as the raw material of plastic and polyester productions, was highly suspected to exist in the wastewater in this treatment plant; If a cleaning process is performed in electric and electronic industries, as the degradation products, alkylphenols could be detected in the wastewater from such kinds of industries. Rattanakosin wastewater treatment plant was chosen as a sampling site to check the contamination of target endocrine disrupting chemicals in official commercial and tourism zones without the disturbance of industrial activities.

Klong Dan (a canal) is selected based on the considerations that though it is located in a suburb area, it is the downstream canal of Bangplee industrial zone, where glue, plastic, foam and paint industries are settled. Klong I of Pathum Thani is the place highly suspected of domestic wastewater contamination. Shabby houses were observed on both sides. Domestic wastewater was directly discharged into this canal. Domestic wastes were floating above the surface. Klong San Sap is located in an urban area with high population density (central Bangkok city). It is a canal highly contaminated by urban pollutions.

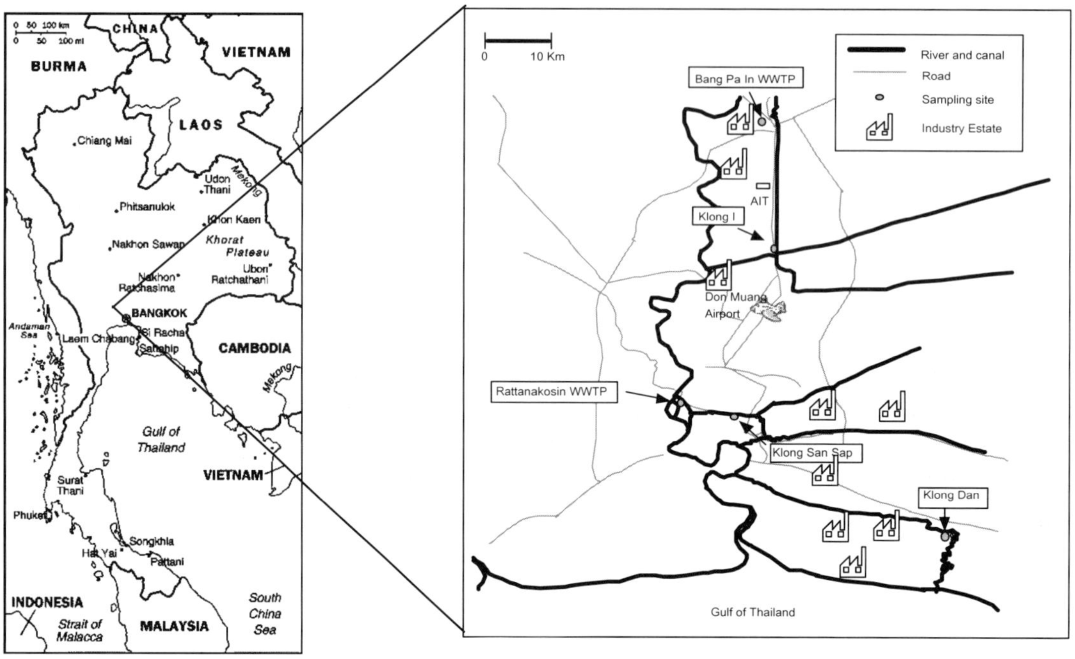

Figure 1. The maps of sampling sites in Bangkok area

Extraction and silylation method

The analysis of the target compounds was performed based on the method suggested by the Environment Agency, Japan (Yasuhara, 1998). A 500-ml water sample was placed into a 1-l separation funnel. 1M H_2SO_4 was added to make the pH around 3, and 120 g of NaCl was added. The sample was extracted with 25ml of dichloromethane under vigorous shaking twice. The extracts were collected, combined, and concentrated by a rotary to 5ml under 30°C. A 100μl of 200ng/μl internal standard (2, 4, 6-tribromophenol) was added to the sample and the sample volume was reduced to 0.5ml by blowing with nitrogen gas. The extract was transferred to a 1-ml vial, and added with 50 μl BSTFA reagent. The extract was heated at 75 °C for 1 hour for silylation.

As for the samples with significant amounts of suspended solids, centrifugation was applied before liquid-liquid extraction. The solid phase was extracted with small amounts of acetone several times by applying ultrasound. The extract was concentrated to 3ml and then added to the supernatant.

GC/MS analysis

The gas chromatography/mass spectrometry (GC/MS) analysis was conducted with a GC/MS system HP5890+HP5971 (Hewlett Packard, USA). A 1.0μl aliquot of the silylated sample was injected to the GC/MS with the splitless injection mode. The injector and the detector temperatures were 250°C and 280°C, respectively. The column oven temperature was 60 °C (5 min) - 230°C at 20°C/min - 238°C at 1°C/min- 280°C at 5°C/min 5 min- 280°C (5min).

The mass spectrometery was performed with the electron impact ionization method at 70eV. The selected ions for quantification and confirmation were: m/z 357 and 372 for silylated BPA; 207,222 for silylated BP; 179,292 for NP and 387,398 for internal standard (2,4,6-tribromophenol).

3. Results and discussion

3.1 Detection Limits and Quantification Limits

Unill now there is no agreement about how to get the detection limit and quantification limit from the standard deviation in gas chromatography, in contrast with other analytical techniques (Zafra *et al.*, 2003). In this work, detection limits were estimated, using the definition proposed by Ho-Sang Shin (Shin *et al*, 2001), in which it is defined by a minimum signal-to-noise ratio of 3. The quantification limit was estimated based on the definition of minimum signal to noise ratio of 9 (Baronti *et al*, 2000). The detection and quantification limits are listed in Table 1.

Table 1. Detection limits and quantification Limits for target compounds

	BP	NP	BPA
DL (μg/ml)	0.0097	0.0051	0.0109
LOQ (μg/ml)	0.0266	0.0159	0.0414

3.2 Recovery

A given water sample was spiked with 100ng/ml of each target compound. The same analysis procedure was applied. The recovery was calculated by comparing the amount recovered as a result of sample preparation with the added amount. The mean recoveries on BP, NP and BPA were 86%, 85% and 104% respectively.

Different sample matrix may vary the recovery ratio in real water sample analysis. In this study, however, the difference of recovery ratio in canal water samples was not significant. The relative errors of each target EDCs recovery in different water samples were 3% of BP (n=3), 4% of NP (n=2) and 5% of BPA (n=3).

3.3 Contamination levels of target EDCs in selected water bodies

The average concentrations of BP, NP and BPA in selected canals, treated and untreated wastewaters are shown in Table 2.

BPA was detected in all three selected canals with high similar concentrations around 1 μg/l, which implies both domestic and industrial wastewater could be the significant contributors to the BPA inputs in selected canals. Both BP and NP were detected in two out of three canals, which was either located downstream of an industry zone (Klong Dan) or in the city center with heavy urban pollution load (Klong San Sap). The concentrations of BP and NP were obviously high in Klong Dan, where it is the downstream canal of an industry zone with plastic, plastic film, glue, foam and rubber industries. As to BP and NP contaminations, the wastewater from these industries may be contributing more to the high concentration levels in the receiving water body.

Regarding the concentrations of target EDCs in treated and untreated wastewater, high concentrations of BPA (8.2 μg/l) and BP (0.32μg/l) were detected in the effluent from Bang Pa-In WWTP, which receives the wastewater from polyester, plastic, electronic and electric industries. Not detectable or considerable low concentrations of target EDCs were found in the effluent from Rattanakosin WWTP, which is located in an office and tourism zone with low potential of target EDCs contaminations.

Table 2. Concentrations of BP, NP and BPA in selected water bodies

Selected water body	4-tert-butylphenol (μg/l)	Nonylphenol (μg/l)	Bisphenol A (μg/l)
Inf. of Bangpa-In WWTP	1.8	0.10	-
Eff. of Bangpa-In WWTP	0.32	N.D.	8.2±0.57
Eff. of Rattanakosin WWTP	N.D.	0.077±0.005	N.D.
Klong I of Pathum Thani	N.D.	-	1.2
Klong San Sap	0.11	0.095	0.92
Klong Dan	0.78±0.038	0.50±0.022	1.3±0.32

The obtained data can be compared with data of NP and BPA contamination levels in surface waters reported by other researchers. In Japan, 95% of rivers and lakes have NP concentration under detection limit; 1% has NP concentration in the range of 0.3-1 μg/l (River Bureau, MLIT of Japan, 2002). Based on these data, it is implied that NP contamination level in Klong San Sap (0.095μg/l) is rather low, and that in Klong Dan (0.50 μg/l) relatively high. However, BPA contamination levels in these three canals were quite high (~1 μg/l), if compared with the contamination level in USA and Japan. In the national wide survey in U.S.A. in 1999-2000, only 5% of the concentrations for these compounds exceeded 1μg/l. In Japan, during 1998-2003, only 1-3% of the rivers and lakes have BPA concentration higher than 0.3μg/l (River Bureau, MLIT of Japan, 2002).

As to the target EDCs concentrations in treated wastewater, much higher concentrations of BPA and BP were detected in Bang Pa-In industrial WWTP than the maximum BPA concentration (0.702μg/l) in Germany WWTPs reported by Fromme and 0.026 μg/l of BP detected in an WWTP effluent in Massachusetts reported by Rudel (1998). The explanations could be the difference of the type of industries, production process and the lack of homogeneity of wastewater.

3.4 Purity checking of peaks observed by GC/MS

With the SIM mode analysis, two ions were monitored for each target analyte, one for quantification, and another for confirmation. The ratio of the abundance of these two monitored ions should be the same for the peaks detected for samples and those detected for standard solutions. If these ratios are significantly different, it means that the detected peak in sample analysis is either not the target compound or overlapped by other unknown compounds. On the other hand, if these ratios are in good agreement, it is strongly supported that the detected peak is for the target analyte.

Table 3 compares the ratios of the mass abundance of confirmation ion to that of quantification ion from samples with those from standards (the value in parentheses). Except for one case for the effluent from Rattanakosin WWTP, the values were in good agreement, meaning that the detected peaks were the target compounds.

Table 3. Relative mass abundance of monitored ions (abundance of confirmation ion over that of quantification ion)

	BP	NP	BPA
Klong I of Pathum Thani	0.12 (0.13)	-	0.15 (0.17)
Klong San Sap	0.15 (0.15)	0.36 (0.36)	0.17 (0.17)
Klong Dan	0.15 (0.15)	0.36 (0.36)	0.17 (0.17)
Effluent of Rattanakosin WWTP	N.D. (0.15)	0.15 (0.29)	N.D. (0.31)
Effluent of Bang Pa-In WWTP	0.17 (0.17)	0.40 (0.36)	0.35 (0.38)
Influent of Bang Pa-In WWTP	0.14 (0.14)	0.39 (0.36)	N.D. (0.32)

*: Values in parentheses are that obtained from standard solutions.

3.5 Possibility of the risk of estrogenic effects

The endocrine disrupting effects of some of EDCs have been reported to have non-mono-relationship of dose to effect. That is, higher concentration can give a smaller effect than at a certain level of dosage. In addition, the sensitivity to EDCs is known to differ very much from species to species. These are the matters that make it difficult to set up clear standards for the control of these chemicals.

As to NP, estrogenic effects have been reported on some aquatic creatures under a low concentration levels. Some critical estrogenic concentrations of NP to specific aquatic creatures have also been reported in previous studies (Warhurst, *et al*, 1995), which were 0.17 μg/k for Sticklebacks, 0.06 μg/k for *Mytilus edulis* and 0.15 μg/k for Macrophylic alga. In the report by River Bureau, MLIT of Japan (2002), the concentration levels of NPs in rivers in Japan are compared with the disorder of testis on Carp. In rivers with the NP concentration around 0.2μg/l, the disorder of testis on Carp was reported to have the high frequency of 40%, though the cause of the disorder cannot simply be attributed to NPs. When these concentration levels are compared with the detected concentrations in this study, the concentrations of NP both in Klong San Sap (0.095μg/l) and Klong Dan (0.5μg/l) could raise a concern about the health of exposed aquatic organisms.

As to BPA, it has been reported that even at very low exposure levels BPA appears to affect the sexual differentiation of frogs (Kloas, *et al*, 1999). Kloas exposed the tadpoles, two or three days after they had hatched, to BPA for 12 weeks. A significant increase in the number of females was found at a BPA concentration of 23μg/l; a smaller, but not significant, increase in the number of females also was found at a concentration of 2.3μg/l. Such an alteration in sex ratio may have an important effect on population levels. Based on the information provided above, with the current BPA contamination level (around 1 μg/l) in three selected canals, the estrogenic effects on local aquatic creatures could not be ruled out.

4. Conclusions

In the present study, at least one or more of endocrine disrupting chemicals, BP, NP, and BPA, were detected in all the six samples examined. High contamination level of BPA (in the order of 1.0 μg /l) was found in all three examined canals, which could be attributed to a number of sources: domestic wastewater, industrial discharges and urban pollutants. Not low concentrations of BP and NP were detected in Klong Dan, a canal downstream of Bangplee industry zone, where the industrial wastewater could contribute higher concentrations of BP and NP in the receiving water body. Low concentrations of BP and NP were also detected in Klong San Sap, a canal located in a heavy-polluted city center area.

BPA concentrations of BPA and BP in the effluent from Bang Pa-In WWTP were found to be much higher than the reported data in the effluent from other WWTPs. Such high BPA residues as 8.2μg/l could contaminate the receiving water body.

Besides the high contamination levels of target compounds in selected water bodies, attention should also be paid to the health of local aquatic organisms exposed to low contamination levels of NP and BPA.

5. Acknowledgement

The authors would like to thank the Southeast Asian Center for Water Environment Technology (Asian Institute of Technology, Thailand, and The University of Tokyo, Japan) for the financial support of the project.

6. References

Baronti, C., Curini, R., Ascenzo, G. D., Gentili A. and Samperi R. (2000). Monitoring Natural and Synthetic Estrogenic at Activated Sludge Sewage Treatment Plants and in a Receiving River Water. Environmental Science & Technology, Vol. 34, pp. 5059-5066.

Blackburn, M. A. and Waldock, M. J. (1995). Concentrations of alkylphenols in rivers and estuaries in England and Wales, Water Res., Vol. 29, pp. 1623-1629.

Colborn, T., Dumanoski, D. and Myers, J.P. (1996). Our Stolen Future: Intelligence, and survival?, Hunan science & Technology Press.

Colborn, T., Saal, F.S.V. and Soto, A.M. (1994). Developmental effects of endocrine Disrupting Chemicals in wildlife and Humans, Environmental Impact Assessment Review, Vol. 14, pp. 469-486.

Fromme, H., Kuchler, T., Pila, K., Muller J. and Wenzel, A. (2002). Occurrence of phthalates and bisphenol A and F in the environment, Water Resource, Vol. 36, pp. 1492-1438.

Japan Environment Agency (2000). SPEED ‘98/ JEA Strategic Programs on Environmental Endocrine Disrupters’98.

Jason, W. B. and John, N.L. (2002). Endocrine disrupters in wastewater and sludge treatment process, Lewis Publisher.

Kloas, W., Lutz, I. and Einspanier, R. (1999). Amphibians as a model to study endocrine disruptors: II. Estrogenic activity of environmental chemicals in vitro and in vivo, Sci Total Environ., Vol. 225, pp. 59-68.

Kolpin, D. W., Furlong, E. T, Meyerm, M. T., Thurman, E. M., Zaugg, S.D., Barber, L. B. and Herbert B. T. (2002). Pharmaceuticals, Hormones, and Other Organic Wastewater Contaminants in U.S. Streams, 1999-2000: A National Reconnaissance, Environmental Science and Technology, Vol. 36, pp. 1202-1211.

Rudel, R. A., Melly, S. J., Geno, P. W., Sun, G. and Brody, J. G. (1998). Identifiation of Alkylphenils and Other Estrogenic Phenolic Compounds in Wastewater, Septage, and Groundwater on Cape Cod, Massachusetts, Environmental Science & Technology, Vol. 32, pp. 861-869.

Shin, H.S., Park, C.H., Park, S.J. and Pyo, H. (2001). Sensitive determination of bisphenol A in environmental water by gas chromatography with nitrogen-phosphorus detection after cyanomethylation, Journal of Chromatography A, Vol. 912, pp. 119-125.

Soto, A.M., Sonnenschein, C., Chung, K.L., Fernandez, M.F., Olea, N. and Serrano F.O. (1995). The E-Screen Assay as a Tool to Identify Estrogens: An Update on Estrogenic Environmental Pollutants, Environ Health Perspect, Vol. 103, pp.113-122.

Warhurst, A. M. (1995). An Environmental Assessment of Alkylphenol Ethoxylates and Alkylphenols, Friends of the Earth Scotland and Friends of the Earth England, Wales and Northern Ireland.

White, R., Jobling, S., Hoare, S.A., Sumpter, J. P. and Parker, M. G. (1994). Environmentally persistent alkylphenolic compounds are estrogenic, Endocrinology, Vol. 135, pp. 175-182.

Yasuhara, A. (1998). Analytical method of phenols. In: Monitoring methods of Environmental Hormones, 24th Seminar of Japan Society of Environmental Chemistry, July 9, 1998, Tokyo, Japan. (in Japanese)

Zafra, A., Olmo, M. D., Suarez, B., Hontoria, E., Navalon, A. and Vilchez, J. L. (2003). Gas Chromatographic-Mass Spectrometric Method for the Determination of Bisphenol A and its Chlorinated Derivatives in Urban Wastewater, Water Research, Vol. 37, pp. 375-742.

Nutrient Budget Analysis and Its Implications as an Indicator of Urban Sustainability: A Case Study on Hanoi, Vietnam

L. N. Cau[1], N. C. Thanh[2] and H. Satoh[3]

1) Environmental Engineering and Management Program, School of Environment, Resources and Development
Asian Institute of Technology (AIT), P.O. Box 4, Klong Luang, Pathumthani 12120, Thailand
e-mail: lengoccau@yahoo.com
2) Asian Institute of Technology Center in Vietnam (AITCV), 21 Le Thanh Torg
GPO Box 136, Hanoi, Vietnam
e-mail: ncthanh@aitcv.ac.vn
3) Environmental Engineering and Management Program, School of Environment, Resources and Development
Asian Institute of Technology (AIT), P.O. Box 4, Klong Luang, Pathumthani 12120, Thailand
e-mail: satohi@ait.ac.th

Abstract

The present study was conducted to estimate the budget of nutrients, phosphorus and nitrogen, in Hanoi city in an attempt to propose an indicator of urban sustainability in terms of nutrients. Hanoi citadel area with a total area of 84.1 square kilometers was selected as the case study. Available data relating to nutrient input and output flows of the target area were collected. By using the obtained data, major nutrient input flows in the forms of water, food, and mineral fertilizer, and output flows in the forms of wastewater and solid waste were estimated and identified.

The estimated results showed that, in the year of 1999, the total N input and output were around 14000 tons and 16000 tons, respectively. Major input was live stock commodities (71% of total input) and crop commodities (25%), and the major output form was as wastewater (88%). The total P input and output were around 2600 tons and 4200 tons, respectively. Major input was again livestock commodities(77%) and crop commodities(17%). The output was 71% as wastewater and the rest as solid wastes. The amount of nutrients recycled through composting was also estimated, which contributed only about 1.3% of nitrogen flow and 4% of phosphorus flow. The difference between input and output still remains uncertain, because the accuracy of the data used in the study is not reliable enough to make such a discussion. But the order of the size of flow and recycling was grasped. The small contribution of recycling by composting indicates that there is room for improvement concerning the budget of nutrients in the study area.

Keywords: Hanoi citadel, nitrogen, nutrient budget analysis, phosphorus, urban sustainability

1. Introduction

In the present world, "sustainability" is one of the key concepts used to realize the long-lasting welfare of human society. However, it is not easy to define "sustainability" or a "sustainable society". Sustainability is a matter that needs to be discussed from different points of view such as energy supplies, materials supplies, the stability of society, and its relationship with other societies. In 1987, the World Commission on Environment and Development developed a definition of sustainability. Under this definition, sustainability is development that "meets the needs of the present without compromising the ability of future generations to meet their needs." Although this definition has become widely publicized, the term sustainability is not limited to one precise definition. Another way to define sustainability is a balancing of economic objectives, social goals, and environmental resources in a way that works for both present and future generations. In the present study, as one of the key issues essential for sustainability, the focus is on the nutrients (nitrogen and phosphorus) budget of urban systems. Here, "nutrients budget" is the inflow, outflow, and accumulation (or reduction) of nutrients within a given geographic region. In addition, the recycling flow inside or around the system would be within its scope.

The present study was conducted with following the objectives: (1) To clarify the nutrients budget of Hanoi citadel; and (2) To evaluate the sustainability of Hanoi citadel from the view point of the clarified nutrient budget.

2. Materials and Methods

Hanoi Citadel

Hanoi, capital city of Vietnam, lies in the Red River Delta. It is situated in a tropical area having a strong monsoon influence. It consists of 7 urban districts and 5 peri-urban districts, with the total area of 1000km^2 and officially around 2.7 million population. The urban area has only 84.1 km^2, but it's population is 1.45 milliion, which is more than half of the total population. While the urban area of Hanoi is now rapidly developing and being rapidly industrialized, it still has a small agricultural activity, with an agricultural area about 19 km^2. On the other hand, the peri-urban area is more agricultural, with around 50% of its area used for agricultural (de Bon and Thi, 2000).

Nutrient Budget Analysis

The outline of the nutrient budget analysis is as illustrated in Figure 1. The method developed here is used for estimating various major nutrients input and output flows of Hanoi citadel mentioned in the model. And the nutrients balance was obtained by comparing the total amounts of nutrients input flows and output flows. The nutrients flows to and from Hanoi city that we considered explicitly are those in drinking water, livestock and crop commodities, mineral fertilizer, atmospheric deposition, solid waste, landfills leachate, compost, and wastewater, as a priori crude analysis indicates that these are by far the most important.

In estimating the amounts of crop and livestock commodities that Hanoi citadel consumes annually, the following two factors were taken into consideration; (1) the percentage of foods supplied to Hanoi citadel from peri-urban areas and from surrounding provinces; (2) the

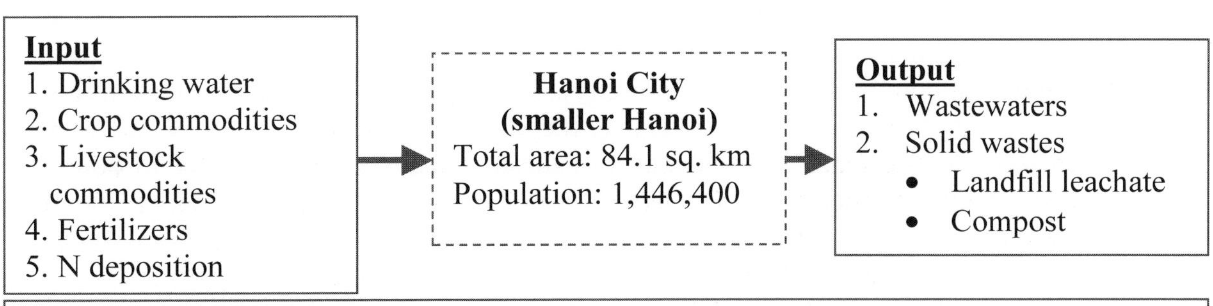

Figure 1. Nutrients inflows to and outflows from Hanoi City

inedible portion of the crop and livestock commodities imported into Hanoi citadel that have inedible parts.

In estimating nutrients contents in vegetable products, in most vegetable products, owing to fundamental biochemical relations, the N/P ratios are quite constant (Walworth and Sumner, 1987). Walworth and Sumner (1987) refer to 16 ratios found in green leaf parts, collected from various literature sources. These ratios range from 8.2 to 15.3. Soybean has a much higher N/P ratio than all other plant types. When soybean is excluded, the average N/P ratio can be estimated to be equal to 10.0.

With regards to the estimation of total P input in livestock and crop commodities, the P content in each kind of livestock and crop commodities as disclosed in Beatonet, 1995. From the disclosed P content and the amount of crop commodities consumed annually in Hanoi citadel, one can estimate P input in crop commodities.

Total N concentration of livestock and crop commodities will be estimated based on published papers, on literature on actual quantities of N contents in livestock and crop commodities. In addition, the total N concentration of a certain livestock/crop commodity can be obtained by dividing total protein content of the commodity by the nitrogen-to-protein conversion factor (% total protein = 6.25 x % total N). For some livestock and crop commodities that their protein or total N content is unknown, the total N content can be estimated from total P content with the assumption that N/P ratio is about 10, and 5 (Færgea et al, 2000) for crop commodities and livestock commodities, respectively.

To justify the methodology of outflow estimation as wastewater adopted here, it is necessary to explain the basic scheme of wastewater flow in Hanoi. In Hanoi, there is no wastewater treatment, except for some industries. Wastewater from households and most industries, and urban runoff in rain are collected by small rivers, and then discharged to outside the urban area of Hanoi. Thus, it is possible to estimate the outflow based on the flow of the rivers in the urban area and their water quality.

The details of the estimation methods are tabulated in Table 1.

3. Results and Discussion

The estimated nutrients input and output flows of Hanoi citadel are shown in Figure 2 for nitrogen and Figure 3 for phosphorus.

The total estimated input of N was 14243 tons per year for the year 1999. The estimated mass balance for N showed that most of N input occurred via domestic or household uses, (import of crop and livestock commodities into Hanoi city). The dominant N input flows are livestock commodities and crop commodities. Especially, the subtotal input amount of N in livestock commodities accounts for approximately 71 % of the total input of N. The second dominant input flow was crop commodities, which accounts for approximately 25 % of the total estimated N input.

The minor N input flows are drinking water, atmospheric deposition and mineral fertilizer used for farming purposes in the agricultural land in Hanoi citadel. Among these minor input

Table 1. Details of the estimation method of N & P input and output flows

Flow	Description	Methods calculation	References
Drinking water	Sources of water: Mainly from Red river and underground water. Methods of distribution: Mainly through water supply bureau. But privately owned wells are also popular.	(1) Total population of urban areas * water demand per capita per year * N or P concentration. (2) Data related to the total amount of drinking water supplied to Hanoi * N or P concentration.	(1) Hanoi water works authority. (2) N & P concentrations were obtained in the present study by measurement of a sample.
Crop commodities	Sources: Around 70 % of crop commodities consumed in smaller Hanoi is from peri-urban areas and surrounding provinces.	- The amount of nutrients in each crop commodity consumed in Hanoi citadel per year (IT) is calculated as follow: IT = (Total urban population)*(food intake per capita per year)*(N/P concentration of each kind of food consumed). - From IT, nutrients in total consumption (CT) of crop commodities which includes inedible part is estimated by following calculation; CT=IT/k, wherein k is ratio of inedible part. - By taking into consideration the crop production inside smaller Hanoi, the actual nutrient input flow in crop commodity is estimated as: IF=0.7 * CT = 0.7 * IT/k.	(1) Tuyen et al, (2000. (2) Beaton et al, (1995). (3) Bon and Thi (2000). (4) Report No.: Report No., 141 – XNNKC, February 11, 2003 (Hanoi Department of Agriculture and Rural Development, 2003). (5) Food Composition Table for Use in East Asia (2003).
Livestock commodities	Sources: Around 70 % of livestock commodities consumed in smaller Hanoi is from peri-urban areas and surrounding provinces.	- The amount of nutrients in each livestock commodity consumed in Hanoi citadel per year (IT) is calculated as follow: IT = (Total urban population)*(food intake per capita per year)*(N/P concentration of each kind of food consumed). - From IT, nutrient in total consumption (CT) of livestock commodities which includes inedible part is estimated by following calculation; CT=IT/k, wherein k is ratio of inedible part. - By taking into consideration the crop production inside smaller Hanoi, the actual nutrient input flow in livestock commodity is estimated as: IF=0.7*CT = 0.7*IT/k.	(1) Tuyen et al, (2000). (2) Beaton et al, (1995). (3) Bon and Thi (2000). (4) Report No.: Report No., 141 – XNNKC, February 11, 2003(Hanoi Department of Agriculture and Rural Development, 2003). (5) Clayton, (2003).

Table 1 (continued)

Mineral fertilizer	Mineral fertilizer imported into Hanoi for farming purposes	Average amount of N and P applied in agricultural lands per hectare * total area of agricultural land in urban districts.	Hanoi URENCO - Hanoi City People's Committee (1998)
Solid waste	Total amount of annually generate solid waste in urban districts of Hanoi	Total amount of generated solid waste per year * N or P concentration of the solid waste.	(1) Hanoi URENCO - Hanoi City People's Committee (1998). (2) Lemna International, Inc., USA, (2002). (3) N & P contents were obtained by analysis of a sample in the present study.
Atmospheric deposition	N atmospheric deposition mainly due to fossil and fuel combustion, and agricultural production	Average amount of N in kg deposited per hectare * total urban area of Hanoi.	(1) Galloway J. N. (1998). (2) Hanoi Statistical yearbook, 2002.
Leachate	Total amount of leachate leached from landfills in Hanoi city	Total amount of leachate * N or P concentration of the leachate.	Hanoi URENCO - Hanoi City People's Committee (1998)
Compost	Compost produced from collected solid waste in a solid waste treatment plant	Total amount of produced compost * N or P concentration of the compost.	Hanoi URENCO - Hanoi City People's Committee (1998)
Wastewater	Source: Domestic wastewater + wastewater from public places + industrial wastewater + storm water	(1) Annual amount of storm water = average annual rainfall in Hanoi citadel* total area of the region. (2) Total amount of discharged wastewater * N or P concentration of the wastewater.	(1) Department of Natural Resources and Environment, and Hanoi Municipal Sewerage and Drainage Company - Hanoi City People's Committee Hanoi (1997) (2) Hanoi statistical yearbook, 2001

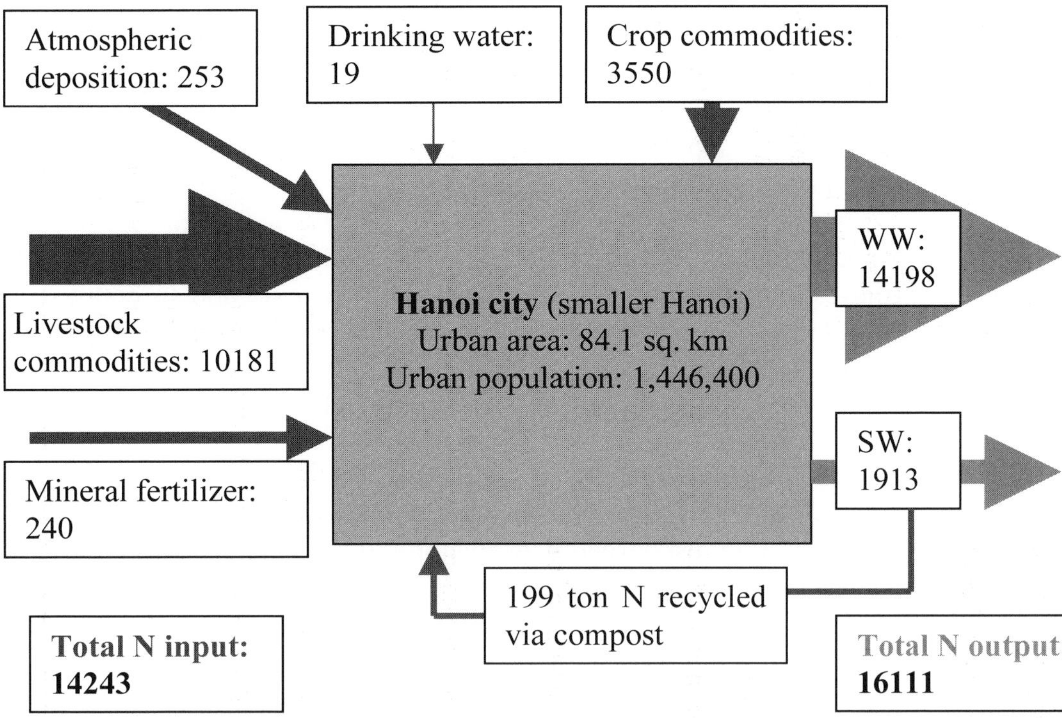

Figure 2. N input and output flows in Hanoi city (ton per year)

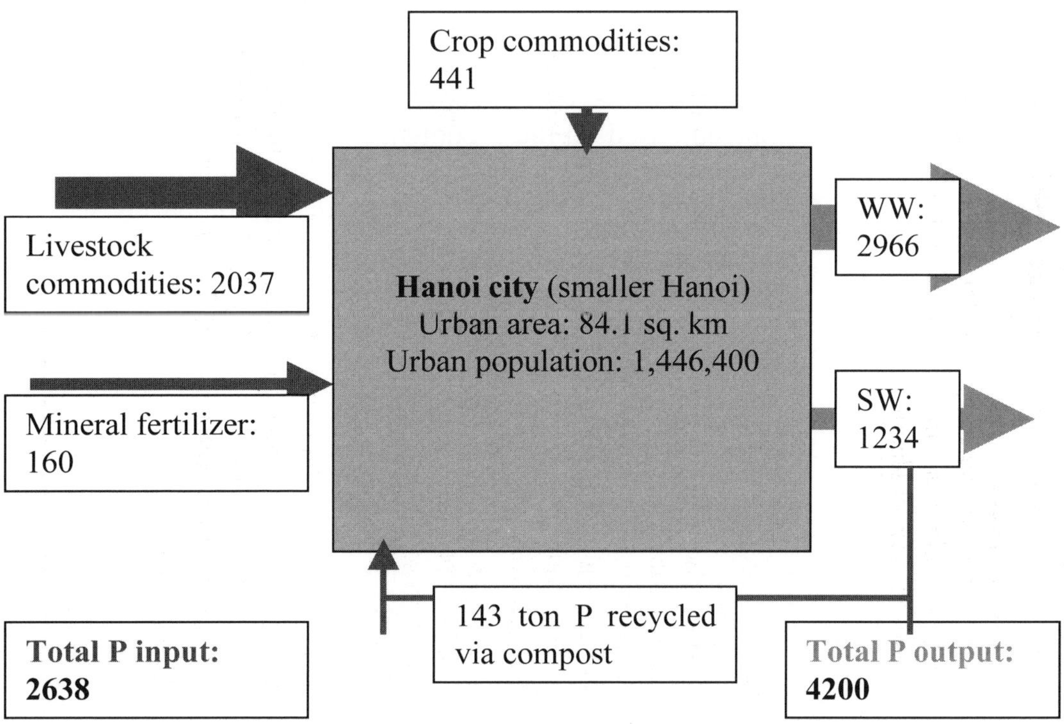

Figure 3. P input and output flows in Hanoi city (ton per year)

flows, drinking water has least affect on total N input flow (estimated N in drinking water accounts for only about 0.2 % of the total N input).

With regards to the input of P, the estimated total input was approximately 2638 tons per year for the year 1999. Similarly to the N input, the P input estimation showed that total P input in livestock commodities is the most dominant input flow of P (total nutrient input of P in livestock commodities accounts for approximately 77 % of the total estimated input of P). The second-most dominated input flow of P is crop commodities (accounting for approximately 17 % of the total estimated input of P).

Regarding the estimated output of N, the estimated total output of N was 16111 tons per year for the year 1999 (the estimated amounts of N in landfills leachate and in compost are not included in the total output). The estimated output amounts of N in discharged wastewater and in generated solid waste account for approximately 88 % and 12 % of the total estimated output of N, respectively. And for N output flows, the dominant one is wastewater.

The estimated total output of P was 4200 tons per year for the year 1999 (the estimated amounts of P in landfills leachate and in compost are not included in the total output). The estimated output amounts of P in discharged wastewater and in generated solid waste account for approximately 71 % and 29 % of the total estimated output of P, respectively.

From the estimated N content in generated solid waste and from the estimated N content in the compost produced from the generated solid waste, one can estimate the percentage of N nutrient recycled via use of compost for farming purposes in agricultural land areas of Hanoi city. Particularly, in the case of Hanoi city the estimated percentage of N recycled in compost was 10.4 % of the N content in the generated solid waste. In comparing with the total N output, the amount of N recycled through composting accounts for a very small percentage portion of the total output (approximately 1.5 % of the total N output).

Similarly to the estimation of the percentage of N recycled via compost, the percentage of P recycled in compost was estimated based on the P content in generated solid waste and on P content in compost produced from the generated solid waste. And the estimated P recycled percentage was 11.5 %, approximately to N recycled percentage. In comparing with the total P output, similar to the case of N, the amount of P recycled through composting accounts for a very small percentage portion of the total output (approximately 3.5 % of the total P output).

The small size of recycling indicates that there is large room for Hanoi to improve its system for better nutrient budget.

4. Conclusions

The estimation methods developed here have successfully clarified the size of input and output flows. The accuracy might be enough to grasp roughly the size of nutrients recycling in comparison to the total input and output flows and to evaluate the sustainability from the view of nutrient recycling. But in order to assess the budget itself, the accuracy still needs significant improvement.

The estimated total N input in Hanoi citadel was approximately 14243 tons per year in 1999. The identified major or dominant N input flows were livestock and crop commodities. The estimated total output of N from Hanoi citadel was approximately 16111 tons per year in 1999. The identified major output N flows were solid waste and discharged wastewater.

The estimated total P input in Hanoi citadel was 2638 tons per year in 1999. The identified major or dominant P input flows were livestock and crop commodities. The estimated total output of P in Hanoi was approximately 4200 tons per year in 1999. The identified major output P flows were solid waste and discharged wastewater.

Nutrients recycling was not well done for both N and P. The estimated results of N and P recycled via uses of compost produced from the generated solid waste showed that approximately 11 % of N content and 12 % of P content in the generated solid waste were recycled. In comparison with the total output, the N recycled amount accounts for approximately 1.5 % and that for P accounts for approximately 3.5 %. This indicated that Hanoi citadel still consumes more nutrients from external sources or it depends much on the external nutrient sources.

The third conclusion indicates that when applying the criteria for using the nutrients budget as an indicator for urban sustainability for the situation of Hanoi citadel, the development of the region was not in the sustainable state, and further improvement was found to be necessary.

5. Acknowledgements

The authors wish to thank Mrs. N. H. Lan; Mr. N. L. Ngoc and Mr. V. V. Dinh for providing data necessary for the nutrient budget analysis.

6. References

Clayton, C.L. (2003). Dead dog food (http://www.flintriverranchpetfood.org/DeadDogFoodArticle.html).

De Bon, H. and Thi, T.K. (2000). The response of peri-urban agriculture to urbanization in Vietnam (http://www.fao.org/ag/ags/agsm/sada/-asia/DOCS/DOC/Debon1.doc), FAO.

Dept. of Natural Resources and Environment, and Hanoi Municipal Sewerage and Drainage Company – Hanoi City People's Committee, Hanoi (1997). Summary report of project on investigation of environmental pollution status of To Lich River and on designing of treatment methods for treating wastewater discharge into the river's cannel systems.

FAO (2003). Food Composition Table for Use in East Asia (http://www.fao.org/DOCREP/003/X6878E/X6878E01.htm).

Hanoi Department of Agriculture and Rural Development (2003). Report No. 141-XNNKC, February 11, 2003.

Beaton, J.D., Roberts, T.L., Halstead, Ed. H. and Cowell, L.E. (1995). Global Transfers of P in Fertilizer Material and Agricultural Commodities, Scope 54 – Phosphorus in the Global Environment.

Færgea, J., Magida, J. and de Vriesba, F.W.T.P. (2000). Urban Nutrient Balance Modeling Parameterized for Bangkok.

Galloway, J.N. (1998). The global nitrogen cycle: changes and consequences, Environ. Pollution, Vol. 102, S1, pp. 15-24.

Hanoi City (2002). Hanoi Statistical Yearbook 2002.

Hanoi City (2001). Hanoi Statistical Yearbook 2001.

Hanoi URENCO-Hanoi City People's Committee (1998). Feasibility study report of the project on treatment of solid waste producing organic fertilizer, Caudien Town, Hanoi.

Keith, J.S., William, S. and John, L. (2002). Nutrient Balance as an Indicator of Sustainable Agriculture, 17th WCSS, 14-21 August 2002, Thailand.

Lemna International, Inc. (2002). Feasibility study report in designing a total solid waste treatment plant for treating urban solid wastes in Hanoi City, Minneapolis, Minnesota, USA.

Tuyen, L.D., Bricas, N., Mai, L.B., Marie, B., Dop, M-C., Chung, N.D. and Khan, N.C. (2000). Trends in food consumption and in the nutritional status in urban Vietnam (http://www.fao.org/ag/ags/agsm/sada/asia/DOCS/DOC/Tuyen1.doc), FAO.

Walworth, J.L. and Sumner, M.E. (1987). The Diagnosis and Recommendation Integrated System, *ADV SOIL SCI*, Vol. 6, pp. 149-188.

Keyword Index

Keywords are indexed by the first page number of the paper in which they appear.